Physical Properties
of Soils

R. E. Means

J. V. Parcher

Oklahoma State University

Charles E. Merrill Publishing Co.   Columbus, Ohio

Copyright © 1963 by *Charles E. Merrill Publishing Company*, Columbus, Ohio. All rights reserved. No part of this book may be reproduced, by mimeograph or any other means, without the written permission of the publisher.

Library of Congress Catalog Card Number: 63-14199

Standard Book Number 675-09821-1

4 5 6 7 8 9 10 11 12 13 14 15-76 75 74 73 72 71 70 69

PUBLISHED IN THE UNITED STATES OF AMERICA

Preface

It is the conviction of the authors that the successful practice of soil mechanics and foundation engineering depends much more on a basic understanding of soil properties—their determination, interpretation, and significance—than on the development of those techniques of engineering mechanics and hydraulics which are used in the mathematical solution of problems encountered in practice. A solution based on a deep understanding of the behavior of soils is much to be preferred to a solution, however mathematically precise, based on a superficial knowledge of the soil properties. In this book, as well as in their teaching, the authors have sought to develop a broad understanding of the nature of soils and an awareness of the necessity for thoughtful and careful interpretation of experimentally determined properties. With such a background the commission of serious errors in practice is likely to be rare.

To broaden the reader's understanding, some topics have been considerably amplified beyond that which can be given detailed class discussion. Laboratory procedures and instructions have been integrated into the text because it has been the experience of the authors that theory and laboratory cannot (and should not) be separated. For convenience, a procedure for carrying out each test is suggested in Chapter XII.

Soil mechanics is a rapidly advancing field in which new concepts, new procedures, and other important contributions are made each year. Our readers are encouraged to retain an open-minded inquisitiveness and to keep abreast of new developments by studying the many excellent papers which appear in the technical publications. The references listed at the ends of the various chapters have been limited to those papers having direct applicability to materials in the text. Graduate students, especially, and others who contemplate doing advanced work in soil mechanics should study these reference papers if at all possible.

At widely separated times, the authors studied at Harvard University where they were profoundly influenced by Professors Karl Terzaghi and Arthur Casagrande. It is inevitable that the influence of those superb teachers is to be found throughout this work. A large share of the credit for whatever is creditable in this text belongs to them. For all else the authors accept full blame.

It is the sincere hope of the authors that students, practicing engineers, and architects will find something of value here. An eagerness on the part of professional men of many years standing to develop a basic understanding of soil behavior is always evident in the technical sessions of the societies. It is believed that the material presented in this book forms the necessary foundation for such understanding.

RAYMOND E. MEANS
JAMES V. PARCHER

Stillwater, Oklahoma
July, 1963

Acknowledgments

The materials presented in this text have been acquired by the authors in a variety of ways—from formal and informal studies and research, from reviews of the literature of the field, and from experience. We have attempted to give specific credit for all materials taken from other publications. Even if we have succeeded in doing this, recognition must still be given to the efforts of countless, untiring workers in the field, all of whom have had a part in developing knowledge to its present state.

This book has, however, been directly and immensely influenced by one person—Arthur Casagrande—in such a way that it has been impossible to indicate, by direct credits, the extent to which his teachings have shaped the entire text. The class notes which provided the basis for this book were in process of development and use over a period of several years, and were modeled after the "Notes on Soil Mechanics—First Semester, compiled by the teaching staff and students during the academic year 1938–39 at Harvard University, and based on the lectures by A. Casagrande." The concepts represented in these notes, and those presented by Professor Casagrande during the many lectures which we were privileged to attend, have become so much a part of our mentalities that an acknowledgement of the general debt of gratitude owed Professor Casagrande is unhesitatingly given. A book such as this should properly have been written by him.

Appreciation is also expressed to Mr. Walter R. Ferris for his help in the preparation of parts of the chapter on Stress-Deformation and Strength Characteristics, and to Professors J. O. Osterberg and Morris Grosswirth, who reviewed the manuscript and made valuable suggestions for its improvement.

To

Arthur Casagrande

Magnificent Teacher—Inspired Engineer

Table of Contents

I. INTRODUCTION

II. SPECIFIC GRAVITY OF SOLIDS

III.   GRAIN SIZE DISTRIBUTION

IV.   PLASTICITY

V.   STRUCTURE

VI.   CLASSIFICATION OF SOILS

IX.   COMPRESSIBILITY AND CONSOLIDATION

X.   SHRINKAGE

XI.   STRESS-DEFORMATION AND STRENGTH CHARACTERISTICS

XII.   LABORATORY TESTING

CHAPTER I

Introduction

1.01 GENERAL

The term soil is used by the engineer and the agriculturist to refer to natural materials made up of separate grains not cemented together into solid rock. The agriculturist is interested in those properties of soils which affect its ability to support plant life. The engineer is interested in those properties which affect its ability to support loads in structures. These properties for soils are the same as for any other material; i.e., time-stress-deformation relationships, ultimate strength, permeability, etc. The agriculturist is primarily interested in organic soils, and the engineer in inorganic soils.

Soil Mechanics may be subdivided for study as follows:

I. Soil Mechanics
  A. Physical Properties of Soils and Testing Methods
    1. Origin and Mineral Constituents of Soils
    2. Specific Gravity of Solids
    3. Grain Size Distribution
    4. Plasticity
    5. Structure
    6. Permeability
    7. Compressibility and Consolidation
    8. Shrinkage
    9. Shearing Resistance
  B. Theoretical Soil Mechanics
    1. Stability of Slopes
    2. Earth Pressures
    3. Carrying Capacity of Soils
    4. Settlement Analysis
    5. Flow of Fluids Through Soils
II. Design of Earth Structures
  A. Application of Theoretical Soil Mechanics
  B. Imperical Knowledge of Soils
  C. Experience, Imagination, and Judgment

Soil Mechanics deals with idealized, homogeneous, and isotropic or uniformly anisotropic granular materials which have not been indurated into solid rock. Such materials may not exist in nature, or may exist in very small volume. There may or may not be a great variation in the subsurface materials beneath an area.

Generally, the properties of soils cannot be determined with as great a degree of accuracy as those of most other construction materials. Even though the properties were known for one sample of soil beneath an area, the properties of the entire soil affected would be only vaguely known, because materials may vary over a wide range in a small area. Experience, imagination, and judgment are of considerably greater importance than mathematical analysis in the design of earth structures. However, it should be borne in mind that in the design of any structure, mathematical analyses are made on simplified hypothetical structures which more or less resemble the structures being designed; and that experience, imagination, and judgment are required on the part of the designer in adapting the results for the hypothetical structures to the design of the real ones. The only difference in the design of earth structures and structures of other materials is in the upper and lower limits for which the properties of the materials are known and the extent of the materials.

1.02   ORIGIN

The rocks that form the earth's surface are classified as to origin
as igneous, sedimentary, and metamorphic. The igneous rocks are
those formed directly from the molten state. If the molten rock cools
very slowly, the different minerals segregate into large crystals forming a coarse grained or granitic structure. The common igneous rocks
of coarse grained structure are granite, which consists principally of
quartz and feldspar, and gabbro, which is composed principally of the
dark ferromagnesian minerals, plagioclase feldspar, augite, horneblende
and olivine. When the solution of minerals is cooled more rapidly,
tiny crystals of the minerals are formed in a vitreous matrix. Such
extremely fine grained rocks are called felsite when they are formed
of the light colored minerals, feldspar and quartz, and basalt when
formed of ferromagnesian minerals. If the solution is cooled very
rapidly, the minerals do not separate into crystals, but solidify as
amorphous glassy rock such as volcanic scoria, pumice, and obsidian.

The igneous and other rocks are disintegrated, either by mechanical
or chemical weathering, into small particles of the original rock or
minerals and into other minerals. These small pieces and new minerals, after being transported by wind or water, are deposited in
layers on the surface of the ground, after which they may be cemented
together by minerals carried in solution in the water which percolates
through the deposits. This cementing together of small particles
produces sedimentary rocks. Some of the minerals formed by the
weathering of igneous minerals which are commonly associated with
sedimentary rocks are: kaolin, bauxite, limonite, hematite, calcite,
gypsum, anhydrite, flint, chert, agate, and opal.

Metamorphic rocks are formed from igneous or sedimentary rocks
by pressure and heat. The extreme pressure and heat recrystallizes
the minerals into new crystals of the same or other minerals. A considerable number of new minerals are formed in the metamorphic
process. The common minerals associated with metamorphic rocks
are: quartz, feldspar, mica, horneblende, actinolite, epidote, garnet,
tourmaline, beryl, and corundum. The minerals orient themselves
in reforming under high pressure forming a flaky structure similar to
fine grained mica, and known as mica schist; or segregate in streaks
or bands of relatively light and dark minerals to form gneiss. Still
others, such as slate, form with perfect cleavage planes so that the
rock can be easily split into thin slabs with comparatively parallel
surfaces. Some of the common metamorphic rocks are: gneiss formed

by the metamorphosis of granite, slate formed by the metamorphosis of shale or clay, mica schist formed by the metamorphosis of slate, and marble formed by the metamorphosis of limestone.

Mechanical weathering, such as frost action, abrasion of the rocks by streams, wind, wave action along the shores of lakes and seas, breaks the rocks into small bulky pieces such as gravels, sands, and silts, or into flat scale-like particles. These uncemented rock particles and minerals deposited by wind and water form the soils with which Soil Mechanics is concerned.

The size of rock and mineral particles carried by a stream of water varies with the velocity of the current. When the velocity of a stream is decreased, heavier particles which have been carried in suspension settle out. As the velocity is still further decreased, smaller and smaller particles are deposited. The variation in sedimentation velocity with size of particle results in segregation into layers of coarse gravels, sands, silts, and clays as the velocity of the stream changes. Gravels are deposited near the source of streams where the gradient is high or during flood stage down stream. Sands are deposited down stream where the gradient is less or as the velocity diminishes after flood stage. Sands are also formed by wave action along the shores of lakes and seas. The very small particles forming silts and clays are often deposited in quiet water of lakes or seas or in back water after flood stage along streams.

Pebbles forming gravels are usually pieces of rock with occasional pieces of quartz, feldspar or other mineral. Sands, generally, consist of grains of quartz and feldspar with occasional grains of other minerals, but may consist of grains of fine grained rock, such as basalt or obsidian. Silts consist of grains of quartz or other minerals which possess a high resistance to chemical weathering. Clays are formed of flat scale-like particles of mica, kaolin, montmorillonite, and other minerals, a majority of which have been broken down into colloidal sizes.

1.03   MINERAL COMPOSITION AND FORM

Minerals are inorganic chemical compounds formed in nature. An identical compound formed in the laboratory is not a mineral. As solids, minerals may occur in an amorphous state or in a crystalline state.

Molecules of minerals are composed of atoms of chemical elements. Although an atom of any element is itself formed of the basic particles, protons, neutrons, and electrons, one might consider atoms as the basic building blocks of all matter. An atom of one of the elements consists of a nucleus having a positive electromagnetic charge or charges and one or more satellite electrons each having a negative electromagnetic charge. The nucleus of an atom consists of one or more protons, each carrying a positive electromagnetic charge, and may or may not contain one or more neutrons carrying no charge. Most of the weight of the atom is concentrated in the nucleus. The electrons, each weighing about $\frac{1}{1840}$ as much as a proton, revolve about the nucleus in comparatively very large orbits and spin about their own axes. Most of the atom is empty space.

The individual elements are determined by the number of protons in the nucleus and the number of electrons attached to the nucleus of their atoms. In every atom of a specific element there are a certain number of positive protons and the same number of negative electrons, making each atom of every element electrically balanced with positive and negative charges.

A certain amount of energy is bound up in the electron as it revolves in its orbit around the nucleus and spins on its axis. An electron following a larger orbit farther away from the nucleus possesses a different amount of energy than one close to the nucleus. Spectra indicate that electrons move at definite energy levels outside the nucleus. There are seven major energy levels in which the electrons are active. The energy level in which an electron or group of electrons moves is often referred to as a shell. Only a limited number of electrons can exist in any one energy level or shell. The first shell closest to the nucleus, called the K shell, can contain only 2 electrons. The second level, L shell, can contain 8 electrons; the third, M shell, 8 electrons; the fourth, N shell, 18 electrons; the fifth, O shell, 18 electrons; and the sixth, P shell, 32 electrons. The seventh level or Q shell is not filled to capacity in any known element but does contain electrons in the rare natural and artificially produced elements. That portion of an atom inside one of the major energy levels is often referred to as a core bearing the name of the inert element formed when a shell is filled to capacity; i.e., Helium core, Neon core, Argon core, Krypton core, Xenon core, and Radon core.

The foregoing discussion is not meant to imply that all the electrons active in a shell have the same energy. Probably no two electrons possess the same energy. Also, in those major shells in which there are more than 8 electrons, there exist slightly lower energy levels or subshells in which some of the electrons of the major shell operate. Within these subshells a high degree of magnetic balance is attained,

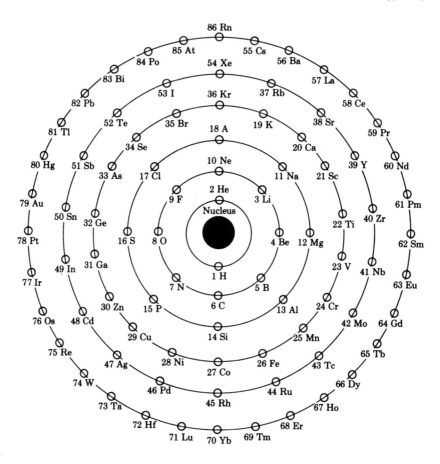

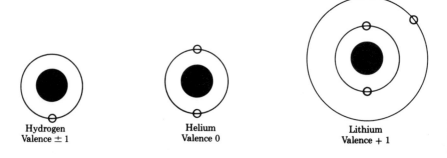

Hydrogen
Valence ± 1

Helium
Valence 0

Lithium
Valence + 1

Figure 1.03a
Diagram Showing Free Electrons in Six Major Energy Levels
(First 86 Elements)

but the inert gases are formed only when the major shells are filled to capacity.

A spinning charged electron forms a magnet exerting a magnetic moment. An electron spinning in the opposite direction exerts a magnetic moment in the opposite direction. Two electrons spinning in opposite directions balance or neutralize their magnetic moments. The two electrons in the first shell of an atom spin in opposite directions thus neutralizing the magnetic moments. In those elements having their shells filled to capacity, the magnetic moments are balanced making those elements chemically inactive. Those elements in which there is an excess or a deficiency of electrons to fill the outer shell are chemically active and unite with other such elements in such proportion as to neutralize the magnetic moments in the combined elements. The number of electrons in excess of the number required to fill the inner shells of an atom is referred to as a positive valence equal to the number of excess electrons. The number of electrons lacking in the outer shell of an atom is referred to as a negative valence equal to the number of electrons required to complete the filling of the outer shell.

Atoms may or may not contain neutrons in their nuclei. But, a specific number of protons and the same number of electrons determines the element regardless of the number of neutrons present in the nuclei. The normal hydrogen atom consists of one proton and one electron with no neutrons in its nucleus. Another form of hydrogen atom contains one neutron in addition to its proton in the nucleus. Still another form of hydrogen atom contains two neutrons in its nucleus. The weight of the neutron is comparable to the weight of the proton. The atomic weight of these atoms containing the neutrons is greater than that of the normal hydrogen atoms. Hydrogen of this type is sometimes called heavy hydrogen. These different forms of the same element are known as isotopes of that element.

In Figure 1.03a are shown the number of electrons and protons for the first 86 elements with the number of electrons in each energy level or shell. The electrons do not occupy the static positions indicated in the diagram. For an example of the use of the diagram, the calcium atom contains 20 electrons and protons with 18 of the electrons in the Argon core and 2 electrons in the next higher level of energy. The electrons above 20 do not exist in the calcium atom.

The 15 most common elements that compose the rocks of the earth's crust or lithosphere are given in a table which follows. The percentages of these elements contained in known matter on the earth including atmosphere, water, and rocks are slightly different from those shown for rocks alone.

15 Most Common Elements in Earth's Crust

Element	% of Earth Crust	No. of Electrons in Atom	Lacks	Valence Excess	Least
O	46.46	8	−2	6	−2
Si	27.61	14	−4	4	±4
Al	8.07	13	−5	3	3
Fe	5.06	26		Varies	
Ca	3.64	20	−6	2	2
Na	2.74	11	−7	1	1
K	2.58	19		1	1
Mg	2.07	12	−6	2	2
Ti	0.62	22		4	4
H	0.14	1	−1	1	±1
P	0.12	15	−3	5	−3
C	0.09	6	−4	4	±4
S	0.06	16	−2	6	−2
Cl	0.05	17	−1	7	−1
Ba	0.04	56		2	2

In solution compounds lose their identity by separating into ions. These ions consist of only one element of the compound or of two or more elements which are not electrically balanced.   The ion is formed when the excess electrons in a shell of an atom take the place of the electrons lacking in a shell of another atom.   This exchange of electrons causes the atom with the excess electrons to lose as many negative charges as were exchanged.   The number of positive protons remains unchanged, leaving the atom electrically unbalanced with an excess of positive charges equal to the excess negative electrons that have left the atom.   The opposite is true of the atom which received the electrons in the exchange.   These electrically unbalanced atoms are called ions.   For example, Na_2CO_3 separates into 2 Na ions each carrying 1 positive charge and into a CO_3 ion carrying 2 negative charges.   Each Na atom has 11 protons with 11 positive charges and 11 electrons with 11 negative charges or an excess of 1 electron above the number required to fill the second shell of the Neon core.   When this excess electron is exchanged, the ion formed contains 11 protons with 11 positive charges and 10 electrons with 10 negative charges, giving, therefore, to the Na ion a positive unit charge.   The carbon atom has 4 electrons excess and the oxygen atom lacks 2 electrons. In the CO_3 ion the carbon atom provides 4 extra electrons and the 3 oxygen atoms are deficient 6 electrons for magnetic balance.   The combined 4 atoms in themselves are in electrical balance but lack 2 electrons for magnetic balance.   The magnetic field pulls the 2 excess electrons from the 2 Na atoms to fill the shells of the combined atoms to form the magnetically balanced ion.   But, the ion is unbalanced electrically by having 2 extra negative electrons. Thus, the CO_3 ion carries 2 negative charges.

The negatively charged ions are called anions and the positively charged ions are called cations.

When the ions are removed from solution, the anions and the cations unite to form the original solid compound. If two compounds are ionized together in solution, the anions from one compound may unite with the cations of the other to form two compounds different from either original. When $CaSO_4$ and Na_2CO_3 are dissolved together, they separate into Ca and Na cations and SO_4 and CO_3 anions. These ions unite to form molecules of $CaCO_3$ and Na_2SO_4.

Many elements do not form ions, yet these elements unite to form compounds. Carbon is probably the most common of these elements which do not form ions. Solutions of these non-ion forming elements or compounds in water are poor conductors of electric current. Atoms of these non-ion forming elements are united by covalent bonding which is discussed later.

Under some conditions when a compound solidifies, its molecules arrange themselves in a haphazard manner so that the molecular bond is the same in all directions. A material with such a haphazard arrangement of molecules is said to be amorphous. Such a material is as likely to fracture along one surface as another. Glass is an amorphous material. Under favorable conditions during solidification the molecules arrange themselves in a definite order with a certain number of molecules forming a unit having a definite shape. This unit is called a crystal. The molecules within the crystal are bound together with strong molecular bonds. This regular orderly arrangement of the molecules produces planes of weakness in certain directions causing the crystals to split or cleave along these planes of weakness. Some mineral crystals, like quartz, have no plane of cleavage but break with a conchoidal or irregular fracture surface. Some minerals have only one plane of cleavage while others may have two, three, or more. Cleavage planes occur parallel to the faces of the crystalline form.

Crystals can be formed in three ways: (1) by solidification from a molten or liquid state, such as the formation of mineral crystals from molten magma as it cools; (2) by precipitation of matter from a solution, such as the formation of salt (halite) crystals after evaporation of sea water; (3) by precipitation from vapor, such as the formation of sulphur crystals from its vapor.

Although there is considerable variation of form within each system, all mineral crystals have been classified as falling in one of six systems.

a. Isometric System

The isometric system includes all crystalline forms having three axes of equal length meeting at right angles.

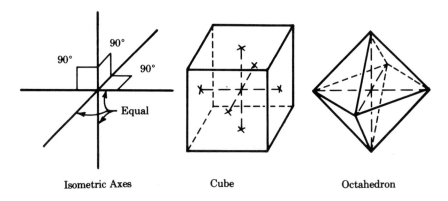

| Isometric Axes | Cube | Octahedron |

Figure 1.03b
Simple Examples of Isometric System

b. Tetragonal System

The tetragonal system includes all crystalline forms having three axes meeting at right angles, two of the axes being the same length and the third either longer or shorter than the other two.

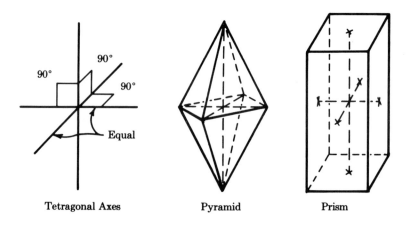

| Tetragonal Axes | Pyramid | Prism |

Figure 1.03c
Simple Examples of Tetragonal System

c. Hexagonal System

The hexagonal system includes all the crystalline forms having three coplanar axes of equal length meeting at 60° and 120° and an axis of different length normal to the plane of the other three.

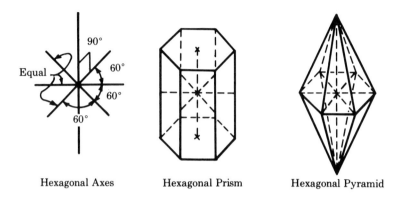

Hexagonal Axes   Hexagonal Prism   Hexagonal Pyramid

Figure 1.03d
Simple Examples of Hexagonal System

d. Orthorhombic System

The orthorhombic system includes all crystalline forms having three axes of unequal lengths meeting at right angles.

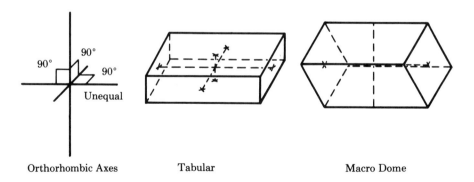

Orthorhombic Axes   Tabular   Macro Dome

Figure 1.03e
Simple Examples of Orthorhombic System

e. Monoclinic System

The monoclinic system includes the crystalline forms having three axes of unequal length, two of which intersect at an acute angle and one at right angles to the other two.

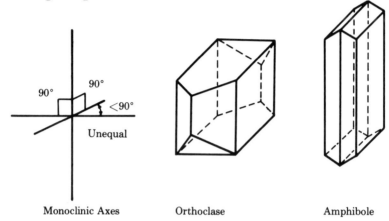

Monoclinic Axes   Orthoclase   Amphibole

Figure 1.03f
Simple Examples of Monoclinic System

f. Triclinic System

The triclinic system includes those crystalline forms having three axes of unequal lengths all intersecting at angles more or less than 90°.

Perfect crystals of minerals are comparatively rare. When the molten magma, which is a solution of several minerals, cools slowly,

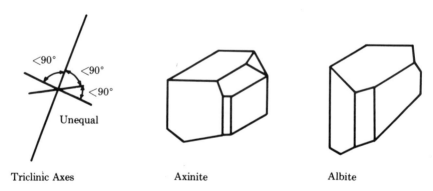

Triclinic Axes   Axinite   Albite

Figure 1.03g
Simple Examples of Triclinic System

some minerals begin to form crystals at a higher temperature than others. Crystals of these early forming minerals are formed while others are still in solution and are free to form perfect crystals unhampered by already formed crystals. The crystals that form later at lower temperatures must conform to the spaces left between those formed earlier. When a magma containing mica, feldspar, and quartz cools, the mica crystals form first and are regular in form. The feldspar crystals form next and are restricted slightly by the already formed mica crystals. The quartz crystals are formed in the spaces between the already formed mica and feldspar crystals making them irregular and imperfect.

As already mentioned, the size of the mineral crystals depends upon the rate of cooling, slow cooling producing large crystals.

1.04   ATOMIC AND MOLECULAR BONDS

To one beginning a study of the nature and structure of atoms and the bonds that unite them into aggregates, it appears that the structure of atoms and their arrangement in aggregates occur in infinite variety. This variety is what one would expect from observation of natural phenomena. It exists in the plant and animal kingdoms, in the distribution of minerals and rocks on the earth, in the planets and satellites of the solar systems, and in the arrangement of the solar systems in the universe. One is impressed with the similarity of the submicroscopic, the microscopic, the macroscopic, and the megascopic.

Scientists have begun to classify the forces that bind atoms and molecules into aggregates after study with X-rays, spectroscope and electron microscope. In reading recently published literature on the nature of chemical bonds, one finds considerable difference of opinion. Some concepts are not much beyond the hypothetical stage. But, in spite of considerable variation in terminology, there is a great deal of agreement in the classification of the principal or primary bonds. Possibly one of the simplest general classification systems for these bonding forces is: (a) Electrostatic or Primary Valence Bond, (b) Hydrogen Bond, and (c) Secondary Valence Bond. Some chemists do not recognize the secondary valence bond as a legitimate bond class. It is, however, the bonding force in which the student of Soil Mechanics is most interested. Because of the variation within each class, one cannot expect that all minerals can be placed in any one distinct class so far as atomic and molecular structure is concerned. Each mineral must be considered as an individual with its own peculiar arrangement of atoms and molecules.

a. Electrostatic or Primary Valence Bonds

This class or type of bonds is known by both these terms. Some authors use one and others use the other. Most specialists in the field of atomic chemistry and physics recognize two types of electrostatic bonds, (1) ionic bond (sometimes called electrovalent), and (2) covalent bond. Some scientists recognize a third type of bond, the heterpolar, which is a combination of the ionic and the covalent. Others apparently do not recognize the heterpolar as a separate bond type, but simply as a combination of the other two.

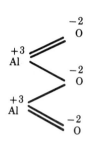

Figure 1.04a
Ionic Bond
Al to O

(1) Ionic Bond

The ionic bond is the simplest and strongest of the bonds that hold atoms together. This bond is made by the exchange of electrons in the union of ions.

In the combination of simple elements as shown in Figure 1.03a, an element having an excess of electrons in a ring or shell can exchange

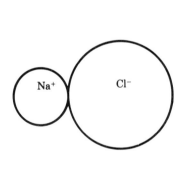

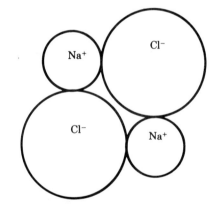

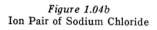

Figure 1.04b
Ion Pair of Sodium Chloride

Figure 1.04c
Two Ion Pairs of Sodium Chloride

the excess electrons with an element lacking electrons to complete its outer ring or shell. This exchange of electrons ties the atoms together with an exceptionally strong electrostatic bond.

An example of such a bond is illustrated by the union of Al and O. Aluminum has an excess of 3 electrons in its outer ring and oxygen lacks 2 electrons in its outer ring. The linking of the ions of these two elements can be indicated as shown in Figure 1.04a. Each of the joining lines in the figure can be considered as a unit electrostatic force bonding the aluminum and oxygen ions into a molecule of aluminum oxide. Another simple ionic bond exists between the sodium ion and the chlorine ion. These two ions in the molecule form an ion pair as shown in Figure 1.04b. Two of these ion pairs attract each other like bar magnets as shown in Figure 1.04c. The ion pairs tied together by these magnetic bonds form the cubic crystalline form of the mineral halite as shown in Figures 1.04d and 1.04e.

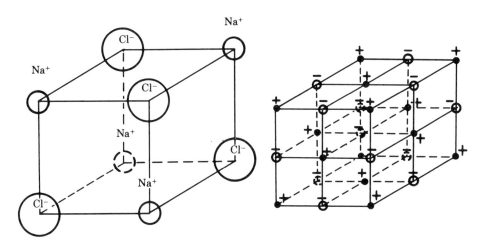

Figure 1.04d
Elemental Halite Crystal

Figure 1.04e
Arrangement of Ions in Halite Crystal

(2) Covalent Bond

Not all combinations of elements can be explained by the simple ionic bond theory. Some combinations seem to be bonded by a sharing of electrons by two atoms. The explanation of the idea that the same electron can be shared by two atoms and thereby create a bond between these two atoms is based upon the theory of resonance.

The electrons attached to the nucleus in an atom do not occupy the static positions shown in Figure 1.03a. Each electron revolves in an orbit about the nucleus and spins on its own axis. Some electrons spin in one direction and others spin in the opposite direction. Not only do the electrons revolve in orbits but they vibrate with a wave motion as they move along their paths. In these particles in motion there is bound a certain amount of energy. The electrons, although acting at a certain energy level, orbit in paths that may remain within their own shells or may penetrate other shells.

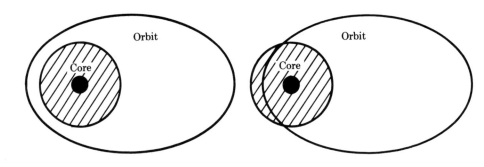

Figure 1.04f
Nonpenetrating Orbit

Figure 1.04g
Penetrating Orbit

When two atoms of like valence having different energies of motion come within a common sphere of influence, one or more pairs of oppositely spinning electrons in the atom of higher energy pass to the atom of lower energy and revolve in the place of a like number of electrons in the second atom. The energy of these electrons is thus transferred from the first to the second atom causing the second atom to possess the greater energy. Then the electrons are transferred back to the first atom, reversing the energy levels. Energy is thus transferred back and forth in a resonating system. This sharing of electrons by atoms of like valence produces a bond known as covalent bond.

Transfer of energy back and forth between atoms in the resonating systems ties the atoms together by a bond that is less strong than the ionic bond. This hypothetical concept explains a great many combinations of atoms which do not conform to the ionic relationship. The sharing of electrons occurs in pairs of oppositely spinning electrons. The covalent bond exists generally in non-electrolytes (elements that do not form ions, such as carbon) and between elements of negative valences.

b. Hydrogen Bond

Hydrogen, because it contains 1 electron in its shell and because only 2 electrons are needed to fill the first shell, may be considered as having 1 excess electron or as lacking 1 electron. It can take 1 electron from another element to form an anion or can give its 1 electron to another element to become a simple positive proton.

The hydrogen atom can enter into chemical combination with other elements in the following ways:

(a) by forming a cation, H^+
(b) by forming an anion, H^-
(c) by forming a normal covalent bond
(d) by forming a 1 electron bond
(e) by forming a hydrogen bond

The ionic bond formed by the cation and anion and the covalent bonds have already been discussed. The 1 electron bond is a rare bond that occurs only between like atoms in which the resonating structures are equal in energy. The molecule-ion $(H \cdot H)^+$ occurs only in discharge tubes.

The hydrogen atom under some conditions is attracted to two atoms instead of only one as suggested by its possession of only 1 electron. This bond between the hydrogen proton and two anions is called the hydrogen bond. Compared to the ionic bond, the hydrogen bond is weak, yet it is strong enough to significantly affect the properties of matter. The hydrogen bond is considered to be largely ionic in character. Its relative weakness causes it to play a part in reactions occurring at normal temperatures.

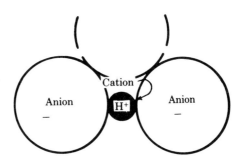

Figure 1.04h
Two Anions Bound by H Proton

Only two atoms can be bonded by the hydrogen ion. The positive hydrogen ion is a bare proton without its electron ring or shell. This very small hydrogen proton can attract two anions which, because each consists of its nucleus with orbital electrons, are much larger than the bare proton. As shown in Figure 1.04h, only 2 anions can approach the H cation close enough to form a stable compound.   A third anion is prevented from approaching the H cation by the 2 anions attached to the cation.

Only the most electronegative atoms are bound by hydrogen bonds. The most negative atoms form the strongest hydrogen bonds.   Fluorine, oxygen, nitrogen, and chlorine form hydrogen bonds of strengths in the same order as named.

The oxygen atoms in ice are tied together by hydrogen bonds.

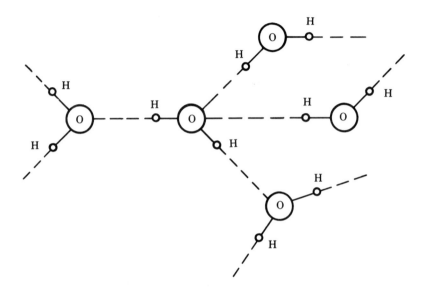

Figure 1.04i
Crystalline Structure of Ice

c. Secondary Valence Bonds, van der Waals Forces

The ionic, covalent, and hydrogen bonds do not account for all the attractive forces bonding atoms and molecules into aggregates, especially is this true of molecules.   The ionic bond accounts for the very strong bond between simple atoms in the formation of molecules as already described.   The less strong covalent bond also accounts for a great many other aggregates of atoms.

About 1873 van der Waals postulated the existence of a common attractive force acting between all atoms and molecules of matter. Later Keesom proposed that this force is caused by two oriented dipoles and calculated the average attractive force between such oriented dipoles.   A common example of such a bond is the attractive force between molecules of water. The water molecule is a permanent dipole similar to a bar magnet. When these dipoles are oriented as shown in Figure 1.04j, the molecules are held by the interaction of the magnetic fields. As the molecules vibrate and constantly change positions, they can assume positions in which there is either repulsion or attraction. But, because the repelling moment tends to throw the dipoles toward positions in which there is attraction, there occurs at any instant more attractive positions than repulsive making the resultant force attractive. The energy holding the molecules together in this manner is known as the orientation energy.

Debye suggested and wrote equations for the attraction between molecules that are permanent dipoles and the magnetic moments induced in adjacent molecules that are not permanent dipoles. This attractive force

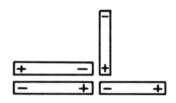

Figure 1.04j
Oriented Dipoles

is similar to that produced when a nonmagnetic bar of iron is brought into the field of a permanent magnet. A magnetic moment is induced in the normally nonmagnetic iron and an attractive force exists between the induced magnet and the permanent magnet. This type of attractive energy is known as induced energy.

In 1930 London conceived of another type of energy holding molecules together which is known as dispersion energy. This type of attractive energy exists between atoms and molecules that are not normally dipolar. As the electrons spin in their orbits, at one instant the negative electrons may be near the side of their orbits farthest away from the positive nucleus, which produces in effect a statistical dipole as

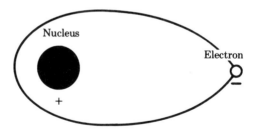

Figure 1.04k
Statistical Dipole

shown in Figure 1.04k. This statistical dipole can induce magnetic moments in adjacent molecules and thus produce the type of attraction known as the London-Van der Waals force.

One need only to refer to the literature in order to realize that the foregoing discussion of chemical bonds is oversimplified and incomplete. It is offered here as an introduction to the field of study which the student of Soil Mechanics must pursue if he is to thoroughly understand the stabilization of soils by chemical means or by electro-osmosis.

1.05 FUNDAMENTAL STRUCTURE AND COMPOSITION OF CLAY MINERALS

Most of the clay minerals have very weak basal cleavage allowing them to be broken down into extremely thin sheets. A few, like attapulgite, break into needle-like crystals and a few, like halloysite, form in thin sheets and roll into tiny open sided tubes. The common clay minerals are hydrous aluminum silicates. In some forms magnesium or iron cations are substituted for some of the aluminum.

These hydrous aluminum silicate minerals are composed generally of two fundamental building blocks, a tetrahedral silica unit and an octahedral hydrous aluminum oxide unit (gibbsite).

The fundamental silica unit is a tetrahedral unit with 4 oxygen ions enclosing a silicon ion in the configuration shown in Figure 1.05a. These tetrahedral fundamental units are bonded together with each of the oxygen ions in the base common to one other tetrahedral unit as shown in Figure 1.05b. In this arrangement there are both exchanged and shared electrons or a combination of ionic and covalent bonds (heteropolar). This strong horizontal bond provided by the sharing of oxygen ions by two molecules produces a sheet structure of silicon ions between two layers of oxygen ions. The oxygens of the bases of the tetrahedrons lie in a common plane. For each silicon

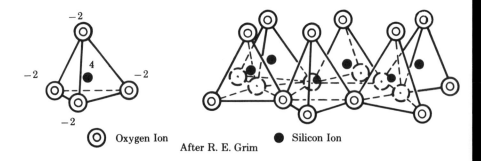

◎ Oxygen Ion   ● Silicon Ion

After R. E. Grim

Figure 1.05a
Fundamental Silica Molecule

Figure 1.05b
Combination of Tetrahedral Silica Units

ion with 4 positive charges, there are 4 oxygen ions with 2 negative charges each. Each of the 3 oxygen ions in the base divides its charges with another tetrahedral unit making a total of 5 negative charges to balance the 4 positive charges, thus leaving a net charge of −1 per unit.

The other fundamental building block for the hydrous aluminum silicate clay minerals is an octahedral unit of 6 hydroxyl ions enclosing an aluminum ion as shown in Figure 1.05c. These octahedral

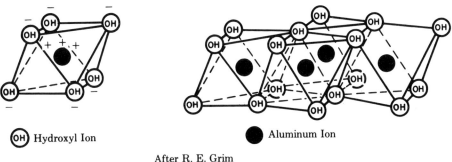

(OH) Hydroxyl Ion   ● Aluminum Ion

After R. E. Grim

Figure 1.05c
Fundamental Hydrous Aluminum Oxide Molecule

Figure 1.05d
Combination of Octahedral Aluminum Hydroxyl Units (Gibbsite)

units are bound together in a sheet structure with each hydroxyl ion common to 3 octahedral units. The Al ion has +3 charges. Each hydroxyl ion divides its −1 charge with 2 other units leaving for each Al ion −2 charges or a net charge of +1 for each octahedral unit. Under some conditions, iron or magnesium may substitute for the aluminum. This substituted magnesium or iron for aluminum is a constituent of the mineral crystal and is not exchangeable without dissolving the crystalline structure. Such a substitution by nature of one element for another in the formation of the crystal without changing the crystalline form is known as isomorphous substitution. For each substitution of a Mg ion with a charge of +2 for an Al ion with a charge of +3, the excess positive charge will not exist in that unit where the substitution is made. This elemental sheet consists essentially of a planar layer of Al ions between two planar layers of hydroxyl ions.

These molecular sheets of silica and gibbsite are bonded together in different combinations to form different clay minerals. The tetrahedral silica units bond to the octahedral gibbsite units with the

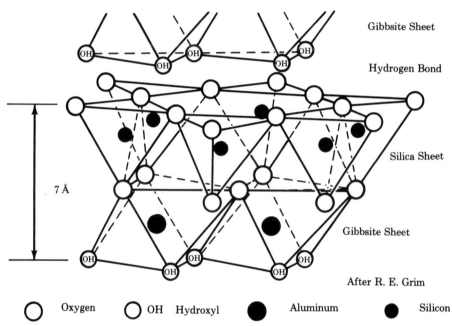

Gibbsite Sheet

Hydrogen Bond

Silica Sheet

7 Å

Gibbsite Sheet

After R. E. Grim

○ Oxygen   ○ OH Hydroxyl   ● Aluminum   ● Silicon

Figure 1.05e
Elemental Silica-Gibbsite Sheet

oxygen ions at the tips of the silica units intergrown with the hydroxyl ions of the gibbsite sheet with relatively strong ionic bond to form an elemental sheet 7 Å thick as shown in Figure 1.05e. Each tetrahedral silica unit carries a -1 charge and each octahedral gibbsite unit has a $+1$ charge. These opposite charges provide a strong ionic bond between the silica and gibbsite sheets. But, because there are more silica units than there are aluminum hydroxyl units in the common silica-gibbsite sheet, there is a surplus of negative charges on the combination sheet. On one side of this elemental silica-gibbsite sheet the shared O ions in the base of the silica sheet are exposed. On the other side of the sheet the hydroxyl ions of the gibbsite sheet are exposed. In kaolinite these elemental silica-gibbsite sheets are bound together with the base of the silica side next to the hydroxyl layer of the gibbsite side with hydrogen bonds as indicated in Figure 1.05e.

If in the formation of the minerals, there is an excess of silica, another silica sheet is bonded to the opposite side of the gibbsite sheet making an elemental sheet consisting of a gibbsite sheet between two

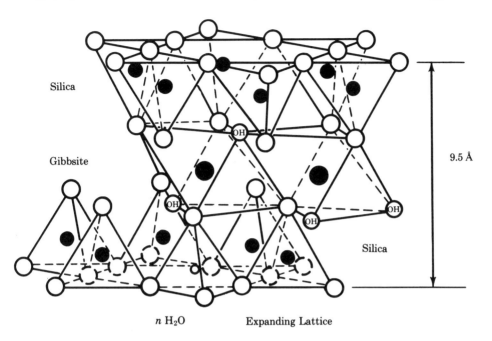

Silica

Gibbsite

Silica

9.5 Å

n H_2O Expanding Lattice

Exchangeable Cations

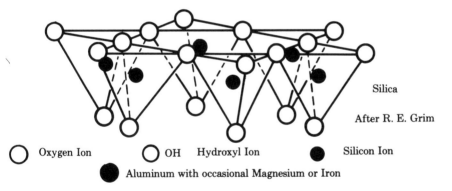

Silica

After R. E. Grim

○ Oxygen Ion   ◯ OH   Hydroxyl Ion   ● Silicon Ion

⬤ Aluminum with occasional Magnesium or Iron

Figure 1.05f
Elemental Structure of Montmorillonite

silica sheets bonded together as shown in Figure 1.05f. In this elemental sheet there are more negative charges because of the additional
silica units than in the elemental silica-gibbsite sheet. The isomorphic
substitution of magnesium for aluminum causes the sheet to carry additional negative charges. These elemental sheets of silica-gibbsite-
silica are about 9.5 Å thick. Both surfaces of the elemental sheets
are the bases of the tetrahedral silica units bearing negative charges

varying in strength with the amount and kind of isomorphic substitution. The negatively charged surfaces of the elemental sheets may be separated by dipolar water molecules oriented with their positive ends toward the negative surfaces of the elemental sheets. The distance between sheets depends upon the amount of water available for attachment to the surfaces. Such a structure in which the distance between sheets is variable depending upon the amount of water available is called an expanding lattice. This is the type of structure that exists in the mineral montmorillonite. These elemental sheets may also be tied together with cations of potassium, calcium, sodium, or other element between the negative surfaces. These cations are exchangeable with other cations without changing the elemental structure of the mineral. Illite consists of alternate elemental silica-gibbsite-silica sheets bonded with potassium cations.

1.06   COMMON SOIL MINERALS

Some of the common minerals which form soil particles are described below.

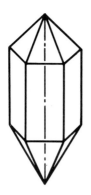

Figure 1.06a
Typical Quartz Crystal

Quartz (Silica) (SiO_2) is a clear transparent mineral formed from the slow cooling of liquid magma. When allowed to form without restraint it crystallizes in the hexagonal system. A common form is illustrated in Figure 1.06a. It is No. 7 in the rock hardness scale and will scratch glass. It is one of the principal minerals in granite and forms the light colored bands in gneiss. Its high resistance to chemical weathering enables it to be broken into small particles by mechanical weathering without change in composition. It is the principal mineral in sands, silts, and rock flour.

Flint, Chert, Agate, and Opal are forms of secondary silica that has been deposited from solution; often replacing other substances, such as limestone, wood, etc. Secondary quartz is one of the end products along with kaolin in the chemical weathering of feldspar. Once out of solution after the chemical reaction, this secondary silica possesses the same hardness and resistance to chemical weathering as primary quartz. The chemical composition of flint, chert, and agate is the same as that of primary quartz. Opal is a hydrous form of silica. Flint and chert are usually white or gray. Agate and opal contain coloring matter.

Feldspars are the most important of the rock forming minerals. All are silicates of aluminum combined with one or more bases of potassium, sodium, calcium, and barium. Feldspars are divided into two groups according to the angle between cleavage planes, (1) Orthoclase —breaking at right angles, (2) Plagioclase—breaking at oblique angles. The most important one of the orthoclase group is orthoclase ($K_2O \cdot Al_2O_3 \cdot 6SiO_2$) which is usually pink in color and occurs in quartz bearing rocks. It is No. 6 in the rock hardness scale and cannot be scratched with a knife blade. It crystallizes in the monoclinic system, often prismatic. Chemical weathering of orthoclase feldspar produces kaolin and secondary silica. It is one of the principal minerals in granite and other rocks containing quartz. Sands produced by rivers rising in the Rocky Mountains contain considerable orthoclase feldspar. The plagioclase feldspars occur in the nonquartz bearing rocks, such as diorite and gabbro. They are usually white or light colored. Sands of ferromagnesian minerals usually contain plagioclase feldspars. Albite ($Na_2O \cdot Al_2O_3 \cdot 6SiO_2$) is a common plagioclase feldspar. It weathers into Na_2CO_3 and secondary SiO_2. It also weathers into one of the kaolin groups of minerals.

Micas crystallize in the monoclinic system and have common perfect basal cleavage which enables them to be split into thin elastic sheets. Muscovite ($K_2O \cdot 3Al_2O_3 \cdot 6SiO_2 \cdot 2H_2O$) is the most common mica. It is colorless and transparent and is one of the primary minerals of granites and gneisses. Secondary muscovite is a principal mineral of schists. Sand size particles of mica are often found in sands. Mica in colloidal sizes occurs in clays.

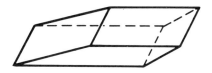

Figure 1.06b
Rhombohedral Block of Calcite

Calcite, which is calcium carbonate ($CaCO_3$), is derived from the chemical weathering of calcium bearing feldspar and other calcium bearing rocks and carried in solution as the bicarbonate of calcium. Evaporation produces crystalline calcite to form limestone. Calcite is No. 3 in the rock hardness scale and can be scratched with a knife blade. It crystallizes in the hexagonal system in a great variety of forms. The cleavage is perfect, yielding rhombohedral blocks. Its color is white to colorless. It is transparent to opaque. Calcite occurs in limited quantities in some igneous rocks. Limestone, chalk, travertine, and marble are almost wholly calcite. Calcite is used in nature as a cementing material in forming calcareous sandstones. A considerable amount of calcium carbonate is present in calichi and a small amount is present in most loess. Its presence can be detected by violent fizzing upon application of a small drop of hydrochloric acid.

Gypsum and Anhydrite are hydrous calcium sulphate ($CaSO_4 \cdot 2H_2O$) and ($CaSO_4$) respectively. They are formed by the action of sulphur fumes in the vicinity of volcanoes with calcium bearing minerals. Calcium sulphate is carried in solution to the sea. Most of the large deposits of gypsum and anhydrite have been deposited when inland seas have dried up.

Hematite is iron oxide (Fe_2O_3) and is dark red in color. It is a cementing material for forming sandstones. Small amounts are present in most red clays.

Limonite, which is hydrous iron oxide ($2Fe_2O_3 \cdot 3H_2O$), is yellow in color and is used by nature as a cementing material in some sandstones and other sedimentary rocks. It imparts a yellow color to sandstones and clays.

Gibbsite or Hydrargillite is aluminum hydrate ($Al_2O_3 \cdot 3H_2O$). It is translucent and white, grayish, greenish, or reddish in color. It crystallizes in the monoclinic system in tabular form. Basal cleavage is good, allowing it to split in thin sheets. Its importance in soil mechanics is due to its presence in the more complicated structure of kaolinite, montmorillonite, and illite, which are principal minerals of clay and give to clay some of its characteristic properties.

Kaolin and the related minerals are hydrous aluminum silicates differing only in the proportions of Al_2O_3, SiO_2, and H_2O with some isomorphic substitution of magnesium and iron. They are soft and break down by basal cleavage into thin plates of colloidal submicroscopic sizes. Such colloidal sizes of these minerals form the clays with which Soil Mechanics is concerned.

Kaolinite ($Al_2O_3 \cdot 2SiO_2 \cdot 2H_2O$) is derived from the chemical weathering of feldspar and other aluminum bearing rocks. It crystallizes in the monoclinic system as thin rhombic, rhombohedral, or hexagonal scales. The structure of the crystal consists of alternate layers of silica and gibbsite as shown in Figure 1.06c. The hydrogen bond between the base surface of the silica layer and the gibbsite layer is strong but less so than the very strong ionic and covalent bond between the apex side of the silica layer and the gibbsite layer. The relatively weak hydrogen bond allows the kaolinite crystals to be split into extremely thin platelets. The basic kaolinite crystal exists as a six sided plate with basal cleavage planes 100 to 200 Angstroms apart. The kaolinite particles occur in clay as platelets from 1000 Å to 20,000 Å wide by 100 Å to 1000 Å thick. Because of the relatively strong hydrogen bond between the elemental silica-gibbsite sheets, the

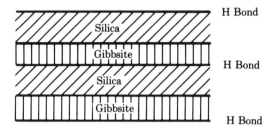

Figure 1.06c
Structure of Kaolinite

particles of kaolinite do not occur as single elemental sheets, but consist of clusters similar to that shown in Figure 1.06d. The platelets carry negative electromagnetic charges on their flat surfaces which attract thick layers of adsorbed water thereby producing plasticity when the kaolinite is mixed with water. China clay is almost pure kaolinite. Nacite and dickite are other forms of kaolin having the

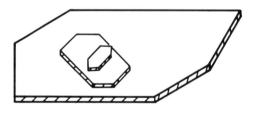

Figure 1.06d
Kaolinite Platelets

same chemical composition but having a different origin and physical properties.

Montmorillonite ($Al_2O_3 \cdot 5SiO_2 \cdot 5-7H_2O$) is hydrous aluminum silicate of the proportions shown and in addition usually contains CaO or MgO. Like kaolinite, the crystals are formed of silica and gibbsite layers. But in montmorillonite, each gibbsite layer lies between two silica layers leaving contact surfaces between the silica layers as shown

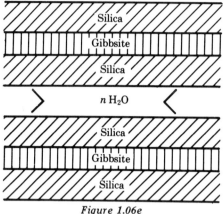

Figure 1.06e
Structure of Montmorillonite

in Figure 1.06e. The negatively charged surfaces of the silica layers attract water which occupies the space between silica layers in varying amounts. The spacing between the elemental silica-gibbsite-silica sheets depends upon the amount of water available to occupy the space. For this reason montmorillonite is said to have an expanding lattice as opposed to a fixed lattice of kaolinite, illite and the other hydrous aluminum silicates. Under some conditions, enough water may be present to separate the montmorillonite into ultimate platelets of elemental silica-gibbsite-silica sheets having a thickness of about 10 Å. Each thin platelet has the power to attract to each flat surface a layer of adsorbed water approximately 200 Å thick thus separating the platelets a distance of 400 Å, assuming zero pressure between the plates. As occurring in clay, montmorillonite platelets may have lateral dimensions of 1000 to 5000 Å and thicknesses from 10 to 50 Å.

Illite is hydrous aluminum silicate similar to montmorillonite, except that the adjacent silica layers are bonded with potassium ions instead of water. Potassium ions bear positive charges which bond the two

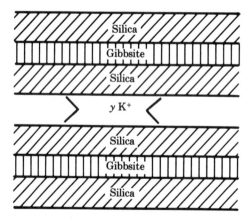

Figure 1.06f
Structure of Illite

negative surfaces of the silica layers. Because the cation bond of illite is weaker than the hydrogen bond of kaolinite, the tendency for illite to split into ultimate platelets consisting of a gibbsite layer between two silica layers is greater in illite than in kaolinite. Because of the stronger cation bond of illite than the water bond of montmorillonite, the tendency for forming ultimate platelets in illite is less than for montmorillonite. The lateral dimensions of illite clay particles are about the same as montmorillonite, 1000 to 5000 Å, but the thickness of illite particles is greater than that of montmorillonite particles, 50 to 500 Å.

Halloysite ($Al_2O_3 \cdot 2SiO_2 \cdot 2H_2O$ + water) is much like kaolinite but contains more water and is amorphous or submicroscopic. The submicroscopic crystals that occur in soils appear to be curled into tubes with open ends and one open side as shown in Figure 1.06g. Halloysite is often associated with other minerals of the kaolin group and is a mineral of minor importance in most clays.

Attapulgite crystallizes in needle-like shapes and is a mineral of minor importance in most clays.

Pyrophyllite ($Al_2O_3 \cdot 4SiO_2 \cdot H_2O$) crystallizes in the orthorhombic system and occurs as foliated, radiating, and laminar plates. The basal cleavage is perfect. Thin plates can be bent and remain so. Pyrophyllite is similar to montmorillonite, except that there is no isomorphous substitution in pyrophyllite. Pyrophyllite is, therefore, more nearly electrically neutral than montmorillonite and illite and less subject to volume change due to change in water content.

Bentonite is a hydrous aluminum silicate consisting mostly of montmorillonite with other minerals of the kaolin group. It contains a greater

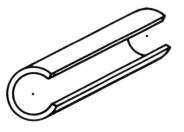

Figure 1.06g
Halloysite Crystal

percentage of silica and the oxides of magnesium, calcium, potassium, and sodium than other minerals of the kaolin group. Most bentonite is derived from the disintegration of volcanic ash or scoria. It usually occurs in seams and beds from a few inches to a few feet thick in clays. It has a soapy feel, is highly plastic, smooth and free from grit, and clings to the tongue. Bentonite clays are highly plastic and subject to large volume change with change in water content.

1.07 SYMBOLS

In order to express relationships efficiently it is necessary to use symbols with appropriate definitions. The symbols and their definitions used in these notes are the same as given in Manual No. 22 of the Am. Soc. C. E., "Soil Mechanics Nomenclature."

Symbol	Property	Dimension
Å	Angstrom	1×10^{-7} mm
A	Area	cm^2
a_v	Coefficient of Compressibility, de/dp	cm^2 g^{-1}
b	Breadth or Width	cm
C	Resultant or Total Cohesion	g
C_c	Compression Index	Dimensionless
C_s	Swelling Index	Dimensionless
C_u	Uniformity Coefficient	Dimensionless
c	Unit Cohesion	g cm^{-2}
c_v	Coefficient of Consolidation	cm^2 sec^{-1}
D	Diameter	cm
$D_d(D_R)$	Degree of Density (Relative Density)	Dimensionless
e	Void Ratio	Dimensionless
F	Total Force	g
f	Force per Unit of Area	g cm^{-2}
G	Specific Gravity	Dimensionless
g	Acceleration of Gravity	cm sec^{-2}
H	Height, Depth, or Thickness	cm
h	Head	cm
I	Index (Subscript indicates particular)	
i	Hydraulic Gradient	Dimensionless
J	Resultant Seepage Force	g
j	Unit Seepage Force, $j = i\gamma_w$	g cm^{-3}
k	Coefficient of Permeability	cm sec^{-1}
L	Distance or Length	cm
n	Porosity	Dimensionless
P	Total Force or Load, Pressure	g
p	Unit Pressure	g cm^{-2}
Q	Total Quantity	cm^3
q	Quantity per Unit of Time	cm^3 sec^{-1}
R	Reading	
r	Radius	cm
S	Degree of Saturation	Dimensionless
s	Unit Shear	g cm^{-2}
T	(a) Temperature	Degrees
	(b) Time Factor	Dimensionless
T_s	Surface Tension	g cm^{-1}
t	Time	sec
$U\%$	Average Degree of Consolidation	Dimensionless
u	Hydrostatic Excess Pressure	g cm^{-2}
V	Volume	cm^3
v	Velocity	cm sec^{-1}
W	Weight	g
w	Water Content	Dimensionless
α, β, θ	Angle	
γ	Unit Weight	g cm^{-3}
Δ	Increment	
ξ	Base of Natural Logarithm	
ϵ	Strain (Unit)	Dimensionless
η	Coefficient of Absolute Viscosity	g sec cm^{-2}
ν	Coefficient of Kinematic Viscosity	cm sec
σ	Normal Unit Stress	g cm^{-2}
τ	Shearing Unit Stress	g cm^{-2}
ϕ	Angle of Internal Friction	
ρ	Mass Density, γ/g	g sec^2 cm^{-4}

Note: Except when a symbol designates a specific unit, given dimensions are to designate their character and are not necessarily in the most convenient or common units.   g indicates weight, cm indicates distance, sec indicates time.   Special symbols using subscripts; such as w_L meaning water content at liquid limit, I_p plastic index, etc.; will be defined when first used in the text.

1.08   DEFINITIONS

In general, definitions will be presented as required.   However, in order to establish basic relationships, it is necessary to become familiar with a few definitions in the beginning.   Some of these common definitions are:

Void Ratio e = ratio of volume of voids to volume of solids in a given mass of soil. V_v/V_s.

Porosity n = ratio of volume of voids to the total volume of a given mass of soil. V_v/V.   (Sometimes expressed as percentage).

Water Content w = ratio of the weight of the water to the weight of the solids in a given mass of soil. W_w/W_s.   (Usually expressed as percentage).

Degree of Saturation S = ratio of the volume of water in a given mass of soil to the volume of voids. V_w/V_v.   (Usually expressed as percentage).

Specific Gravity G = ratio of unit weight in air of a material to the unit weight in air of a reference material at a stated temperature. The most common reference material for determining specific gravity is distilled water at 4 degrees centigrade, γ_0.   Hydrometers are commonly calibrated to read specific gravity referred to distilled water at the calibration temperature.

1.09   RELATIONSHIPS

In order to be able to determine the physical properties of soils from laboratory tests, it is necessary to know, or to be able to work out, the relationships which exist between the different properties.   In working out these relationships, it is convenient to represent the solids and voids in the given volume by means of a sketch.   If the area be

Volume   Weight

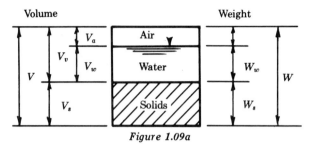

Figure 1.09a

assumed as unity, or other constant, the height will represent the volume. For convenience, the volumes may be shown on the left hand side of the sketch and the weights on the right hand side as shown in Figure 1.09a

$$e = \frac{V_v}{V_s} \qquad n = \frac{V_v}{V}$$

$$w = \frac{W_w}{W_s} \qquad S = \frac{W_w}{V_v \gamma_w}$$

$$W_s = V_s \cdot G_s \cdot \gamma_0$$

If the relationship between water content, void ratio, and specific gravity of solids is desired, the volume of the solids may be represented as 1, in which case the volume of voids is equal to the void ratio, e. If the voids are completely filled with water, the weight of the water is equal to the volume of water, e, times the unit weight of water, γ_w; and the weight of the solids is equal to the specific gravity of the solids, G_s, times the unit weight of water at reference temperature, γ_0, times the volume of solids, 1.

Volume   Weight

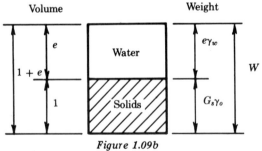

Figure 1.09b

These relationships are illustrated in Figure 1.09b.

$$w = \frac{e\gamma_w}{G_s \gamma_0} = \frac{e}{G_s} \text{ or } e = wG_s$$

γ_0 = unit weight of water at reference temperature, 4° C.
γ_w = unit weight of water at given temperature.
For most purposes the ratio γ_w/γ_0 may be considered as 1.

In relationships involving the use of porosity, n, a sketch may be used in which the total volume equals 1, in which case, by definition, the volume of voids is equal to the porosity, n, and the volume of solids is $1 - n$. From these sketches, Figure 1.09b and Figure 1.09c,

Volume   Weight

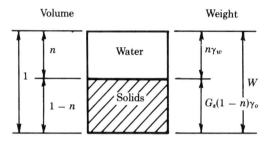

Figure 1.09c

one can readily arrive at the relationships between void ratio and porosity because the total volumes are proportional to the volumes of the voids and to the volumes of the solids in different volumes of the same soil, or they can be determined directly from the definitions.

$$\frac{1}{1 + e} = \frac{n}{e} = \frac{1 - n}{1}$$

$$n = \frac{e}{1 + e} \qquad e = \frac{n}{1 - n}$$

$$w = \frac{n\gamma_w}{G_s(1 - n)\gamma_0} = \frac{n}{G_s(1 - n)}$$

The unit weight of a given soil, dry, saturated, or submerged, can be determined from the void ratio and the specific gravity of the solids in the following manner:

As seen from Figure 1.09d, the dry weight of a volume of soil, $1 + e$, is equal to the weight of the solids, $G_s \cdot \gamma_0$. The unit weight of the dry soil is then equal to $\frac{G_s\gamma_0}{1 + e}$. If the soil is completely saturated, the

Volume   Weight

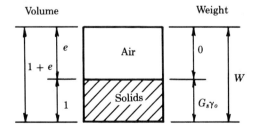

Figure 1.09d

total weight is $e\gamma_w + G_s\gamma_0$, and the unit weight is $\dfrac{e\gamma_w + G_s\gamma_0}{1 + e}$. If the soil is submerged, the unit weight of the submerged soil is equal to the unit weight of the saturated soil reduced by the unit weight of water; i.e., $\gamma_{\text{sub}} = \gamma_{\text{sat}} - \gamma_w$.

Similar relationships can be determined for partially saturated soils by imagining the solids, water, and gas separated as shown in Figure 1.09e. Water content can be determined by weighing a sample of the

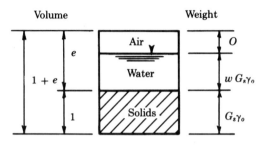

Figure 1.09e

soil with its entrapped water, then drying the soil in a drying oven until all the free water has been driven off, and then weighing the dry soil. The water content is equal to the weight of the water (difference in wet and dry weights) divided by the weight of the solids (dry weight). If the value of e is known, the degree of saturation S can be determined from the water content as shown in Figure 1.09e.

$$\text{Wt of water} = wG_s\gamma_0$$

If the voids were completely filled with water, the weight of the water would be $e\gamma_w$. From definition,

$$S = \frac{wG_s\gamma_0}{e\gamma_w} = \frac{wG_s}{e}.$$

It is sometimes desirable to express the ratio of the volume occupied by gas to the total volume of the soil, which is similar in definition to porosity, n. This ratio is usually referred to as air porosity and is designated n_a.

$$n_a = \frac{V_a}{V} = \frac{V_v - V_w}{V}$$

$$n_a = \frac{n - wG_s(1 - n)}{1}$$

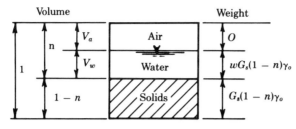

Figure 1.09f

Needed relationships should be worked out directly from basic definitions without the use of previously developed general formulas. The foregoing examples are illustrations of the use of phase diagrams as aids to an understanding and statement of specific relationships.

The following example illustrates the use of the phase diagram and basic definitions to the solution of a specific problem.

Example: In a compaction test the following data were obtained:

$$\begin{aligned}
&\text{Volume of compaction cylinder} \dots \dots & 970 \text{ cm}^3 \\
&\text{Weight of compacted soil} \dots \dots \dots & 1986 \text{ g} \\
&\text{Water content } w \dots \dots \dots \dots \dots & 8.6\% \\
&\text{Specific gravity of solids} \dots \dots \dots & 2.71
\end{aligned}$$

a. Determine the dry weight of the soil per unit of volume.
b. Determine the porosity and the void ratio of the soil as compacted.
c. Determine the degree of saturation.

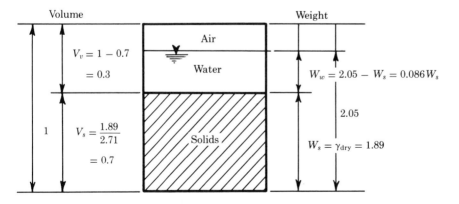

Figure 1.09g

Solution:

a. Unit weight, wet, $\gamma_{wet} = \frac{1986}{970} = 2.05$ g cm^{-3}

By definition, water content, $w = \frac{W_w}{W_s} = 0.086$

Therefore, $W_w = 0.086 W_s$

W_w is also equal to $W - W_s = 2.05 - W_s$

Then, $2.05 - W_s = 0.086 W_s$

From which $W_s = 1.89$ g and since the total volume is unity,

$W_s = \gamma_{dry} = 1.89$ g cm$^{-3} = 1.89 \times 62.4 = 118$ lb ft^{-3}.

A simple method of determining the unit weight of any substance in any other units than the known unit weight is by proportion with the known unit weights in the two units of a reference material. Water weighs 1 g cm^{-3} and 62.4 lb ft^{-3}, therefore n g cm$^{-3} = 62.4n$ lb ft^{-3}.

b. The porosity and void ratio can be determined directly from the diagram.

Volume of solids $V_s = \frac{W_s}{G_s \gamma_0} = \frac{1.89}{2.71} = 0.70$ cm^3.

Volume of voids $V_v = 1 - 0.70 = 0.30$ cm^3.

Porosity $n = \frac{0.3}{1} = 0.3$.

Void ratio $e = \frac{0.3}{0.7} = 0.43$.

c. Volume of water $V_w = \frac{2.05 - 1.89}{1} = 0.16$ cm^3.

Degree of saturation $S = \frac{V_w}{V_v} = \frac{0.16}{0.30} = 0.53$ or 53%.

Similar relationships can be readily worked out with the help of appropriate sketches.

1.10 CLASSIFICATION OF SOILS

In general, soils are classified as cohesionless and cohesive. By cohesion is meant shearing resistance inherent in the material which does not have to be developed by normal pressure or other outside influence.

Cohesionless soils possess no shearing resistance, except as developed by normal pressure between the grains. Sand, a cohesionless soil, has no shearing resistance on the surface, but at some depth beneath the surface where it is subjected to the pressure of the overburden, the frictional resistance to movement of the grains relative to each other is great. Such confined sands will support heavy loads comparable to those that can be carried by solid rock.

A fine grained cohesionless soil, such as fine sand or silt, will show an apparent cohesion when moist, due to capillary forces. Such a moist soil will stand as a short column, or can be cut into vertical or even overhanging embankments, but when dried out or submerged the shearing resistance is eliminated and the column or embankment collapses. Such shearing resistance is due to frictional resistance produced by pressure between the grains and is not true cohesion.

A fat plastic clay possesses shearing resistance practically independent of normal pressure and is classed as a cohesive soil. Most soils are mixtures of cohesionless and cohesive materials and fall between the two extremes represented by cohesionless clean sands and cohesive fat plastic clays.

Classification of soils in detail is discussed in the chapter on "Classification." Cohesion is discussed in detail in the chapters of "Plasticity" and "Shearing Resistance."

REFERENCES

Brown, G. I., *Modern Valence Theory*, London, New York: Longmans, Green and Company, 1953.

Dapples, Edward C., *Basic Geology*, New York: John Wiley and Sons, Inc., 1959.

Dennen, William H., *Principles of Mineralogy*, New York: The Ronald Press Company, 1959.

Conference, *Clays and Clay Minerals*, Proc. 9th National Conference on Clays and Clay Minerals, Vol. 9, Lafayette, Indiana, Purdue University, 1960.

George, Russell D., *Minerals and Rocks*, New York: D. Appleton-Century Company, Inc., 1943.

Grim, Ralph E., "Physico-Chemical Properties of Soils," *Journal Soil Mechanics and Foundations Division*, Proc. Am. Soc. C.E., Vol. 85, No. SM2 (April, 1959).

Grim, Ralph E., *Applied Clay Mineralogy*, New York: McGraw-Hill Book Company, Inc., 1962.

Hough, B. K., "Appendix I," *Basic Soils Engineering*, New York: The Ronald Press Company, 1957.

Ketelaar, J. A., *Chemical Constitution*, Amsterdam, New York: Elsevier Publishing Company, 1958.

Lambe, T. William, "The Structure of Compacted Clay," *Journal Soil Mechanics and Foundations Division*, Proc. Am. Soc. C.E., Vol. 84, No. SM2, (May, 1958).

Pauling, Linus, *The Nature of the Chemical Bond*, Ithaca, New York: Cornell University Press, 1960.

PROBLEMS

1.1 Using the customary diagrammatic sketch of a three phase soil system define the following terms:
(a) void ratio, e
(b) porosity, n
(c) water content, w
(d) degree of saturation, S

1.2 Assuming the soil to be saturated derive an expression for the quantity in column I in terms of those in column II. The unit weight of water, γ_w or γ_0, or the specific gravity of water G_w, should be used also, when required in the derivation. Make use of appropriate sketches.

	I	II
(a)	w	$G_s,\ e$
(b)	w	$G_m,\ e$
(c)	e	$w,\ G_s$
(d)	e	$G_s,\ G_m$
(e)	e	$G_s,\ \gamma_{dry}$
(f)	G_m	$G_s,\ n$
(g)	G_m	$G_s,\ e$
(h)	n	$G_s,\ w$
(i)	n	$G_s,\ G_m$
(j)	G_s	$w,\ n$
(k)	G_s	$w,\ G_m$
(l)	γ_{dry}	$w,\ G_s$
(m)	γ_{sat}	$w,\ G_s$
(n)	γ_{sub}	$w,\ G_s$

1.3 A saturated specimen of undisturbed inorganic clay has a volume of 17.4 cc and weighs 29.8 g. After drying at 105°C, the volume is 10.5 cc and weight 19.6 g. For the soil in its *natural state*, compute
(a) the water content
(b) the specific gravity of solids
(c) the void ratio
(d) the saturated unit weight.

1.4 The moisture content of a specimen of saturated, undisturbed silty clay is found to be 60 per cent. Make a reasonable estimate of the natural void ratio of this material. Do not use any formula without first deriving it.

1.5 A cube of dried clay having sides 3 cm long weighs 46 g. The same cube of soil, when saturated *at unchanged volume*, weighs 56.5 g. Determine the average specific gravity G_s of the mineral particles.

1.6 In terms of the specific gravity of the solids, the degree of saturation, and the void ratio, derive an expression for the unit weight γ_m, of a soil. Assume $\gamma_w = \gamma_0$.

1.7 Derive an expression for the degree of saturation S of a soil in terms of w, G_s, G_w, γ_{dry}, and γ_0.

1.8 In terms of G_s, S, w, G_w, and γ_w, derive an expression for γ_m, the unit weight of the material.

1.9 A sample of moist soil has a volume of 40.5 cc and weighs 59.2 g.   After drying, it weighs 48.3 g.   The specific gravity of the solid matter is 2.70. Compute the degree of saturation, the water content, and the porosity, all in per cent, and the void ratio.

1.10 A 50 cc specimen of moist soil weighs 95 g.   Its dry weight is 75 g and the specific gravity of the solids is 2.68.   Compute the water content, void ratio, porosity, specific mass gravity, and degree of saturation of the soil.

1.11 A soil specimen having a volume of 80 cc contains mineral grains having an average specific gravity of 2.70.   The specimen has a natural weight of 140 g and an oven dried weight of 108 g.   For the natural soil determine: (a) w, (b) e, (c) n, and (d) S.

1.12 The volume of a moist sample of soil is 85 cc, and its weight is 140 g. After oven drying the specimen weighs 115 g.   Assume a value for the specific gravity of solids of 2.75, and determine the water content, the degree of saturation, and the porosity of the soil.

1.13 The unit weight of a certain quartz sand is found to be 100 lb/cu ft when the water content is 10 per cent.   Determine its void ratio and degree of saturation.   (Estimate $G_s = 2.63$).

1.14 A small cylinder, weighing 270 g and having a volume of 300 cc, is pressed into an embankment of moist, fine sand, filling the cylinder. The filled cylinder weighs 800 g.   The weight of the soil, after drying, is 500 g.   Determine the void ratio and degree of saturation of the sand specimen.

1.15 A field test of a fine sand deposit shows that the sand has a natural (moist) unit weight of 110 lb/cu ft at a moisture content of 10 per cent. Determine the void ratio and degree of saturation of this sand.

1.16 A 60 cc specimen of silty clay weighs 74.4 g when wet (at natural water content) and 54.4 g dried.   Determine its natural water content, and estimate the void ratio and the degree of saturation of the specimen.

1.17 Water was thoroughly mixed with pulverized loessial soil to give an average water content of 15 per cent.   A quantity of this soil sufficient to fill the mold was then uniformly compacted into a 10 lb steel mold having a volume of 0.1 cu ft.   The weight of the mold and soil was found to be 21.5 lb.   Determine the approximate degree of saturation of the soil.

1.18 The water table in a certain area is at a depth of 10 feet below the ground surface.   To a depth of 37 feet the soil consists of very fine sand having an average void ratio of 0.8. Above the water table the sand has an average degree of saturation of 50 per cent.

Determine (*a*) the average unit weight (lb/cu ft) of the soil above the water table,

  (*b*) the saturated unit weight of the soil below the water table,

  (*c*) the submerged unit weight of the soil below the water table, and

  (*d*) the *effective* pressure (lb/sq ft) on a horizontal plane at a depth of 30 feet below the surface.

1.19 Determine the overburden pressure acting on a horizontal plane at a depth of 14 feet beneath the surface of a deposit of sand for which the average natural void ratio is 0.62. The water table is at a depth of 6 feet, and the average degree of saturation of the sand above the water table is 46 per cent.

1.20 For the given soil profile compute the effective pressure at a depth of 30 feet.

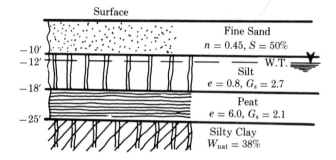

1.21 For the subsurface profile shown, determine the effective stress (in units of lb/sq ft) along the surface of contact between the clay and the bed rock.

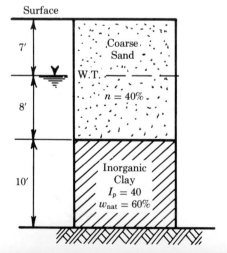

CHAPTER II

Specific Gravity of Solids

2.01 GENERAL

Specific gravity is the ratio of the unit weight in air of a given material to the unit weight in air of a reference material. Unless otherwise stated, specific gravity is generally referred to distilled water at 4° C.

The specific gravity of the solids making up soils is useful mainly for deriving other needed properties of soils. For practical purposes it is the average or overall specific gravity of the mixture of different mineral grains, rather than the specific gravity of individual grains, that is needed in Soil Mechanics.

The values of the specific gravity of some of the common minerals of soils are about as follows:

		Illite	2.60
Quartz	2.60	Kaolinite	2.60–2.63
Feldspars		Bentonite	2.13–2.18
Orthoclase	2.57	Augite	3.20–3.60
Plagioclase	2.62–2.76	Hornblende	3.00–3.47
Micas	2.70–3.10	Gypsum	2.30
Calcite	2.80–2.90	Gibbsite	2.30–2.40
Magnetite	5.17–5.18	Talc	2.70–2.80
Hematite	4.90–5.30	Anhydrite	3.00
Siderite	3.83–3.88	Olivene	3.27–3.37

Because the range of these values for individual minerals is quite small, the overall specific gravity of soil solids consisting of these minerals varies over about the same small range, from about 2.60 to 2.80. Clean light colored sand of quartz and feldspar has a specific gravity of about 2.64; dark colored sand of quartz, feldspar, hornblende, augite, etc. has a somewhat higher value; sandy or silty clays about 2.72; fat clay of mica and kaolinite and other clay minerals about 2.75 to 2.80.

With a little experience one can estimate the specific gravity of soil solids from a visual inspection with sufficient accuracy for most practical purposes of Soil Mechanics.

Generally, the only use made of the specific gravity of the soil solids is for determining other needed properties.

2.02   DETERMINATION OF SPECIFIC GRAVITY

The specific gravity of soil solids can easily be determined with a calibrated flask or pycnometer bottle from the weight of the flask

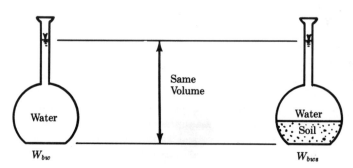

Figure 2.02a
Determination of Specific Gravity with Flask

filled with deaired distilled water, W_{bw}; the dry weight of a sample of soil solids, W_s; and the weight of the flask filled with the sample and deaired distilled water, W_{bws}.

The relationship for determining the specific gravity can be derived as follows:

Assume a bottle of volume V_b filled with deaired distilled water which weighs W_{bw}. When a sample of soil of volume V_s and weight W_s is placed in the bottle filled with water, it displaces a volume of water equal to the volume of solids V_s and weight $V_s \gamma_w$. From the sketches in Figure 2.02b, one can see the following relationships:

$$W_{bws} = W_{bw} + W_s - V_s \gamma_w.$$

By definition,
$$V_s = \frac{W_s}{G_s \gamma_0}$$

$$W_{bws} = W_{bw} + W_s - \frac{W_s \gamma_w}{G_s \gamma_0}$$

$$G_s = \frac{W_s}{W_{bw} + W_s - W_{bws}} \cdot \frac{\gamma_w}{\gamma_0}.$$

Because the unit weight of water at room temperature is very little less than at 4° C., the error in calling γ_w / γ_0 (the specific gravity of water) equal to unity is negligible for practical purposes.

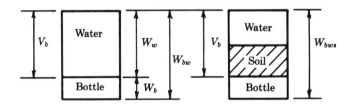

Figure 2.02b

In this derivation it is assumed that W_{bw} and W_{bws} are determined at the same temperature, because a change in temperature changes the volume of the bottle and the unit weight of the water. The exact volume of the bottle is not important but it should be the same for the two weighings for any one determination. It is convenient, there-fore, to use water that has been stored in the room long enough to assume the room temperature in order to avoid temperature changes during the determination. Ordinary tap water can be used instead of distilled water without appreciable error.

In order to determine the specific gravity of the soil particles, it is necessary that the grains be thoroughly dispersed. Cohesionless materials may be oven dried and W_s determined before the test and without special dispersion. Cohesive materials must be dispersed before placing in the bottle or flask. Dispersion can be achieved for cohesive materials by mixing the soil sample with water in an evaporating dish with a spatula to form a thin batter or slurry and then mixing in a mechanical mixer until the soil mass is broken down into individual particles. The value of W_s for cohesive materials is found by evaporating the water and oven drying the sample after the value of W_{bws} has been determined.

Any air dissolved in the water and trapped in the soil in the bottle or flask must be removed for accurate determination of the specific gravity. The air can be removed by applying a vacuum of about 65 cm mercury for about 15 minutes during which time the flask should be gently rolled to release the air bubbles from the soil. Until the air has been removed, the flask should be filled only to a point slightly below the neck to prevent boiling over during application of the vacuum. The air can also be removed by boiling, but the temperature must be brought to the same value for both weighings.

2.03 ACCURACY OF MEASUREMENTS

Generally, it is desirable to know the specific gravity of soil solids to three significant figures; i.e., two figures to the right of the decimal. Because the denominator of the fraction expressing G_s will be in the range from 20 to 50, an error of 0.1 g in weight of W_s will cause an error of 0.005 to 0.002 in the value of G_s. If the values of W_s, W_{bw}, and W_{bws} are determined within 0.1 g, the determined value of G_s should be accurate to three significant figures.

The unit weight of water changes approximately 0.0005 g for each degree C change in temperature. This would cause a change in weight of 500 cc of water about 0.25 g per degree C change in temperature and a change in the volume of the flask of about

$$500 \times 1° \times 0.1 \times 10^{-4}$$

or 0.005 cc. For an accuracy in the value of W_{bw} and W_{bws} of not more than 0.1 g the difference in temperature of the water for the two weighings should be less than 0.5° C. An accuracy of 0.2° C is sufficient.

2.04 TEST PROCEDURE

Detailed instructions for determining specific gravity of the solids in a sample of soil are given in Chapter XII, Laboratory Testing, Section 2.04L.

2.05 CALIBRATION OF FLASK

One of the weighings in the determination of specific gravity, W_{bw}, can be eliminated by preparing a calibration curve for the flask showing the relationship between W_{bw} and temperature. Using such a calibration curve and measuring the temperature of the water in the flask with the sample, the value of W_{bw} can be read from the curve. This procedure eliminates the necessity for determining W_{bw} for a particular temperature during the performance of the test.

Each flask must have its own calibration curve.

The relationship between W_{bw} and temperature can be determined as follows:

Change in temperature affects both the volume of the flask and the unit weight of the water

$$
\begin{aligned}
\text{Let}\quad V_b &= \text{Vol of flask at } T° \text{ C}\\
\gamma_T &= \text{Unit weight of water at } T° \text{ C}\\
\Delta V_b &= \text{Change in Volume from } T_1 \text{ to } T\\
T_1 &= \text{Temperature corresponding to } V_{b_1}\\
\alpha_v &= \text{Coefficient of volume expansion of flask.}
\end{aligned}
$$

Then
$$\Delta V_b = \alpha_v(T - T_1)V_{b_1}$$

and $\Delta W_{bw} = (\gamma_T - \gamma_{T_1})V_{b_1} + \alpha_v(T - T_1)V_{b_1}\gamma_T = W_{bw} - W_{bw_1}.$

Over a 10° C calibration temperature range, which includes all ordinary laboratory determinations, the change in V_b will be in the order of $500 \times 10 \times 0.1 \times 10^{-4} = 0.05$ cc. Calling V_{b_1} the calibration volume, 500 cc for a 500 cc flask, instead of the actual value of V_{b_1} at T_1 and calling $V_{b_1}\gamma_T = V_b$ instead of its actual value produces a maximum error in W_{bw} for 10° C change in temperature of

$$(.99568 - .99823)0.05 + (0.1 \times 10^{-4} \times 10)0.05 = 0.0001325 \text{ g.}$$

The error introduced by these approximations is much less than the error due to weighings made to the nearest 0.1 g so that they may be used with no loss in accuracy of determinations to three significant figures.

CHAPTER III

Grain Size Distribution

3.01 GENERAL

Early in the development of the science of Soil Mechanics, it was believed by many that most of the physical properties of soils could be related to the grain size distribution. It is now known that most of the physical properties of soils are relatively little affected by the grain size. Permeability, and therefore the rate of deformation of saturated soils, is dependent upon the grain size, but other factors are not so directly influenced. Because of this early belief, considerably more attention has been given to mechanical analysis and grain size distribution than its importance justifies. Only for coarse grained soils is the grain size distribution indicative of physical properties. Fine grained soils are so much affected by structure, shape of grain, geologic origin, and other factors that their grain size distribution alone tells little about their physical properties.

However, regardless of its importance, it is necessary that the engineer be familiar with the methods of mechanical analysis in order to interpret the results of a grain size analysis intelligently.

3.02 MECHANICAL ANALYSIS

By mechanical analysis is meant the separation of a soil into its different size fractions, usually expressed as percentage of the whole by weight of particles smaller than a given diameter.

a. Sieve Analysis

Fortunately, grain size distributions for coarse grained soils, which are the only ones of significance, can be made with a simple sieve analysis.   Grains larger than 0.074 mm diameter, which is the size of the No. 200 sieve, are easily separated into any desired fractions by shaking or washing a sample of soil through a nest of sieves of the proper sizes.   Cohesionless material may be oven dried and shaken through the sieves dry, making sure that all lumps are broken into individual grains.   Cohesive soils must be dispersed before passing through the sieves.   Dispersion of cohesive soils can be accomplished by mixing with water in an evaporating dish with a spatula to form a homogeneous batter or slurry.   The batter should be rinsed into the cup of a mechanical mixer with about 250 cc of water and mixed for about 15 minutes.   This suspension should be poured and rinsed into the top sieve of the nest of sieves and washed through until the different sizes have been retained on the different sieves.   The fraction retained on each sieve is dried out and weighed, after which the per cent by weight passing each sieve can be computed.

The Tyler standard screen scale is arranged so that the area of each opening in any size is double that of the next finer sieve and half that of the next coarser sieve.   The diameter just passing one sieve is then 2 times the diameter just passing the alternate sieve.   By omitting sieves, the ratio of one size to the next can be made 2, 3, or 4 to 1; or fractions of these size ratios can be taken directly from an analysis without calculation.

The sieve sizes are given in terms of the number of openings per inch.   The number of openings per square inch is equal to the square of the sieve size.   The standard sizes with the diameter of a sphere that will just pass through an opening are given in the following table.

Mesh	200	150	100	65	48	35	28	20	14	10	8
Opening in mm	0.074	0.104	0.147	0.208	0.295	0.407	0.589	0.833	1.168	1.651	2.362

The sieve method of mechanical analysis is superior to others because it separates the soil into fractions which will and will not pass a given opening independent of temperature or other conditions which prevail during the test. The sieve analysis is a true grain size determination. The wet analysis, as will be seen from the following discussion, is not a true grain size distribution.

b. Wet Mechanical Analysis

Grain sizes smaller than 0.074 mm in diameter must be separated into fractions by some method other than sieving. The methods most commonly used depend upon the sedimentation of soil grains from suspension in a liquid. Water is used as the suspending liquid for wet mechanical analyses of soils.

The velocity at which grains settle out of suspension, all other factors being equal, is dependent upon the shape, weight, and size of the grains. If it be assumed that all grains are perfect spheres and have the same specific gravity, then the velocity of sedimentation depends upon the diameter of the grains. The following relationship between velocity of fall through fluid, diameter of sphere, unit weight of solid sphere, and the unit weight and absolute viscosity of the suspending fluid was determined and expressed by the English physicist, G. G. Stokes, (1850) and is known as Stokes' Law.

$$v = \frac{2}{9}\left(\frac{\gamma_s - \gamma_f}{\eta}\right)\left(\frac{D}{2}\right)^2$$

In which

$v =$ velocity of fall through fluid in cm sec^{-1}
$\gamma_s =$ unit weight of solid sphere in g cm^{-3}
$\gamma_f =$ unit weight of suspending fluid in g cm^{-3}
$\eta =$ absolute viscosity of suspending fluid in g sec cm^{-2}
$D =$ diameter of sphere in cm

Solving for D the relationship becomes

$$D = \sqrt{\frac{18\eta v}{\gamma_s - \gamma_f}}.$$

It has been found that Stokes' Law is not valid for mineral grains larger than about 0.2 mm in diameter because of excessive turbulence accompanying their settlement. It is also not valid, of course, for grains smaller than 0.2 micron (0.0002 mm) which are affected by the Brownian movement. Because the sieve analysis covers a range of sizes down to 0.074 mm diameter, there is a considerable range between the lower limit of the sieve analysis and the upper limit of the wet mechanical analysis for which both methods are available. This overlap is of great convenience in determining the grain size distribution of soils for which both methods must be used after separating into fine and coarse fractions.

Because Stokes' Law applies only to a single sphere settling out of suspension at a velocity which produces laminar flow, grain sizes determined by the use of Stokes' Law are the diameters of perfect spheres that would settle at the same rate as the actual grains. A large percentage of the grains of fine grained soils are thin platelets which do not settle out of suspension in the same manner and at the same rate as smooth spheres.

A further discussion of Stokes' Law is included in the chapter on Hydraulics.

3.03 PIPETTE METHOD OF WET ANALYSIS

Points on the grain size distribution curve can be found by taking samples of a soil suspension with a pipette at intervals during continuous sedimentation. The suspension should be made from a properly dispersed sample of soil in a liquid that will prevent flocculation of colloidal particles. Silts having practically no colloidal particles may be suspended in plain water. Soils, like clay, containing colloidal particles may be suspended in water with the addition of a dispersing agent, such as sodium silicate.

The suspension should be placed in a deep vessel such as a 1000 cc glass cylinder. At the start of the test the top of the cylinder should be covered (the palm of the hand can be used) and the position of bottom and top reversed about every two or three seconds for about one minute to obtain a uniform suspension. As the cylinder is set upright, a stop watch is started for measuring the time of settlement. After a short interval, depending upon the largest grain size, a pipette is inserted into the suspension and a sample taken out. A record must be kept of the time since beginning of settlement, the distance from the surface to the point from which the sample was taken, and the volume of the sample taken in the pipette. One point on the grain size distribution curve can be determined from each sample taken in this manner.   A similar procedure makes use of one or more draw off points on the side of the cylinder.

Because all grains of the same size settle at the same rate, grains of a given size where they exist at all are in the same degree of concentration as at the beginning.   Therefore, all grains smaller than a given size will be present in the pipette sample in the same degree of concentration as at the beginning, and all grains larger than this size will have settled below the depth at which the sample was taken and are not present in the sample.

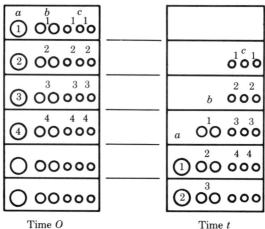

Time O   Time t

Figure 3.03a
Settlement of Different Sizes

The weight smaller than the given size in the entire suspension, W_D, is equal to the dry weight of the solids in the pipette sample, $W_s{}'$, times the ratio of the total volume of suspension, V, to the volume of the pipette sample, V'.

$$W_D = W_s{}' \cdot \frac{V}{V'}.$$

The percentage by weight smaller than the given size is

$$W_D\% = 100\,\frac{W_s{}'V}{W_s V'}$$

in which W_s is the dry weight of solids in the total suspension.

The diameter of the largest grain present in the pipette sample can be obtained from Stokes' Law, $D = \sqrt{\dfrac{1800\eta v}{\gamma_s - \gamma_f}}.$ In this relationship, D is in mm when the velocity, v, is the depth in cm from which the pipette sample was taken divided by the time in seconds since the beginning of settlement. The unit weight of the solids, γ_s, has the same value in the cgs system as specific gravity and the unit weight of the suspending fluid at the temperature of the suspension can be taken from a table of densities. When the suspending fluid is water, γ_f may be considered as unity for this purpose. The absolute viscosity η at the temperature of the suspension may be taken from a table of viscosities.

These two values D and W_D per cent establish one point on the grain size distribution curve. Additional points can be determined by taking other samples with the pipette, provided that each sample is taken from a point not lower than that at which earlier samples were taken.

The density of the suspension should be great enough that pipette samples will have enough solids to make accurate weighings, but not so great as to cause interference in the settlement of grains as individual units.

The pipette method was developed independently in three different countries about 1922. These developments were based on work done earlier by Sven Oden in 1915 and Wiegner in 1918.

3.04   HYDROMETER METHOD OF WET ANALYSIS

In 1926 Goldschmidt in Norway proposed a method of mechanical analysis based upon the specific gravity of a suspension as determined with a hydrometer. This method was developed independently in the United States in 1927 by George Bouyoucos of the Michigan Agricultural Experiment Station.

Bouyoucos developed a hydrometer graduated in grams of soil per liter of suspension existing at the center of immersion. A correction is made for temperatures different from that for which the hydrometer is calibrated. The hydrometer may be used with a standardized cylinder of such diameter and volume that the corrected hydrometer reading at the end of 40 seconds is the weight in grams of soil smaller than sand (0.05 mm dia.) in a liter of suspension, and the reading at one hour is the weight in grams of soil smaller than silt (0.005 mm dia.) in a liter of suspension. The percentages of sand, silt, and so-called clay size particles in the sample can be easily determined from these corrected hydrometer readings and the total weight of the suspended solids.

So many serious errors result from the use of the Bouyoucos method that, while with the Bureau of Public Roads, A. Casagrande made a study of the hydrometer method for the purpose of eliminating or reducing the errors.   Some of the errors in the Bouyoucos method will be apparent after a study of the theory and application of the hydrometer method of analysis which is given below.

The hydrometer method of analysis is based upon the assumption that Stokes' Law, $D = \sqrt{\dfrac{1800\eta v}{\gamma_s - \gamma_f}}$, is valid for the particular case of settlement of soil grains in a suspension.

If the test is started with a uniformly dispersed suspension of fine grains in a liquid, in time t any grain of a certain diameter D' will

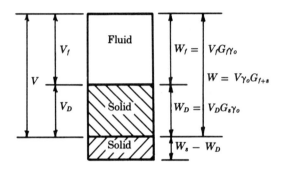

Figure 3.04a

have settled a distance H below the original position, so that at distance H from the top, the suspension will consist of grains smaller than D' only and each size which is present at all will occur in the same degree of concentration as in the original mixture.   If time interval, t, and depth, H, are known, the velocity $v = \dfrac{H}{t}$ can be computed; and with the viscosity and the unit weight of the suspending fluid and the unit weight of the solid particles known, the diameter of the largest grain in suspension at depth H after time t can be determined from Stokes' Law.   The solution of Stokes' Law can be accomplished quickly by means of a nomographic chart.   Such a chart was designed by A. Casagrande.   A copy of the chart is included in the chapter on Laboratory Testing.

The percentage of solid matter in suspension at depth H can be determined from the density as measured by the hydrometer as follows:

Consider the relationship as shown in Figure 3.04a in which

W_s = Total weight of solids in entire suspended sample

W_D = Total weight of solids in entire sample smaller than D'

G_{f+s} = Specific gravity of suspension

V = Total volume of suspension including all fluid and fraction of sample smaller than D'

W = Weight of suspension represented by volume V

$$G_{f+s} = \frac{W}{V\gamma_0} = \frac{W_D + V_f G_f \gamma_0}{V\gamma_0}$$

$$= \frac{W_D + (V - V_D)G_f \gamma_0}{V\gamma_0}$$

$$= \frac{W_D + \left(V - \dfrac{W_D}{G_s \gamma_0}\right) G_f \gamma_0}{V\gamma_0}$$

$$= \frac{W_D}{V\gamma_0} + G_f - \frac{W_D G_f}{V G_s \gamma_0}$$

$$G_{f+s} = G_f + \frac{W_D}{V G_s \gamma_0}(G_s - G_f). \qquad \text{(Eq. 3.04a)}$$

The relationship of Equation 3.04a is simply a mathematical statement which follows directly from definitions of terms.

At the calibration temperature, the hydrometer measures the specific gravity of the suspension referred to the unit weight of water at the temperature at which the hydrometer was calibrated. Therefore, in order to obtain the specific gravity referred to water at 4° C, the hydrometer reading made at calibration temperature must be multiplied by the specific gravity of water at the calibration temperature.

As measured by the hydrometer $G_{f+s} = r_c G_c$, wherein r_c is the hydrometer reading at calibration temperature and G_c is the specific gravity of water at calibration temperature.

Unity can be subtracted and added to each factor without affecting its value

$$G_{s+f} = r_c G_c = [(r_c - 1) + 1][(G_c - 1) + 1].$$

Expanding this relationship produces

$$G_{s+f} = (r_c - 1)(G_c - 1) + (r_c - 1) + (G_c - 1) + 1$$

r_c and G_c are both nearly equal to 1, so the first term can be neglected without affecting the results within the limits of accuracy available. Neglecting this term the relationship becomes

$$G_{f+s} = r_c + G_c - 1 \qquad \text{(1st approximation).} \quad \text{(Eq. 3.04b)}$$

The hydrometer at some other temperature than the calibration temperature reads differently than at the calibration temperature, because it expands with an increase in temperature thereby displacing more liquid and floating higher in the suspension, and vice versa.

Neglecting that part of the stem that is immersed, the expansion of the bulb is $\alpha_v(T - T_c)V_b$, in which α_v is the coefficient of volumetric expansion of the bulb, T is the test temperature, T_c is calibration temperature, and V_b is the volume of the bulb at calibration temperature. The volume of the bulb at test temperature T is

$$V_T = V_b + V_b\alpha_v(T - T_c).$$

The readings are proportional to the volumes displaced, so when r_T is the reading at any temperature T,

$$\frac{r_c}{r_T} = \frac{V_b}{V_T} = \frac{V_b}{V_b + V_b(T - T_c)\alpha_v} = \frac{1}{1 + \alpha_v(T - T_c)}.$$

Performing the indicated division, the following series is obtained

$$\frac{r_c}{r_T} = 1 - \alpha_v(T - T_c) + \alpha_v{}^2(T - T_c)^2 - \alpha_v{}^3(T - T_c)^3 + \cdots$$

Because α_v is a very small number compared to 1, near 1×10^{-5}, and $(T - T_c)$ is usually less than 10, $\alpha_v{}^2(T - T_c)^2$ has a value approximately equal to $(1 \times 10^{-5})^2 \times 10^2 = 1 \times 10^{-8}$, and because all subsequent terms become smaller and smaller and are alternately added and subtracted, the summation of all terms above the first power can for practical purposes be neglected. Neglecting these terms, r_c becomes $r_c = r_T[1 - \alpha_v(T - T_c)]$.

Inserting this value of r_c in Equation 3.04b produces

$$G_{f+s} = r_T[1 - \alpha_v(T - T_c)] + G_c - 1 \qquad \text{(2nd approximation).}$$
$$\text{(Eq. 3.04c)}$$

Equation 3.04c shows the approximate relationship between specific gravity of the suspension at the center of immersion and the hydrometer reading. Equating the value of G_{f+s} in Equation 3.04a to its value in terms of the hydrometer reading as expressed in Equation 3.04c produces the following relationship

$$G_f + \frac{W_D}{VG_s\gamma_0}(G_s - G_f) = r_T - r_T\alpha_v(T - T_c) + G_c - 1$$
$$\frac{W_D}{VG_s\gamma_0}(G_s - G_f) = r_T - 1 - r_T\alpha_v(T - T_c) + G_c - G_f.$$

Because the last three terms immediately above are dependent upon temperature, they may be considered a temperature correction and may be called $m_T \times 10^{-3}$, in which case

$$-r_T\alpha_v(T - T_c) + G_c - G_f = m_T \times 10^{-3}.$$

Because r_T is nearly 1 and, because m_T is little affected by small changes in r_T, the value of r_T as used in the temperature correction may be used as 1 without appreciable error. This approximation will allow the use of a temperature correction chart which is independent of r_T. After applying this approximation, Equation 3.04c becomes

$$\frac{W_D}{VG_s\gamma_0}\,(G_s - G_f) = r_T - 1 + m_T \times 10^{-3} \qquad \text{(3rd approximation)}.$$

If for convenience $(r_T - 1)$ be called $R_H \times 10^{-3}$, the numbers on the hydrometer may be used as whole numbers with the 1 omitted; i.e., if $r_T = 1.021$, $R_H = 21$; and the relationship becomes

$$R_H + m_T = \frac{W_D \times 10^3}{VG_s\gamma_0}\,(G_s - G_f).$$

In the derivation of Equation 3.04a, V was assumed to be the volume of the suspending fluid and the volume of the fraction of solids below D' in diameter, 'and represented by V_D. Each hydrometer reading indicates a different W_D, so that during the test the part of the volume V represented by V_D is variable. But if the volume of solids is small compared to the volume of suspending fluid, the volume of solids larger than D' will be a small proportionate part of the total volume and V may be assumed to be constant and equal to the total volume of suspension without appreciable error. (4th approximation). Then, if the total volume of the suspension be made 1000 cc and V assumed to be 1000 cc, and, if the cgs system be used so that $\gamma_0 = 1$,

$$R_H + m_T = \frac{W_D}{G_s}(G_s - G_f) \quad \text{or} \quad W_D = \frac{G_s}{G_s - G_f}\,(R_H + m_T). \quad \text{Then the}$$

per cent smaller than D' at depth H is

$$W_D\% = \frac{100}{W_s}\frac{G_s}{G_s - G_f}\,(R_H + m_T).$$

If distilled water at not more than ordinary room temperature is used as the suspending fluid, G_f can be called 1 without appreciable error, because the change in G_f from 4° C to room temperature is less than 0.01 and the value of G_s is not generally known closer than the nearest hundredth. (5th approximation). Then for a standard test at room temperature, using 1000 cc of suspension with distilled water as the suspending fluid,

$$W_D\% = \frac{100}{W_s}\frac{G_s}{G_s - 1}\,(R_H + m_T).$$

Thus the percentage by weight smaller than a given diameter in a suspension of soil can be determined directly from the hydrometer readings with a temperature correction.

In order to prevent flocculation of some clays, it is necessary to add a dispersing agent to the water in which the clay is suspended for a hydrometer grain size analysis.   This dispersing agent changes the value of G_f and causes the hydrometer to float higher in the suspension than if distilled water is used.   The change in G_f required to prevent flocculation is usually in the range from 0.0002 to 0.0005. For values of G_s accurate to three significant figures, this change does not affect the value of $\dfrac{G_s}{G_s - G_f}$.   The reading R_H, however, would be changed in the range from 0.2 to 0.5.   This error in readings ranging from 10 to 30 is an appreciable error.   This error is constant for all readings taken during a single test, so may be determined at the beginning of the test and applied as a correction for density, c_d. Applying this correction, the relationship becomes

$$W_D\% = \frac{100}{W_s} \frac{G_s}{G_s - 1}\ (R_H + m_T - c_d).$$

The value of H for the application of Stokes' Law for the determination of the diameter D' may be related to the hydrometer reading by calibrating the hydrometer and cylinder by direct measurement.

3.05 CALIBRATION OF HYDROMETER AND CYLINDER

First the distance in cm from the center of immersion to the calibration marks on the stem of the hydrometer must be determined. This may be done fairly accurately as described in Section 3.05L in the chapter on Laboratory Testing.

These values of H in terms of hydrometer readings can also be determined by locating the center of immersion on the bulb of the hydrometer, laying the hydrometer on a sheet of paper, projecting the graduations on the hydrometer stem to the paper, and measuring the marked distances from the center of immersion to the readings. The center of immersion can be located by inserting the hydrometer in a graduated cylinder of small diameter, noting the displaced volume, and withdrawing the hydrometer until exactly one-half this volume has been displaced.

Because the settlement, except for the first 2 to 4 minutes, takes place with the hydrometer out of the water and the readings are made with the hydrometer inserted, it is necessary to make a correction to this measured H' in order to determine the distance of settle-

ment H. Assume that during settlement with the hydrometer out of
the suspension the surface of the suspension is at level A as shown in
Figure 3.05a, and that the level at which the density is to be measured
is at B. The distance through which grains of diameter D' have
settled is H. When the hydrometer is inserted to determine the
density, the surface of the suspension rises in the cylinder. The volume of water between the new and old surfaces is equal to the volume
of the bulb. The rise is V_b/A_c, when A_c equals the area of the cylinder.
A volume of water below B equal to half the volume of the bulb is
displaced when the hydrometer is inserted and the suspension at B
rises to B', a distance equal to $V_b/2A_c$. The relation between H and

H' is $H = H' - \dfrac{V_b}{2A_c}$. The volume

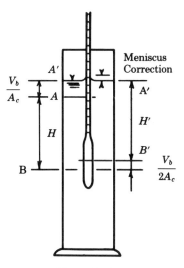

of the bulb can be determined by
displacement in a graduated cylinder or by weight. The area of the
cylinder can be determined by
pouring a known volume of water
into the cylinder and measuring the
distance the water surface rises in
the cylinder. The area of the cylinder is the volume of water placed
in the cylinder divided by the rise.

Having determined the value of
H for all of the measured values
of H', a scale may be made which
will allow the values of H to be
taken directly from the readings of
the hydrometer. A different scale
must be prepared for each hydrometer. These relationships between

Figure 3.05a
Hydrometer in Cylinder

H and R_H may be shown on the nomographic chart for the solution
of Stokes' Law so that the grain diameter can be read directly from
the hydrometer reading R_H.

Because of the opaqueness of the suspension, it is necessary to read
the hydrometer at the top of the meniscus which forms around the
stem. Because hydrometers are calibrated to be read at the surface
of the liquid (bottom of meniscus), readings taken at the top of the
meniscus must be corrected by adding the height, in terms of R_H,
that the meniscus climbs the stem. The meniscus correction c_m can
be determined by placing the hydrometer in clear water and reading
the top and bottom of the meniscus. The meniscus correction is a
constant for each hydrometer.

3.06 PRECAUTIONS IN THE USE OF HYDROMETER METHOD

Certain precautions must be observed in the use of the hydrometer for grain size determinations. The principal precautions are as follows:

1. The hydrometer should be removed from the suspension between readings and wiped off with a clean dry cloth. Grains of soil settle on the hydrometer bulb causing it to settle too low if left in the suspension between readings.
2. The hydrometer should be inserted into and removed from the suspension very slowly in order to prevent excessive disturbance. It should require about 10 to 15 seconds to insert the hydrometer and the same time to withdraw it.
3. The cylinder should be fairly large in diameter compared to the diameter of the hydrometer bulb in order to prevent excessive disturbance when the hydrometer is inserted and withdrawn.
4. Care should be taken to prevent convection currents in the suspension produced by placing the cylinder in air currents or in the sunlight so that one side is heated more than the other.
5. The temperature should not change more than a few degrees during the progress of the test because such change will affect the viscosity of the water making the rate of settlement variable. This error is slight for only a few degrees change, because the hydrometer reading is corrected each time for the temperature at which the reading was made. The influence of temperature changes during long intervals between readings is not evaluated and should not be allowed to occur. Variation in temperature during an analysis causes irregularities in the grain size distribution curve.
6. The hydrometer stem should be clean and free of grease, oil or other material that will prevent the development of a complete meniscus. Hydrometers are calibrated for correct reading with a fully developed meniscus. If the meniscus is not fully developed around the stem, the hydrometer will float too high in the suspension indicating too high a specific gravity.
7. The volume of the solids in suspension should be small in comparison to the volume of the suspending fluid. This precaution will be observed if the density of the suspension at the beginning of the test is not great enough to float the hydrometer so high that the lowest mark on the stem is above the surface.

3.07　ACCURACY OF WET MECHANICAL ANALYSIS METHODS

From a study of the wet mechanical analysis methods which are dependent upon the rate of settlement, it is readily apparent that the actual grain size is not determined. Theoretically, the diameter determined is that of a sphere that would settle at the same rate as the actual grain. Also, because not all the grains are the same shape nor the same unit weight, the size distribution determined by the wet analysis is not correct. The accuracy is also dependent upon the validity of Stokes' Law which was derived for a single sphere settling out of suspension. For soils of rather uniform gradation, an error in the hydrometer analysis is introduced by the assumption that the density of the suspension at the center of immersion of the hydrometer is adequately represented by the average density over the height of the bulb. For soils of normal gradation (most natural soils) the assumption is satisfactory, but for very uniform soils an appreciable error results.

However, these errors are of little consequence, because the grain size distribution for soils passing the 200 mesh sieve has little significance, and the distribution of the larger grain sizes can be more accurately determined by a sieve analysis.

3.08　TEST PROCEDURE FOR HYDROMETER ANALYSIS

Detailed directions for determining the grain size distribution for fine grained soils by means of a hydrometer analysis are given in Section 3.08L in the chapter on Laboratory Testing.

3.09　GRAPHICAL REPRESENTATION OF GRAIN SIZE DISTRIBUTION

The grain size distribution is commonly shown by a curve with the per cent by weight smaller, W_D per cent, plotted as the ordinate to an arithmetic scale and the diameter of grain, D, to a logarithmic scale as the abscissa. The analysis of the fine and coarse fractions as determined by the hydrometer and by sieves is usually combined into a curve covering the entire range. Such a curve is shown in Figure 3.09a.

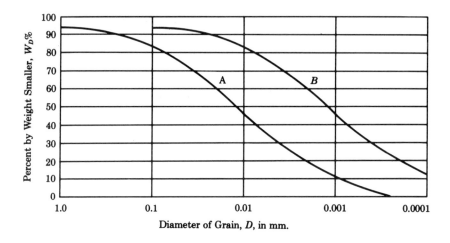

Figure 3.09a
Grain Size Distribution Curve
(Semilogarithmic)

There are several advantages to showing the grain size distribution curve on a semilogarithmic plot.

First, the scale for the smaller sizes is extended so that the distribution of the smaller sizes can be shown. If the grain size distribution curve Figure 3.09a were plotted to arithmetic scales, it would appear similar to Figure 3.09b. As seen by reference to this figure, more than 50 per cent of the soil by weight is represented by less than one-twentieth of the horizontal scale when the diameter is plotted to an arithmetic scale. The distribution of the fine fractions is not adequately represented by such a curve. This defect in the mathematical plot, as can be seen by comparison of the curves of Figure 3.09a and 3.09b, is eliminated by plotting the diameter to a logarithmic scale.

Second, the semilogarithmic plot for grain size distribution allows easy comparison between the gradation of soils of different sizes. On a semilogarithmic plot, soils of the same gradation are represented by parallel distribution curves. If two soils are so similarily graded that the diameter of one is one-tenth the diameter in the other for every percentage by weight, all points on the curve will be shifted by one cycle of the logarithmic scale and the two curves will be parallel.

Two such soils are represented by curves A and B in Figure 3.09a. It is obvious that such similarly graded soils of different sizes would not be represented by parallel distribution curves when plotted to arithmetic scales.

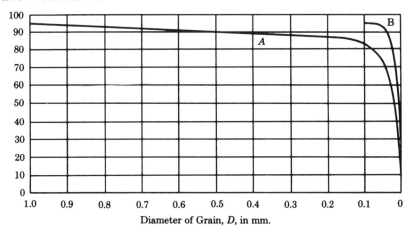

Figure 3.09b
Grain Size Distribution Curve
(Arithmetic)

3.10 EFFECTIVE GRAIN SIZE AND UNIFORMITY COEFFICIENT

A soil made up of grains of uniform size may have the same permeability as a graded soil of larger and smaller sizes. The diameter of the grains of a uniform soil that would have the same permeability as the graded soil is referred to as the effective grain size. Allan Hazen determined the effective grain size as approximately the 10 per cent size. The symbol for effective grain size is D_e.

The general slope of the grain size distribution curve indicating the gradation may be represented by two widely separated points on the curve. Allan Hazen used the ratio of the 60 per cent size to the effective diameter to express this general gradation and applied the term uniformity coefficient, designated C_u. Applying this definition, $C_u = \dfrac{D_{60}}{D_e}.$ Using the 10 per cent size as the effective diameter, the uniformity coefficient becomes $C_u = \dfrac{D_{60}}{D_{10}}.$

A soil of uniform grain size is represented by a vertical line and a well graded soil by a curve extending diagonally across the diagram.

The uniformity coefficient for a soil of uniform grain size is 1. For a graded soil the value of C_u is greater than 1, becoming larger with better gradation. The term is really a gradation or nonuniformity coefficient rather than a uniformity coefficient. The uniformity coefficient for natural sands is seldom less than 2.

Another coefficient sometimes used to indicate gradation is called

$$\text{coefficient of gradation } C_g = \frac{(D_{30})^2}{(D_{60})(D_{10})}.$$

3.11 GRAIN SIZE CLASSIFICATION

There are several classifications based upon grain size only. In all these classifications the terms gravel, sand, silt, and clay are used to indicate grain sizes. Such use of these terms, especially the terms silt and clay, may be misleading and should not be used without stating specifically that the term is used in this sense. In referring to soil particles of a certain size, it is better to state the size and reserve the use of such terms as silt and clay for soils having the properties associated with these soils. Whenever any of these terms are used to indicate size it is necessary to state the classification used. Instead of saying "per cent clay sizes according to U.S. Bureau of Soils Classification," it is easier and more meaningful to say "per cent smaller than 0.005 mm."

The three grain size classifications in common use are, U.S. Bureau of Soils, International, and M.I.T.

The Bureau of Soils Classification was developed about 1900 and was the one in common use in the United States for some time. In this classification the division points between gravel, sand, silt, and clay are the diameters in mm of 1.0, 0.05, and 0.005. Sand is subdivided into coarse, medium, fine, and very fine. Silt and clay are not subdivided.

The International Classification was proposed for general use at the International Soils Congress at Washington, D.C. in 1927, and was adopted by all countries except the United States. In addition to sand, silt, and clay, a size range between sand and silt, called mo, is included. Before adoption for international use, this classification was known as the Swedish classification.

The M.I.T. Classification was proposed by Professor Glennon Gilboy of Massachusetts Institute of Technology as a simplification of the Bureau of Soils Classification. It is built upon the numbers 2 and 6 alternating with the decimal point moved one place with each alternation. The sand, silt, and clay divisions are each subdivided into coarse, medium, and fine. This classification is simplest, most logical, and easiest to remember of the three and will probably become more popular with use.

Grain Diameter in mm.

2.0	1.0	.5	.25	.1	.05	.005

Fine Grav.	Coa.	Med.	Fine	V.F.	Silt	Clay
		Sand				

U. S. Bureau of Soils Classification

2.0	1.0	.5	.2	.1	.05	.02	.006	.002	.0006	.0002

V.C.	Coa.	Med.	Fine	Coa.	Fine	Coarse	Fine	Coarse	Fine	Ultra Clay
	Sand			Mo.		Silt		Clay		(Colloids)

International Classification

2.0	.6	.2	.06	.02	.006	.002	.0006	.0002

Coarse	Med.	Fine	Coarse	Med.	Fine	Coarse	Med.	Fine
	Sand			Silt			Clay	

M. I. T. Classification

Figure 3.11a
Grain Size Classifications

These three classifications are shown for comparison in Figure 3.11a with the diameters plotted to a logarithmic scale.

3.12 EFFECT OF GRAIN SIZE UPON PHYSICAL PROPERTIES

The permeability or velocity of seepage through soils depends almost entirely upon the grain size. The smaller the grain size, the smaller will be the pores between the grains and, therefore, the lower will be the permeability.

The consolidation rate of a saturated loaded soil is also dependent upon the grain size. Volume changes can occur in a saturated soil only as water is squeezed out, and because the rate at which water is squeezed out depends upon the permeability, the rate of settlement of a loaded area is dependent upon the grain size.

Because factors other than grain size affect permeability and factors other than permeability affect the rate of settlement, definite relationships between grain size and settlement cannot be worked out successfully for fine grained soils. One should, therefore, concern himself with measuring permeability and other properties directly rather than depending upon relationships which are, at best, only indicative.

The liquefaction of loose, saturated cohesionless soils is also dependent upon grain size. Deformation of bulky grained soils in a loose state is accompanied by a decrease in volume. When the voids are completely filled with water, the volume change cannot take place until some water escapes. When the deformation occurs, the tendency to decrease the voids places the incompressible water in compression causing it to take part or all of the load off the soil grains. If the deformation occurs suddenly and the resistance to flow of water out of the voids is great, at the instant of deformation the water takes practically all the load and the pressure between the grains becomes almost zero. Because the shearing resistance of cohesionless soil is due entirely to the pressure between the grains, the soil becomes liquid. This liquefaction will occur only in fine grained, cohesionless soils which are loose and saturated.

3.13   USE OF GRAIN SIZE DISTRIBUTION

The grain size distribution of soils smaller than the 200 mesh sieve is of little importance in the solution of engineering problems. The grain size distribution of larger sizes and the percentage of material smaller than the 200 mesh sieve have several important uses.

The grain size distribution affects the void ratio of soils and provides useful information for use in the design of cement and asphaltic concretes. Well graded aggregates require less cement paste to fill the voids and, because there is less water per unit of volume of concrete, produce denser concrete that is less permeable and more resistant to weathering than one made with uniform aggregates. The grading of aggregates is of particular importance in asphaltic concretes. Asphalt is a highly viscous liquid and a large amount of asphalt in a unit volume of the concrete allows bleeding and excessive flow under load. A well graded aggregate requires much less asphalt per unit of volume, thus preventing a great deal of bleeding and reducing the plastic flow or creep under load.

The very fine particles in soils are usually cohesive and act as a binder or low grade cement when mixed with coarse grained cohesionless materials. Therefore, a knowledge of the amount of this fine material and the gradation of the coarse particles is useful in making a choice of materials for base courses under highways, airports, and other structures carrying moving loads on the surface. The percentage of this fine grained material also affects the stability of soils for use in embankments.

Silt and sand grains may be moved by seepage forces as water emerges at the surface producing a condition known as piping. These grains can be held in place by means of a filter of coarser material. This filter material must be coarse enough to reduce the velocity of the water to such an extent that the seepage forces do not move the larger particles of the filter. At the same time it must not be so coarse as to allow the finer particles to pass through the pores of the filter. A grain size distribution curve may be used as an aid in the design of such filters.

Another use of a grain size distribution curve is for estimating the coefficient of permeability of coarse grained soils, such as very coarse sands and gravels. Darcy's Law which applies to laminar flow through fine grained soils does not apply to these very coarse grained materials.

In studying the action of frost in soils, the per cent of particles smaller than about 0.02 mm in diameter is considered to be a significant indicator of the probability of excessive frost heave.

REFERENCES

Brown, H. H., and W. G. O'Hara, "Suggested Method for Test for Amount of Material Finer than the No. 200 Sieve in Aggregates and Soils," *Procedures for Soil Testing*, Philadelphia: ASTM Committee D-18, April, 1958.

Casagrande, A., "The Hydrometer Method for Mechanical Analysis of Soils and Other Granular Materials," Cambridge, Massachusetts: June, 1931.

Havens, J. H., "Suggested Method for Separation and Fractionation of Clays and Associated Materials from Soils," *Procedures for Soil Testing*, Philadelphia: ASTM Committee D-18, April, 1958.

Johnston, C. M., "Suggested Method for Determining Center of Volume of Soil Hydrometers," *Procedures for Soil Testing*, ASTM Committee D-18, April, 1958.

Keen, B. A., "The Physical Properties of Soil," *The Rothamsted Monographs on Agricultural Science*, London: 1931.

Krumbein, W. C., "A History of the Principles and Methods of Mechanical Analysis," *Journal of Sedimentary Petrology*, London: 1931.

Lambe, T. William, *Soil Testing for Engineers*, New York: John Wiley and Sons, Inc., 1951.

Standard Test, "Grain Size Analysis of Soils, ASTM Designation: D 422–61T," *ASTM Standards, Part 4*, Philadelphia: American Society for Testing Materials, 1961.

Standard Test, "Density of Soil in Place by the Sand Cone Method, ASTM Designation: D 1556–58T," *ASTM Standards, Part 4*, Philadelphia: American Society for Testing Materials, 1961.

PROBLEMS

3.1 (a) Illustrate by means of a rough sketch (on a semi-logarithmic plot) a grain size distribution curve for a well-graded soil having an effective grain size of 0.01 mm and a coefficient of uniformity of 10. (Label this curve a.)

  (b) If, instead of being well-graded, the soil described above contained *no* particles in the size range 0.03–0.08 mm sketch the grain size distribution curve. What name would you apply to such a soil?

3.2 Explain why it is preferable, in plotting a grain size distribution curve, to plot the grain diameter on a logarithmic rather than an arithmetic scale.

3.3 Briefly discuss the relationship between the grain size determined during any sedimentation process and the actual size of the particles.

3.4 Due to inherent limitations of the method, the range of grain sizes which may be determined by means of the hydrometer analysis is restricted. State the maximum and minimum sizes which may be determined and give reasons for these limitations.

3.5 A combined mechanical analysis of a certain soil shows a substantial discrepancy in the grain size distribution curve over the range where the sieve analysis and the hydrometer analysis overlap. Explain the probable cause of the discrepancy.

3.6 What is the purpose of adding a small quantity of sodium silicate to the water with which a soil suspension is made?

3.7 Consider the hypothetical case of a *perfectly* uniform fine-grained soil composed of spherical particles of equal weight and density.

  (a) Will the results of a hydrometer analysis provide data which will yield a correct grain size distribution curve? Explain.

  (b) Sketch the curve which would be obtained from the hydrometer analysis.

3.8 A quantity of dry soil having a specific gravity of 2.74 and weighing 133.7 g is uniformly dispersed in water to form 1000 cc of suspension. After standing for 2 min and 30 sec, 10 cc of the suspension was removed from a depth of 21 cm beneath the surface. The dry weight of the soil in the 10 cc of suspension removed was found to be 0.406 g. The temperature of the suspension was 20° C. Determine by means of Stokes' Law one point on the grain size distribution curve for this soil.

3.9 22 g of dry soil having a specific gravity of 2.7 is uniformly dispersed in 400 cc of *water*. Determine the specific gravity of the suspension.

3.10 54 g of dry soil particles having an average specific gravity of 2.7 are placed in uniform suspension to form a total volume of suspension of 1000 cc. What should be the initial reading (at t = 0) of a hydrometer placed in the suspension if the temperature of the suspension is the same as the calibration temperature of the hydrometer?

3.11 A sample of soil weighing 160 g is separated by decantation into a coarse fraction weighing 121.7 g and a suspended fraction weighing 38.3 g. The following data were obtained from a combined mechanical analysis:

Sieve Analysis:	Sieve No.	Size in mm	Fraction Passing
	4	4.70	108.0
	8	2.36	99.5
	14	1.17	90.0
	28	0.59	77.9
	48	~~0.295~~ .295	58.6
	65	0.208	49.0
	100	0.147	34.0
	150	0.104	24.1
	200	0.074	13.8

Hydrometer Analysis:	Time in Min	Hydrometer Reading, R_H
Note: All readings taken	0.5	21.6
with suspension at	1	19.8
temperature of	2	17.6
23.4° C.	5	12.6
	15	7.0
	45	3.3
	120	1.4
	300	0.3
	600	−0.2

Specific gravity G_s = 2.87   No deflocculent was used.
Hydrometer characteristics:
    Depth of center of immersion below 1.000 = 23.0 cm.
    Depth of center of immersion below 1.030 = 10.3 cm.
    Volume of bulb = 65 cc.
    Area of graduate = 30 cm^2
    Calibration temperature = 60° F = 15.6° C.
    Meniscus correction c_m = +0.4.
    Coefficient of cubical expansion of glass = 25 × 10^{-6} cc./°C.

REQUIRED:

(*a*) Computations of grain diameters and percentage weights; grain size distribution curve plotted on 4-cycle semi-log paper; and the effective diameter and uniformity coefficient.

(*b*) Determination of the "m_T" and "B" values from the nomographs; solution of grain diameters by means of the nomographic charts.

CHAPTER IV

Plasticity

4.01 DEFINITION

In soil mechanics, plasticity is defined as that property of a material which allows it to be deformed rapidly, without rupture, without elastic rebound, and without volume change. The term plasticity is often used differently to define any non-elastic deformation, such as workability, creep, or yield. Clay may be mixed with water to form a plastic paste which can be molded into any form by pressure. The addition of water reduces the cohesion making the clay still easier to mold. Further addition of water reduces the cohesion until the material will no longer retain its shape under its own weight, but flows as a liquid. Enough water may be added until the soil grains are dispersed in a suspension. When water is evaporated from the plastic clay, the cohesion is increased and at low water contents the clay begins to rupture when deformed rapidly. After the clay is dried out, it becomes a hard and brittle solid. The clay may exist as a brittle solid, as a plastic material, or as a liquid. The transition from one state to another is gradual so that one cannot say at exactly what point the material changes from one state to another.

4.02   CAUSE OF PLASTICITY

The Swedish agriculturist Atterberg, during the first decade of the 20th century, found by experimentation that quartz ground into particles 2 microns and less in diameter has no plasticity, while mica

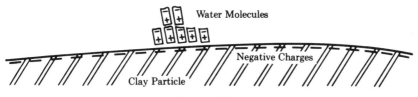

Figure 4.02a
Surface of Clay Particle with Attached Water

ground to the same size particles exhibits plasticity which increases with decreasing grain size. A little later Goldschmidt advanced a generally accepted theory explaining the cause of plasticity. According to the Goldschmidt theory, the plasticity is due to the presence of thin, scale-like particles, which carry on their surfaces electro-magnetic charges. The molecules of water being bi-polar, orient themselves like tiny magnets in the magnetic field next to the surface of the particles as shown in Figure 4.02a.

Those molecules of water adjacent to the particle are bound by this electro-magnetic field in an immobile state forming a thin layer of solid water adhering to the particle. Farther from the particle surface, the water becomes highly viscous, like asphalt, growing less viscous with distance from the particle until at some distance ordinary water exists. Therefore, with enough

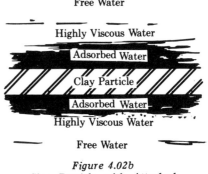

Figure 4.02b
Clay Particle with Attached Water

water present, the particles are separated by molasses-like water which allows the particles to slip past each other to new positions without any tendency to return to their former positions, without change in volume of voids, and without impairing the cohesion.   As more water is evaporated, the more viscous water separating the particles increases the cohesion and reduces the plasticity until eventually, when the clay is dried out, only solid or highly viscous water remains, cementing the particles together in a solid state.

Some minerals possess stronger electro-magnetic fields than others and attract thicker layers of attached water than others.   Some minerals break into extremely thin flaky particles not as thick as the attached water film.   The total mass of soils of these minerals may consist of more attached water than solid matter.   The correctness of Goldschmidt's theory for the cause of plasticity is evidenced by the fact that clay does not become plastic when mixed with a liquid of non-polarizing molecules like kerosene.

The film of solid water attached to the soil particles is often called adsorbed water and the viscous water outside the adsorbed water is sometimes called double layer water.

4.03   ATTERBERG LIMITS

About 1911, the Swedish agriculturist Atterberg divided the entire cohesive range from the solid to the liquid state into five stages and set arbitrary limits for these divisions in terms of water content as follows:

    a. Upper limit of viscous flow above which the clay and water flow like a liquid.

    b. Liquid limit or lower limit of viscous flow above which the soil and water flow as a viscous liquid and below which the mixture is plastic.

    c. Sticky limit above which the mixture of soil and water will adhere or stick to a steel spatula or other object that can be wet by water.

    d. Cohesion limit (Zusammenhaftbarkeitsgrenze) is the water content at which crums of the soil cease to adhere when placed together or in contact.   This limit is seldom mentioned in English literature but is encountered in German writings.

    e. Plastic limit or lower limit of the plastic range is the water content at which the soil will start to crumble when rolled into a thread under the palm of the hand.

f. Shrinkage limit or lower limit of volume change at which there is no further decrease in volume as water is evaporated.   This limit is discussed further under "Shrinkage."

Atterberg used the amount of sand that could be mixed with a plastic soil at the liquid limit before making the soil completely lose its plasticity as a measure of the plasticity.   The difference in the water content at the liquid limit and the plastic limit expressing the plastic range (plastic index) was suggested by him as a numerical classification for soils as to plasticity.

The Atterberg limits which are most useful for engineering purposes are liquid limit, plastic limit, and shrinkage limit.   These limits are expressed as per cent water content.

4.04   APPARATUS FOR LIQUID LIMIT

Atterberg determined the liquid limit as follows: A groove about 2 mm wide at the bottom was cut with a spatula in a soil cake mixed to an even consistency in a small evaporating dish.   The bottom of the dish with the soil pat was struck lightly on the palm of the hand a sufficient number of times to cause the bottom of the groove to just close but not flow together.   The water content at which the groove closed after a certain number of blows was defined as the liquid limit.   As can readily be imagined, such a determination was subject to great variation even when made by persons of considerable experience.   A liquid limit determination made in one laboratory could not be accurately compared with one made in another laboratory.   Consequently, in 1927 A. Casagrande, at the suggestion of K. Terzaghi, designed a mechanical device which eliminates the personal variation and which has standardized the liquid limit determination.

The apparatus consists of a brass cup which is dropped onto a hard rubber base by a cam operated by a crank, Figure 4.04a.   The cup is semi-spherical, 2 mm thick, 54 mm in diameter inside, and should weigh about 200 g.   It is hinged at the back hanging over a pin supported on a brass standard at the back of the hard rubber base.   A cam mounted on a shaft through the standard lifts the cup by pushing against the flat surface of a block at the back of the cup and rotating the cup about the supporting pin.   As the cam is rotated, the cup is lifted and let fall upon the hard rubber base.   The height of fall is adjustable by moving the supporting pin closer to or farther away from the cam.

A standard groove is cut in the sample in the cup with a grooving tool as shown in Figure 4.04a and 4.04b. The trapezoidal groove is 2 mm wide at the bottom, 11 mm wide at the top, and 8 mm high. The tool is about $\frac{3}{16}$ in. thick with the cutting edge formed by rounding

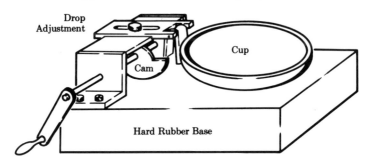

Figure 4.04a
Casagrande Liquid Limit Device

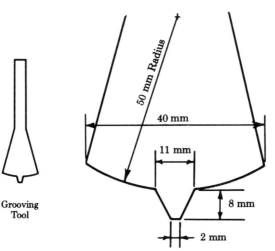

Figure 4.04b
Cutting Edge of Grooving Tool

one side only. Shoulders on the grooving tool form a groove of constant standard depth as well as width.

4.05 DETERMINATION OF LIQUID LIMIT

A test for determining the liquid limit of a sample of soil is made by mixing the sample of soil with water to form a smooth paste of uniform consistency in the cup of the liquid limit device.   The paste is smoothed with a spatula pulled along the front edge of the cup, filling the front three-fourths of the cup to a depth of about 12 mm at the center of the sample.   The groove is cut by holding the grooving tool perpendicular to the surface of the cup with the rounded edge to the front as it is pulled through the sample from back to front in line with the supporting pin at the back.   The cam is rotated by turning the handle at a rate of about 2 revolutions per second.   The height of the drop is adjusted to exactly 1 cm.

Because a large number of liquid limit determinations had been made by an experienced technician with the Bureau of Public Roads and these consistent results were in the laboratory files, it seemed desirable to calibrate the new device to give comparable results.   In consultation with several experienced technicians and engineers, Casagrande set the standard liquid limit as the water content at which the standard groove would close $\frac{1}{2}$ inch along the bottom after 25 blows in the mechanical device.

In some soils, especially nonplastic and organic soils, the grooving tool described above may not cut smooth surfaces when pulled through the sample in the cup.   In such cases the groove may be cut with a clean spatula and carefully checked for size and shape with the grooving tool.   Because of this difficulty in cutting the groove in some soils, a type of grooving tool has been developed and adopted by A.S.T.M. which may be pressed into the soil in the cup to form the groove.   It is curved to fit the inside surface of the cup.   During a test there is a tendency for the two sides of the sample to slide together along the surface of the cup rather than for the groove to close by deformation of the soil.   In this case, the closing of the groove is not a measure of the shearing resistance of the soil.   Also this tool does not control the depth of the groove.   For these reasons, the results of tests made by using this tool may be different from those of tests in which the grooves are cut and the depth controlled.

4.06 FLOW CURVE

Experience has shown that when the water content is plotted to an arithmetic scale and the corresponding number of blows required to close the groove for $\frac{1}{2}$ inch is plotted to a logarithmic scale, the points fall along a straight line.   Such a semilogarithmic plot showing the relationship between water content w and number of blows N is called the flow curve.

The equation for the flow curve can be written by referring to the flow curve in Figure 4.06a.

$$w - w_1 = -I_F(\log N - \log N_1) \text{ and } w = -I_F \log \frac{N}{N_1} + w_1,$$

wherein I_F is the slope of the flow line, always taken positive for convenience, and is called the flow index. If N_1 is 1 blow, w_1 is the water content at which the groove will be closed by 1 blow. Using these values for N_1 and w_1, the equation becomes $w = -I_F \log N + w_1$. The slope of the flow line I_F is indicative of the rate at which a soil loses shearing resistance with increase in water content.

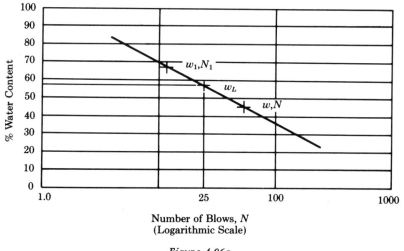

Figure 4.06a
Flow Curve

Because such a linear relationship exists between w and $\log N$, it is more convenient to find the liquid limit from the flow curve than by adjustment of the water content until the exact value is found by trial. The flow curve may be plotted from 4 to 6 determinations of w and N made on a single sample. For greater accuracy, the points should be grouped so that 2 or 3 of them fall between 28 and 35 blows and 2 or 3 fall between 8 and 15 blows. A line drawn between these two widely separated groups will be more nearly correct than one drawn as the average of 4 to 6 points scattered along the entire length of the flow curve. The water content for 25 blows, which is the liquid limit w_L, can be taken directly from the flow curve.

It will be observed that I_F is equal to the difference in water content over one cycle of the logarithmic scale, $w_1 - w_{10}$ or $w_{10} - w_{100}$.

4.07   PROCEDURE FOR LIQUID LIMIT TEST

A detailed description of the procedure for performing the liquid limit test is given in the corresponding section in the chapter on Laboratory Testing.

4.08   PLASTIC LIMIT

Atterberg defined the plastic limit as the water content at which a sample of soil begins to crumble when rolled into a thread under the palm of the hand.   In order to standardize the test, Terzaghi set the diameter of the thread at 3 mm or $\frac{1}{8}$ inch.

Several attempts have been made to devise a mechanical device for rolling out the threads for the plastic limit test, but none has been successful up to the present time.   There is no need for such a mechanical device for this purpose, because the water content at which the thread crumbles is very definite and easily determined.   Anyone with a little experience can check his own or other operator's values for the plastic limit by the method outlined above within 1 or 2 per cent.   Rolling the threads by hand has the advantage that the pressure of the hand can be adjusted to suit the toughness of the sample and the toughness of the soil can be estimated by the pressure required to roll the thread.

4.09   PROCEDURE FOR PLASTIC LIMIT TEST

A detailed description of the procedure for performing the plastic limit test is given in the corresponding section in the chapter on Laboratory Testing.

4.10   TOUGHNESS INDEX

The number of blows required to close the groove in the liquid limit device is dependent upon the shearing resistance of the soil.   The straight line relationship of the flow curve does not extend over the entire plastic range.   But, if it be assumed that the flow line is straight between the liquid and plastic limits and that the shearing resistance

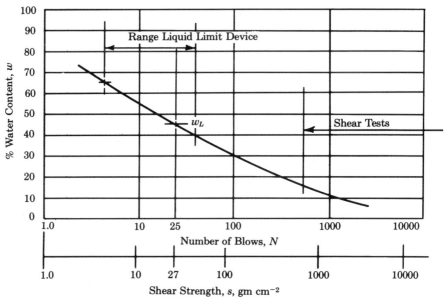

Figure 4.10a
Approximate Relationship Between Shear Strength and Number of Blows

is directly proportional to the number of blows, one may write for the liquid limit and plastic limit the following relationships:

$$ks_L = N_L \quad \text{and} \quad ks_P = N_P,$$

in which k is a constant, s_L is shearing resistance at the liquid limit, s_p is shearing resistance at the plastic limit, N_L is number of blows at the liquid limit, and N_p is number of blows at the plastic limit. For the liquid limit

$$w_L = -I_F \log N_L + w_1 = -I_F \log ks_L + w_1.$$

For the plastic limit

$$w_P = -I_F \log N_P + w_1 = -I_F \log ks_P + w_1.$$

Subtracting these relationships for w_L and w_P

$$w_L - w_P = -I_F(\log ks_L - \log ks_P) = I_P$$

$$I_P = -I_F \log \frac{s_L}{s_P} \text{ and } \log \frac{s_P}{s_L} = \frac{I_P}{I_F}.$$

Because the unit weight of soils varies little at the liquid limit and the number of blows is constant for all soils at the liquid limit, the shearing resistance of all soils at the liquid limit is approximately constant. Early in the development of soil mechanics, A. Casagrande estimated the shear strength at the liquid limit as 27 g cm^{-2}. Later evidence indicates that the shear strength at the liquid limit may be somewhat less than this value. Then the ratio of the shearing resistances at the liquid and plastic limits may be taken as a measure of the resistance to deformation at the plastic limit, or toughness.

Log s_P/s_L is indicative of the toughness at the plastic limit and is defined as the toughness index I_T. $I_T = \dfrac{I_P}{I_F}$.

The shear strength of a soil at the plastic limit can be estimated from the relationship $\log s_P = \dfrac{I_P}{I_F} + \log 27$.

The approximate relationship shown in Figure 4.10a applies to strictly plastic soils only.

4.11 LIQUIDITY INDEX

The liquidity index I_L of a soil is defined as

$$I_L = \frac{w_n - w_P}{w_L - w_P} = \frac{w_n - w_P}{I_P}$$

in which w_n is the natural water content of the soil and the other terms are as previously defined. The quantity is also occasionally referred to as the water-plasticity ratio. The liquidity index should be computed as a matter of routine procedure when the index properties are determined. The liquidity index would be unity for soils whose natural water content is equal to the liquid limit and zero for soils having a natural water content equal to the plastic limit. The term is applicable only to plastic soils.

4.12 ACTIVITY NUMBER

Activity is a term applied to plastic soils in reference to their propensity for undergoing change in volume in the presence of varying moisture conditions. The more active is a soil, the greater, in general, will be its change in volume when passing, for example, from the liquid limit to the shrinkage limit. The quantity of water involved in effecting this change is dependent largely upon the type and quantity of colloidal clay particles present. These same factors also determine the plasticity index of the soil.

A. W. Skempton has proposed the use of an activity number, A, as an indication of the activity of a soil. The activity number is defined as

$$A = \frac{I_P}{\% < 0.002 \text{ mm}}$$

in which the numerator is the plastic index, and the denominator is the per cent by weight of particles having an equivalent diameter smaller than 0.002 mm.

4.13 SIGNIFICANCE AND RANGE OF INDEX PROPERTIES

One should keep in mind that the plasticity limits and indexes are made on remolded soil and are at best only indicative of other physical properties of the remolded soil and of the properties of the undisturbed soil. Structure, geologic origin, or other physical condition of the undisturbed soil affect the actual stress-deformation characteristics to such an extent that these limits and indexes cannot be used to take the place of direct measurements of stress-deformation-time relationships. These properties which are not generally of primary interest to the engineer but which are indicative of the stress-deformation-time relationship in which the engineer is primarily interested are often called index properties.

The liquid limit is dependent upon the mineral constituents of the soil, the intensity of the surface charges and the thickness of the attached water, and the ratio of surface area to volume or shape of the particles. In general, those minerals having the strongest surface charges break into extremely thin scale-like particles. The stronger the surface charge and the thinner the particles the greater will be the proportion of attached viscous water and, therefore, the higher will be the liquid limit. Because the proportion of these thin scale-like particles affects the compressibility of the soil, the liquid limit is indicative of compressibility. The liquid limit of inorganic clays is seldom greater than 100. Clays of volcanic origin and those containing organic matter may have liquid limits considerably in excess of 100.

Bentonite, which is a mineral formed from weathering of volcanic ash or scoria, has a liquid limit of about 400 per cent. Clays containing considerable bentonite have high liquid limits. The liquid limits of some clays, especially those containing organic matter, are affected by drying. In some cases the liquid limit is increased by drying, although usually it is decreased. A liquid limit of 50 per cent and above probably indicates the presence of montmorillonite. Clays with liquid limits of 50 per cent and less probably consist principally of the mineral kaolinite and associated minerals.

The liquid limit for cohesionless soils is meaningless, even though a value can be found for a fine grained cohesionless soil. The apparent cohesion, due to capillarity may permit the formation of a groove. However, if the soil is in a loose state, liquefaction may occur when the cup is dropped. If a value for the liquid limit is to be obtained, the fine sand or silt must be in a dense state in order for the shearing resistance to persist during the deformation caused by dropping the cup. In such a case the shearing resistance is not cohesion and the water content is not a measure of plasticity.

High toughness at the plastic limit indicates a high percentage of colloidal clay particles in the soil. It also indicates that the clay probably contains a high percentage of montmorillonite or other highly active colloidal clay particles. Weakness and lack of cohesion in the thread at the plastic limit indicates an inorganic clay of low plasticity probably containing a considerable amount of kaolinite. Organic clays are weak and have a spongy feel at the plastic limit.

The plastic index is a measure of the plastic range. The smaller the flaky particles making up a plastic soil, the easier the soil can be rolled into a thin thread without crumbling. Because the grain size determines the capillary stress in the pore water, in general, soils with a high plastic index are subjected to high pressure when dried out. A high plastic index coupled with a high liquid limit is indicative of a soil that will shrink a great deal when moisture is evaporated, and that is capable of swelling against heavy loads when water is applied to the dry soil. For given liquid limits, organic soils have lower plasticity indexes than inorganic; but, a low index may be due to coarse flaky grains rather than to organic matter.

The toughness index is a measure of the shearing resistance at the plastic limit. The value of the toughness index varies from 0 to 3 for most clays but may be as high as 5. A value less than 1 indicates that the soil is friable at the plastic limit.

For plastic soils the value of the liquidity index I_L is indicative of the stress history of the soil. Normally consolidated clays usually have liquidity indexes of about unity, while over-consolidated clays have liquidity indexes approaching zero. All intermediate values are possible and are frequently encountered. In exceptional cases the liquidity index may substantially exceed unity, as in the case of extra sensitive clays; or it may take on negative values, as in the case of heavily over-consolidated or desiccated clays.

Higher values of the activity number A are associated with soils containing the more active clay minerals. The more active clay minerals, such as montmorillonite, can produce a given value of the plastic index I_P, when in much smaller quantities than would be the case if the same I_P were produced by the presence of a less active mineral such as kaolinite. Soils of the same geologic origin tend to be composed of the same minerals (perhaps in different proportions) wherever encountered over a more or less restricted geographic area. For a soil of specific origin, higher percentages of clay particles in samples of the soil will result in a corresponding increase in the plasticity index. Thus, the value of A will remain approximately constant (though I_P may vary widely) for samples from a given geologic stratum. Consequently, the value of A may be the most reliable means of ascertaining the existence of different geologic strata encountered in a boring. Clays for which A is less than 0.75 are considered relatively inactive; normal activity is associated with values of A between 0.75 and 1.5, while values greater than 1.5 indicate clays progressively more active. Since for a given clay I_P may be greatly influenced by the type of dissociated ions available in the pore water, base exchange activities or leaching of a soil may produce marked effects upon the activity number.

4.14   USE OF INDEX PROPERTIES

The index properties are useful primarily for comparison purposes. In investigating deposits of soils of the same geologic origin, structure, etc., such as might be found underlying a construction site upon which foundations are to be placed, it is impractical to investigate the entire volume of soil by direct measurements on undisturbed samples. By means of index properties, many disturbed samples taken from bore holes can be compared with a few undisturbed samples for which these and other needed properties have been determined by direct measurement.

When used for the purpose outlined above, liquid limit determinations should be made on the single soil as it exists and not on a mixture of adjacent soils. A so-called average value as determined on a mixture of the different soils making up a layer has little meaning, because such a value may be quite different from the average for the separate soils and not representative of the properties based upon a summation of the separate materials making up the layer. As an example, the compressibility of a layer of alternate thin layers of clay and silt is dependent upon the clay as it exists and the silt as it exists. The compressibility of the layer made up of these thin layers of clay and silt is quite different from the compressibility of the layer if the clay and silt were mixed together and not in separate layers. Therefore, the liquid limit of the mixture is not indicative of the separate soils as they exist. This may be further illustrated with two fairly uniform sands, one coarse and the other fine, with enough water added to each separately to make each moist but not saturated. When the two moist sands are mixed together, the mixture will be saturated, possibly with excess water flowing out, although no water has been added. This phenomenon indicates that the void ratio of the mixture is different from the average of the two separate sands.

4.15 SIMPLE FIELD TESTS FOR PLASTICITY

Several simple field tests are available for estimating the plasticity of soils.

a. Shaking Test

This test is sometimes called the dilatency test. A small sample of soil about $\frac{3}{4}$ inch in diameter is remolded with enough water to make the soil soft but not sticky. Gravel or coarse sand particles should be removed from the sample. The pat of soil is placed in the palm of one hand and shaken horizontally or by striking the back of the hand containing the sample with the other hand. If the soil reacts to this test, the surface of the sample will become shiny with the appearance of water which comes to the surface when treated in this manner. The pat of soil is then deformed by folding the palm of the hand containing the soil. In a positive reaction to the test the surface water will be drawn into the sample when deformed and the surface of the pat will become dull.

Fine clean sands react quickly and distinctly to the shaking test. Very fine grained cohesionless soils, like silt and rock flour, react to the shaking test but somewhat more slowly than fine sands. Plastic clays do not react to this test. Cohesionless soils when moist do not stain the hands when rubbed.

b. Surface Test

Cohesive soils can be identified by rubbing a small ball of remolded moist soil with a clean knife blade or fingernail. A small sample of the soil at about the plastic limit is kneaded into a ball about 1 inch in diameter and rubbed with a clean knife blade or fingernail. If the knife blade leaves a shiny surface, the soil is highly plastic.

c. Dry Strength

The dry strength test requires two pieces of undisturbed soil about $\frac{1}{2}$ inch in each dimension. One of the pieces of soil is remolded and rolled into a ball. The undisturbed and the remolded samples are allowed to stand at room temperature for two or three days until they are thoroughly air dried. An attempt should then be made to crush these two specimens of dried soil between the fingers, noting whether they crush easily or with great difficulty. If they cannot be crushed with the fingers, an attempt should be made to crush them under the ball of the hand with the aid of the weight of the body. If they cannot be crushed in this manner, perhaps they can be crushed under foot. The hardness should be tested further by crushing a small piece between the fingers or the fingernails.

Dry strength increases with increasing plasticity. A high dry strength indicates a high plasticity. A large difference in the dry strength between the undisturbed and the remolded specimens indicates that the soil in the undisturbed state is probably quite loose. Silty fine sands and silts have about the same very slight dry strength. Fine sand feels gritty whereas typical silt has the smooth feel of flour.

d. Toughness Test

This simple field test furnishes a rough means of determining the amount and type of colloidal materials in the soil. By toughness is meant high strength coupled with a large deformation before rupture.

Following exactly the same procedure as used for the laboratory test for plastic limit, a small remolded sample of the soil is rolled into a thread about $\frac{1}{8}$ inch in diameter, repeating until the thread breaks. The pressure required to deform the thread at the plastic limit should be noted. For a discussion of the significance of toughness, see Section 4.13.

4.16 TERMS FOR DESCRIBING PLASTICITY

Some terms which can be used to describe plasticity are as follows:

Degree of Plasticity when wet

Non-plastic.......... Has no cohesion. Cannot be rolled into a ball.

Trace of Plasticity... Barely holds its shape when rolled into a ball.

Medium Plastic...... Has considerable cohesion. Can be molded into a ball and will withstand considerable deformation without rupture.

Highly Plastic....... Can be kneaded like dough without trace of rupture.

Dry Strength

Very slight.......... Will barely hold together in small pieces under its own weight.

Slight............... Can be easily broken between the fingers.

Medium............. Can be broken between the fingers with considerable pressure.

High............... Can barely be broken under palm of hand with weight of body applied.

Very High.......... Cannot be broken under palm of hand.

REFERENCES

Atterberg, A., "Über die Physicalische Bodenuntersuchung und über die Plastizität der Tone," *Internationale Mitteilungen fur Bodenkunde*, Vol. 1. (1911).

Lambe, T. William, "The Structure of Inorganic Soil," *Proc. Am. Soc. C.E.*, October, 1953, Sep. No. 315.

Lambe, T. William, *Soil Testing for Engineers*, New York: John Wiley and Sons, Inc., 1951.

Standard Test, "Liquid Limit of Soils, ASTM Designation: D 423-61T," *ASTM Standards, Part 4*, Philadelphia: American Society for Testing Materials, 1961.

Standard Test, "Plastic Limit and Plasticity Index for Soils, ASTM Designation: D 424-59," *ASTM Standards, Part 4*, Philadelphia: American Society for Testing Materials, 1961.

PROBLEMS

4.1 Define the term *plasticity*, as used in soil mechanics.

4.2 Formulate precise definitions for the terms *liquid limit* and *plastic limit*.

4.3 Define the following terms and state what significance may be attached to their numerical values.
(*a*) Liquidity Index   (*b*) Activity Number

4.4 What is the principle use of the liquid and plastic limits for engineering purposes?

4.5 State whether the *shape* or the *size* of the mineral grains is the more important factor contributing to clay its conspicuous, distinguishing characteristics.   Justify your answer.

4.6 Two cohesive soils, A and B, have the same liquid limit and identical flow indices.   Soil A has a w_p of 30 and soil B a w_p of 20.
(*a*) Which is the tougher soil?   (*b*) Explain very briefly what is meant by toughness.   (*c*) State reason for your answer to part (*a*).

4.7 The shear strength of clay at the liquid limit is about 25 g/sq cm.   The liquid limit of a fat clay was found to be 80 and the plastic limit 20. At 10 blows the water content was 90 per cent.   Estimate conservatively the shear strength of the fat clay at the plastic limit.

4.8 The following data were obtained from a liquid limit test of a highly plastic clay from Seven Sisters, Manitoba:

No. of Blows	28	26	17	13
Water Content	105.1	105.7	110.9	112.2

Two plastic limit determinations gave water contents of 34.5 per cent and 35.2 per cent.   Plot the flow curve and determine the liquid limit, the plasticity index, the flow index, and the toughness index.   Assuming $G_s = 2.82$, compute the void ratios of the saturated soil at the liquid and plastic limits.

CHAPTER V

Structure

5.01 GENERAL

By structure is meant the arrangement of the grains in relation to each other. Terzaghi has classified soils with regard to structure as single grained, honeycomb, and flocculent. The structure is dependent upon the size of the grains and upon the shape as well as upon the minerals of which the grains are formed. Orientation and stratification affect the properties of soils. The structure of clays is affected by the minerals of the grains and by the conditions prevailing at the time of deposition.

5.02 SINGLE GRAINED

Grains of sand and silt larger than 0.02 mm in diameter settle out of suspension in water as individual grains independently of other grains.

The weight of the grain causes them to settle and roll to positions of equilibrium among the other grains practically independent of other forces. This arrangement of single grains is called single grained structure.

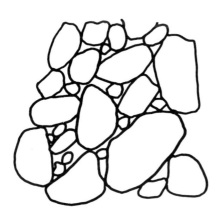

These single grains may be deposited in a loose state having a high void ratio or in a dense state having a low void ratio. In the loose state the grains are in contact about as shown in Figure 5.02b. In the dense state the grains are nested in the hollows between grains about as shown in Figure 5.02c.

For comparison purposes it is useful to know that for equal spheres in the densest state $e_D = 0.35$ and in the loosest state $e_L = 0.91$. The void ratios in both the densest and loosest states for well graded sands and silts are lower than for equal spheres.

Figure 5.02a
Single Grained Structure Sand
(Diameter >0.02 mm)

Loose sand when deformed decreases in volume, the grains sliding or rolling down into a denser state. On the other hand, deformation of dense sand is accompanied by an increase in void ratio. This change

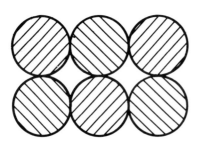

Figure 5.02b
Loose High Void Ratio

Figure 5.02c
Dense Low Void Ratio

in void ratio accompanying deformation has a marked effect upon the stability of cohesionless materials. Structures built on or above layers of loose sand and which have stood for years without appreciable settlement have experienced large sudden settlements when vibrated by earthquake, dynamite blasts, etc.

The effect of deformation upon loose and dense sand can be illustrated with two rubber bulbs, similar to ear syringes, with glass tubes inserted in the openings. One bulb is filled with loose sand and then with water until it shows in the glass tube. The other bulb is filled with dense sand (compacted during filling by vibrating the bulb) and with water well up into the glass tube. When the loose sand is deformed by squeezing the bulb, the decrease in void ratio will be shown by the water rising in the tube. When the dense sand is deformed, the increase in void ratio will be shown by water being drawn into the sand and falling in the tube.

Fine sand in a loose state and saturated is a very unstable material, especially in embankments. When a saturated, loose, fine-grained, cohesionless material carrying a static load is deformed by sudden shock or vibration, the void ratio tends to decrease and the water filling the voids, being relatively incompressible, is stressed in compression. Because of the low permeability of the fine grained soil, the water cannot escape immediately so the load is momentarily carried by the water causing the entire mass to become liquid. The liquefaction of fine grained saturated cohesionless soils is discussed further in Chapters IX and XI. *Has to do with sand.*

The degree of density of a sample of soil has been expressed by Terzaghi as $D_d = \dfrac{e_L - e}{e_L - e_D}$, in which e is the void ratio of the sample, e_L is the void ratio in the loosest state, and e_D is the void ratio in the densest state. Degree of compaction, relative density, and per cent compaction are terms sometimes used to mean the same as degree of density. Degree of compaction is defined in Manual No. 22, "Soil Mechanics Nomenclature" of the Am. Soc. C. E. as $\dfrac{e - e_D}{e_L - e_D}$. According to this definition 100 per cent would indicate the loosest state or no compaction, which appears to the writers as misleading. Because of the possibility of misunderstanding, the term should be used with caution.

The term relative density has been approved and suggested for use in describing the relationship $\dfrac{e_L - e}{e_L - e_D}$ instead of degree of density by the Committee on Definitions and Standards of the Soil Mechanics and Foundations Division of ASCE in their Progress Report "Nomenclature of Soil Mechanics" published in the *Journal of the Soil Mechanics and Foundations Division*, June 1962. This report proposes that relative density be designated D_R.

Relative density can easily be determined for clean coarse sand in either the dry or submerged state. Because of bulking, inaccurate results will be obtained if the sand is moist. Fine grained soils absorb enough hygroscopic moisture from the atmosphere, when allowed to stand for some time after being removed from a drying oven or desiccator, to make an appreciable error in the determination of the void ratio. The void ratio in the densest state can be found by compacting oven dried sand in a cylinder of known volume and weight by rodding and vibrating in thin layers to the densest state possible. After the cylinder has been filled to overflowing, the top should be struck off with a straight edge and the weight of the sand W_s determined. If the specific gravity of the sand particles is known, the volume of the solids is easily determined and the value of e_D can be computed. The void ratio in the loosest state can be determined in the same way, except that the sand should be gently placed in the cylinder through a funnel, care being taken not to compact the sand by vibrating.

The density of a soil may be expressed either by means of the void ratio or the unit weight, γ_{dry}. The unit dry weight is often referred to as dry density. The relationship between dry weight and void ratio has already been shown as $\gamma_{dry} = G_s \dfrac{\gamma_0}{1 + e}$. Because the specific gravity of soil solids varies over a narrow range, the void ratio and unit weight (dry density) are almost proportional. Methods for determining the void ratio and unit dry weight (dry density) are given later in this chapter.

5.03 HONEYCOMB STRUCTURE

Grains of silt or rock flour smaller than 0.02 mm diameter and larger than 0.0002 mm (dust particles when dry) settle out of suspension more or less as single grains but are so small that the molecular forces at the contact areas, as grains come in contact at the bottom, are large enough compared to the submerged weight to prevent the grains from rolling down immediately into positions of equilibrium among the grains already deposited. Other grains coming in contact are held until miniature arches are formed, bridging over relatively large void spaces and forming a honeycomb structure somewhat as shown in Figure 5.03a.

In this way a loose deposit is built up having a high void ratio and capable of carrying relatively heavy loads without excessive volume

change. Such soils appear to have a critical loading below which
settlements are comparatively small and above which there is a very
great increase in settlement. When a load great enough to break down
the arches is applied, a more or less single grained structure results with

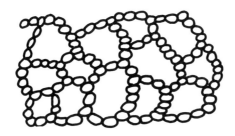

Figure 5.03a
Honeycomb Structure
(Diameter <0.02 mm >0.0002 mm)

a large decrease in void ratio. The structure might also be broken
down, with a resulting volume decrease, by driving piles into a deposit
of silt having a honeycomb structure.

5.04 FLOCCULENT STRUCTURE

In order to understand the flocculent and other structures of clay,
it is necessary to know something of the nature of the minerals which
form clay.

The principal clay minerals are members of the kaolin group as
described in Section 1.04 "Common Soil Minerals." Kaolinite is
probably the most common of the clay minerals, some clay being
almost entirely kaolinite. Others contain illite, montmorillonite, and
bentonite in varying quantities. Halloysite, attapulgite, pyrophyllite,
micas, bauxite, and other minerals are present in some clays but usually not in great enough quantities to affect materially the physical
properties of the clay. As already described, these minerals are composed of layers of silica and gibbsite sheets. The properties of these
minerals depend largely upon the type of bond between the layers of
these sheets.

All the clay minerals possess weak bonds between the thin layers
causing them to break down into thin platelets or needle-like particles

of microscopic and submicroscopic sizes. These mineral platelets bear electro-magnetic charges (usually negative) on their surfaces which attract bipolar water molecules.

Single grains of soil smaller than 0.0002 mm in diameter (colloids) do not settle out of suspension in pure water, but move about in the suspending water, the grains avoiding each other as they move about. The molecules of the liquid water are relatively far apart and are vibrating at a high velocity (depending on the temperature). At any instant more molecules of water strike one side of a soil particle than the other producing a resultant force in one direction. The mass of these colloidal particles is so small that this resultant force drives the particle in the direction of the force. The particles, being all charged alike, repel each other when they come close together, preventing grains from forming in groups large enough to settle out. This constant bombardment of the colloidal particles keeps the particles in constant motion so that they never settle out. This phenomenon is known as the Brownian movement.

Clay may exist as a sol or as a gel. A sol is a solution or suspension of colloidal particles of a solid dispersed in a homogeneous medium which may be gas, liquid, or solid. Sols are classified as extrinsic or intrinsic. Extensic sols consist of electrically charged particles in the suspending medium. The particles can be precipitated by electrical charges or by the addition of concentrations of multivalent ions of opposite sign charge to the charges carried by the particles. Extrinsic sols are not stable. Intrinsic sols can be precipitated by another solvent or high concentration of salts. Intrinsic sols are in equilibrium in the sol state and tend to swell in a suitable medium or solvent to form a stable sol. A sol always takes the shape of the containing vessel.

Sols are often transformed into gels. A gel is a colloidal system consisting of a solid and a liquid phase. Gels may have a fairly high or a low solid content. The attractive forces which form the gel hold the particles together in the same relative position so that they retain the shape of the vessel in which they were formed and behave as elastic solids. Gels when dried to form an apparently solid state are called xerogels. In a suitable solvent these gels sometimes swell and redisperse to form a sol. Clay gels may be transformed into sols by shaking. Many of these clay sols upon standing quietly for some time revert to gels. This transformation from a sol into a gel is known as thixotropy.

The phenomenon can be explained by the character of the double layer water attached to the soil particles. In the gel state the clay particles are held together by the oriented molecules of water attached to the particles. When the gel is disturbed enough to break the orientation of the water molecules, free liquid water develops between the particles and the bond is broken and a sol is formed. When left to stand for considerable time, the water molecules reorient themselves between the soil particles causing the sol to revert back to a gel. Because this orientation of the water molecules is a crystallization process, considerable time is required for the transformation to take place. The regain in strength is considerable for montmorillonite and bentonitic clays, moderate for illite, and practically none for kaolinite. The thixotropic regain in strength seems to be greatest for water contents at about the liquid limit.

When a salt is dissolved in the water in which clay particles are suspended, the positive ions of the solution attach themselves to the negative surfaces of the soil particles, neutralizing them. When these discharged particles come close together as they move about in the suspension, they no longer repel each other but are attracted by gravity forces of the

Figure 5.04a
Flocculent Structure of Clay

masses to form loose flocks or aggregations of randomly arranged particles. These large flocks are too heavy to be affected by the Brownian movement and settle out of suspension to form a deposit of clay having a loose flocculent structure.

A clay having such a flocculent structure has a high void ratio. When pressure is applied to the deposit, high concentration of stress exists at the points of contact causing the particles to be bent and to slip along the contact surfaces to positions of greater stability and producing a denser arrangement of the particles with a resulting decrease in volume. Remolding such a soil breaks down the structure into a more or less single grained structure with a considerably reduced void ratio.

Another and possibly more common type of flocculent structure is formed when a stream carrying silt grains and colloidal clay particles discharges into a sea or salty lake of quiet water. In this case flocks of the colloidal particles form immediately and settle out of suspension together with the silt grains, forming arches of the silt grains and the loose honeycomb flocks. This type of honeycomb structure, which is characteristic of marine clays, might be considered as a honeycomb within a honeycomb structure. As the weight of the overburden increases in a deposit of clay having such a flocculent structure, a concentration of stress is produced in the arch rings, compressing the flocks between the silt grains to a high degree of density.

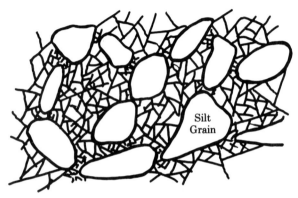

Figure 5.04b
Flocculent Structure of Marine Clay

A. Casagrande has called this dense clay "bond clay." The flocks under the arches, being protected from pressure, remain in a relatively loose state, even under heavy loads. Such a structure is shown in Figure 5.04b.

5.05 DISPERSED STRUCTURE

When colloidal clay particles are carried into a fresh water lake of quiet water, they do not flocculate and settle along with silt particles as they do in salt water. There is always enough ionized salt in fresh water to cause the colloidal particles to flocculate after a considerable time and settle. Remolding of the flocculent structure either by pressure or kneading causes the flaky particles to slip to more nearly parallel positions. As the angle between particles becomes acute, the attached water molecules between the two adjacent particle surfaces turned with their negative ends toward each other, produce a repulsion force which tends to push the particles apart. The greater pressure is at the end where the flakes are closer together. This greater pressure tends to separate the closer ends and to draw water to this area (osmotic pressure) until the platelets are pushed to a parallel position separated by the double layer water. This oriented arrangement of particles is known as dispersed structure. In a dispersed structure

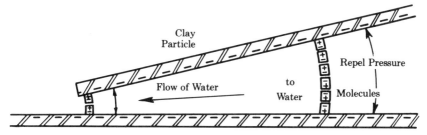

Figure 5.05a
Pressure Between Clay Particles at Acute Angle

the clay particles are supported in a viscous water matrix. Remolding and compacting clays tends to produce a dispersed structure. Loading of a flocculent structure consisting entirely of clay particles also tends to produce a dispersed structure. Pressure applied to a dispersed structure forces some of the double layer water out from between the flakes until they are separated by water with a high enough viscosity to support the applied load.

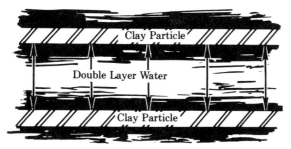

Figure 5.05b
Clay Particles Separated by Double Layer Water
(Dispersed Structure)

Under zero pressure kaolinite has attached to its surface a layer of double layer water about 400 Å thick. Two kaolinite platelets would be separated by 800 Å of water. As reported by T. William Lambe, a kaolinite crystal 1000 Å (1×10^{-4} mm) thick may have attached to its surface about 10 Å of solid (adsorbed) water and 400 Å of viscous (double layer) water. A dispersed structure of kaolinite flakes under zero pressure and ordinary temperature and with sufficient water available is 44 per cent water and 56 per cent solids by volume.

Montmorillonite crystals under some conditions may be only 10 Å thick with 10 Å of adsorbed water next to the surfaces and 200 Å of double layer water. Montmorillonite having a dispersed structure under this condition is 97 per cent water.

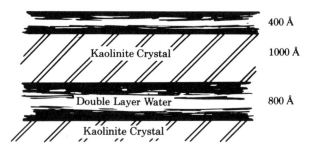

Figure 5.05c
Kaolinite Crystals Separated by Attached Water

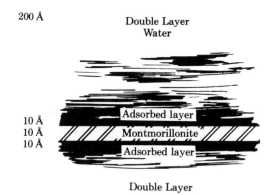

Figure 5.05d
Montmorillonite Crystal with Attached Water

This attachment of water to the surface of clay minerals is the same phenomenon as the wetting of the surface of glass or other matter with water.   The extreme thinness of the clay particles causes a much larger percentage of attached water on them than on thick substances.   The ratio of surface area in square meters to the weight of matter in grams is called specific surface.   The specific surfaces of three common clay minerals are: montmorillonite $800 \, \mathrm{m^2 \, g^{-1}}$, illite $80 \, \mathrm{m^2 \, g^{-1}}$, kaolinite $10 \, \mathrm{m^2 \, g^{-1}}$.

5.06 OTHER STRUCTURES

One can easily imagine the great number of variations and combinations of arrangement of grains when one takes into consideration the processes of sedimentation, deposition by wind, and the formation of soil by disintegration of the parent rock without being transported. Solid particles heavier than water settle out of suspension with a velocity which is dependent upon diameter, shape, and specific gravity of the grains. Large particles settle faster than small ones and spherical shapes faster than flat ones. Because the specific gravity of soil solids varies over a narrow range, its variation little affects rate of settlement. Where streams have a high gradient (in general near the source) the high turbulent velocity keeps grains of fairly large diameter in suspension; but, as the velocity decreases smaller and smaller particles settle out. During flood stage streams carry a heavy load of sediment of all sizes below the largest that the stream will hold in suspension. As the laden water spreads out over a large area and the velocity decreases, particles of a given size settle out. A further reduction in velocity allows still smaller particles to settle out. Thus, the streams tend to segregate materials into layers of the same size and shape; i.e., to cause stratification, first depositing a layer of coarse then a layer of finer materials with each flood stage. The next flood stage lays down other layers of coarse and fine, but never quite the same. This stratified arrangement is excellently exemplified in the varved clays of glaciated regions. During warm weather when the ice melted rapidly, a thin layer of silt was deposited in the glacial lakes by the rapidly flowing melt water. During cool weather when the ice melted more slowly, a thin layer of clay was deposited above the previously deposited layer of silt. Each year a new deposit of silt and clay was formed. This alternation repeated many times built up thick deposits consisting of alternate thin layers of silt and clay.

5.07 ORIENTATION

The orientation of the grains in a soil affects its properties to a large extent. If the grains are deposited in an irregular pattern so that the relative position of the grains is the same in all directions, the resulting material will be isotropic. The bulky grained soils are generally approximately isotropic, although sands of grains varying considerably from spherical in shape may show a difference in permeability in two directions. Probably, most natural deposits are anisotropic to some extent.

Flaky grains may be deposited with most of the particles lying with their flat surfaces parallel to each other. Such an orientation produces an anisotropic material with different properties in two directions. The permeability is greatly affected by orientation, being, in general, greater in a direction parallel to the flat surfaces than in a perpendicular direction. Compressibility and other properties are also affected by orientation.

5.08 STRATIFICATION

As pointed out above, soils may be stratified by the deposition of different materials in layers. Stratification greatly affects the properties in different directions. The permeability of a varved clay is much greater parallel to the layers than perpendicular to them. Stratification should not be confused with orientation.

5.09 COMPACTION

In the remolding of soils in backfills, embankments, and fills, it is often necessary to compact to a certain void ratio or dry density. The engineer is interested in knowing the water content of the soil which will enable the lowest void ratio to be achieved with a given amount of compactive effort, or to know the water content that will enable a certain void ratio or dry density to be achieved with the least effort.

Cohesionless materials are compacted by vibration. Static load produces very little compaction of loose sand. Medium and fine sands do not compact easily when moist because of the shear strength developed by capillary forces. When dry or submerged all sands can be compacted by vibration. The compaction of loose sand by vibration can be illustrated with a glass cylinder filled with sand. If the cylinder of sand is turned upside down and the sand allowed to run slowly back into the cylinder as the cylinder is turned upright, the sand will be in a loose state with the top at a certain height. Heavy pressure applied to the surface of the sand through a plunger which loads most of the area, produces little deformation of the sand. When the cylinder is shaken and vibrated, the compaction of the sand is evident from the subsidence of the surface of the sand.

Clays cannot be consolidated by vibration. Pressure is required to consolidate clay. The compressibility of clay can be illustrated with loose mica flakes in a glass cylinder. A small amount of static pressure produces a large volume decrease of the mica flakes. Shaking the cylinder produces no volume decrease of the mica flakes.

Figure 5.09a
Loose Structure of Clay Before Compaction

The loose clay before compaction generally has a loose random structure as shown in Figure 5.09a. Each clay particle has attached to its surface a thin layer of solid adsorbed water and a thicker layer of viscous double layer water. The thickness of the double layer water depends upon the water content of the soil mass. In compacting the clay, the position of the particles must be changed by forcing the contact points along adjacent surfaces to positions more nearly parallel and with reduced voids as shown in Figure 5.09b. When the clay to be compacted is saturated, compaction can be achieved only as incompressible water is forced out of the voids. Because of the extremely low permeability of clay, considerable time is required for the water to flow out to allow a volume decrease. Under the suddenly applied load, the compacting hammer or device, instead of compacting the soil, forces the saturated soil out and upward from under the loaded area as shown in Figure 5.09c.

When the clay has a high water content less than saturation, a thick layer of double layer water surrounds the particles so that the contact

Figure 5.09b
Dense Structure of Compacted Clay

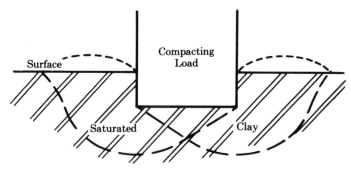

Figure 5.09c
Failure of Saturated Clay Under Compacting Hammer

points are separated by water of low viscosity. Under this condition, only a small amount of pressure is required to force the grains to new positions. But a high degree of compaction cannot be produced with this high water content because the thick layer of double layer water prevents the particles from being forced close together before failure is produced as shown in Figure 5.09c. As the water content is decreased, the thinner layer of attached water allows a higher degree of compaction but more work is required to move the particles in the highly viscous water to the new compact positions.

For a given amount of compactive effort there is a corresponding water content that will produce the densest state. When the water content is very low, a given compactive effort will produce a low degree of compaction because of the high resistance of the highly viscous water. A higher water content of less viscous attached water allows this compactive effort to move the particles to a denser state. As the water content is increased, the effort produces greater density until the water content becomes great enough that the large amount of water with the residual air will not allow further compaction. Beyond the water content that enables this compactive effort to produce the densest state, additional water allows less compaction until at saturation no compaction is produced.

As the compactive effort is increased, the water content at which the densest state will be produced is decreased. Under no circumstances can all the air be driven out of the mixture during compaction so that the mass is completely saturated. Sometimes a comparison is made between the degree of compaction achieved and the maximum possible density for the water content; i.e., a density for which the water content produces saturation.

The water content that will allow the densest compaction under a given compactive effort is known as the Optimum Moisture Content. This optimum moisture content is determined by trial on a sample of the soil by a test devised by R. R. Proctor of the Bureau of Waterworks and Supply of Los Angeles, California, and is generally known as the Standard Proctor Test. The test essentially as devised by Proctor has been adopted as a standard test by the American Association of State Highway Officials and by the American Society for Testing Materials. A modification of the compaction test using a 10 lb hammer falling 18 in. instead of the 5.5 lb hammer falling 12 in. has also been standardized by both AASHO and ASTM. This modified test is often referred to as the modified AASHO test or sometimes as the modified Proctor test.

5.10   STANDARD AASHO (PROCTOR) TEST

The purpose of this test is to determine the relationship between water content and dry density or void ratio of a soil compacted in a standard manner and to determine the optimum moisture content for the soil.

The apparatus consists of a standard brass cylinder mold of the dimensions shown in Figure 5.10a having a volume of $\frac{1}{30}$ cu ft. The mold is provided with a machined base which forms a bottom for compacting soil in the mold.   A removable extension fastens to the top of the mold to hold the soil during compaction.

The compacting is done with a standard hammer having a milled face 2 in. in diameter and weighing 5.5 lbs.   An arrangement is pro-

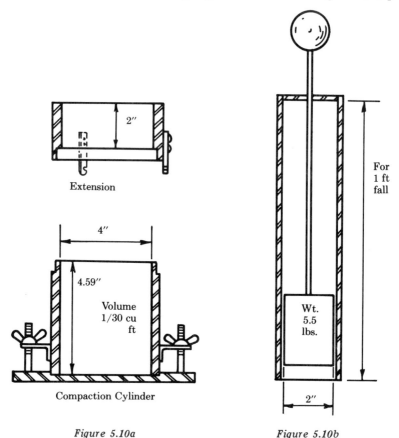

Figure 5.10a
Compaction Cylinder
(Standard Mold)

Figure 5.10b
Standard Compaction Hammer

vided for lifting the hammer exactly 1 foot and allowing a free fall to the surface of the soil.   Several mechanically operated compaction hammers have been devised.

The compaction of the soil in the standard test is made in three lifts under 25 blows for each lift.   Compaction tests may be made using more or less blows per lift than the 25 used in the standard test in order to determine the water content that will enable maximum compaction to be attained with more or less compactive effort than that applied in the standard test.

5.11   TEST PROCEDURE FOR STANDARD COMPACTION TEST

The steps required in performing the standard compaction test are listed in the chapter on Laboratory Testing under Section 5.11L.

5.12   DETERMINATION OF OPTIMUM MOISTURE CONTENT

From the test data recorded on the Data and Computation Sheet the water content, unit weight dry (dry density), and void ratio should be computed and recorded in the proper place on the sheet. These values should then be plotted on cross section paper showing water content as the abscissa and unit dry weight and void ratio as the ordinate. A curve drawn through these points shows the relationship between water content and unit dry weight or void ratio for this constant compactive effort. The top of the curve indicates the maximum density that can be achieved with this amount of compactive effort and the water content which will allow this maximum density to be achieved (optimum moisture content).

For comparison a curve is sometimes drawn on the diagram showing the relationship between greatest possible density and water content. This curve is called a zero air voids curve and represents compaction to the extent that the given water content produces complete saturation; i.e., all the air has been forced out of the sample. The equation for this curve is:

$$\gamma_{\text{Dry}} = \frac{G_s \gamma_0}{1 + w G_s}.$$

An example showing test data and plotted curves are shown in the chapter on Laboratory Testing, Section 5.11L.

As indicated by the two optimum moisture curves for 25 and 15 blows per layer, the maximum density increases with increase of compactive effort. Also the optimum moisture content decreases with increase in compactive effort.

Because the optimum moisture content found by repeatedly compacting the same soil in the different trials is slightly different from that found by compacting a fresh sample for each trial, a more accurate determination can be obtained by mixing a larger sample and compacting the soil at different water contents only once.

5.13   MODIFIED AASHO TEST

It is sometimes useful to study the properties of a soil under a considerably greater compactive effort than that provided by the standard test.   For this purpose the Modified AASHO Test has been devised and standardized by AASHO.   This test is made in the same manner as the standard test, except that the hammer weighs 10 lbs instead of 5.5 lbs and is let fall 18 inches instead of 12 inches, and the sample is compacted in 5 lifts instead of 3.   25 blows per layer are used in both tests.

The modified AASHO test was devised to more nearly duplicate the compactive effort of some of the very heavy compacting equipment used for compacting fills for highways and airfields.

5.14   HARVARD MINIATURE COMPACTION TEST

Apparatus for use in compacting small specimens of soil was designed in 1949 by S. D. Wilson, then at Harvard University.   The apparatus consists of a mold $1\frac{5}{16}$ in. in diameter and 2.816 in. long, having a volume of $\frac{1}{454}$ cu ft.   During compaction, the mold and its extension collar are held firmly to a base.   Compaction of the soil is accomplished by means of a tamper employing a spring loaded plunger.   The size of the spring to be used may be chosen, and its length regulated, so as to produce a desired pressure on the soil at that instant when the spring just begins to deflect.   Twenty and forty pound tamping forces are commonly used.   The soil may be compacted in any desired number of layers in the mold.   After compaction, a special device is used to remove the compacted specimen from the mold.   Usually the entire specimen is dried to determine the moisture content.   The weight of the entire soil specimen, in grams, is numerically equal to the unit weight of the soil in lb ft^{-3}.

The miniature compaction test has the advantage of being able to provide data from a small quantity of soil, thus greatly reducing the labor involved. It has the further distinction of utilizing a kneading action in producing compaction. This type of action probably more nearly resembles that obtained in the field by a sheepsfoot roller than does the dynamic action of the Proctor test.

The test has the disadvantage of not yet having been sufficiently correlated with the standard tests to permit its general use in control of field compaction. However, there is no inherent reason why such procedures could not be developed. A further disadvantage exists in that the test is intended for fine grained soils. Particles larger than about 2 mm in diameter should be removed from the soil by passing the soil through a No. 10 sieve, whereas in the Proctor test only those particles retained on a No. 4 sieve are excluded.

The miniature test has been employed primarily in research investigations. Some care in performing the test is required to obtain consistent results, but in the hands of an experienced operator the apparatus can provide reliable information quickly and economically.

5.15 PROPERTIES OF COMPACTED SOILS

Clean cohesionless soils can be compacted in a dry or submerged state by vibration. The properties of compacted cohesionless soils whether compacted by nature or by man are essentially the same if the conditions of placement and density attained are equal. Backfills of clean sand can be compacted by flooding and vibrating with a vibrator similar to those used for vibrating concrete. Dry sand can be compacted in layers with vibrating rollers. Medium and fine sands are more difficult to compact when damp than when dry or submerged because of the capillary forces retarding displacement of the damp sand grains to positions of greater stability. Loose sand lying beneath the surface can sometimes be compacted in place by means of a process known as "Vibroflotation." A large self contained motor driven vibrator is provided with ports in the top and bottom through which water can be forced at will. In use, the vibrator is started, water is forced through the bottom ports, and the vibrator allowed to sink into the loose sand. At the extreme depth of penetration, the water is cut off at the bottom and forced out the top ports. The vibrator is allowed to operate until the desired density is attained after which the vibrator is raised in short lifts to the surface.

Clay is usually compacted in the field in thin layers by pressure applied with sheepsfoot rollers which apply the pressure over small areas with a kneading action somewhat similar to that produced by the hammer used in the compaction test. When clay is compacted by pressure applied in one direction only, as is the usual case, the compacting process tends to orient the flat, scale-like particles to form a more or less dispersed structure. Because of the greater thickness of the double layer water attached to the thin particles acting as a lubricant, compaction on the wet side of optimum tends to orient the particles with their flat faces perpendicular to the applied force to a greater extent than compaction to the same density on the dry side of optimum.

Permeability of clay is affected materially by compaction. Clays compacted to the same density on the dry side of optimum are usually much more permeable than when compacted on the wet side. The clay compacted on the dry side has a more random or flocculent structure than when compacted on the wet side. Even though the volume of pores is the same in both, the pores are larger in the clay with a random structure than in the dispersed clay.

Compression of clay under pressure can occur: by compression of gas contained in the pores, by particle deformation, by rearrangement of particles, or by change in thickness of the attached double layer water. Under small pressures an increment of load causes a larger volume decrease of clays compacted on the wet side of optimum than those compacted on the dry side. Under heavy pressures an increment of load produces larger volume change in clay that has been compacted on the dry side than in the same clay compacted on the wet side. In clays compacted on the wet side the soil particles are more nearly oriented with their flat faces parallel and have thicker layers of attached water than in those compacted on the dry side. Low pressures force some of the double layer water out from between the oriented particles without having to produce displacement of the particles. In the clays with a random structure and less double layer water heavy enough pressures to cause rearrangement of the particles produce greater volume change than low pressures which do not displace the soil particles.

Expansion of compacted clays is probably due almost entirely to increase in thickness of the double layer water. Clay compacted on the dry side of optimum picks up more water than that compacted on the wet side and swells more when water is made available. A clay compacted on the dry side of optimum to a certain density requires greater pressure to prevent swelling than the same clay compacted to the same density on the wet side.

Clays compacted on the dry side of optimum possess greater shear strengths than when compacted to the same density on the wet side. In clays compacted dry of optimum the double layers are not fully developed. The soil particles trying to attract more water to themselves produce meniscuses and pore water tension in the pore spaces thus increasing the intergranular pressure and producing greater strength in the as molded state. In general, clays compacted dry of optimum and then saturated possess greater strengths than those compacted wet of optimum and then saturated.

5.16 DENSITY OF SOIL IN PLACE

The unit dry weight or so-called dry density of a soil in place can be determined by taking a sample of known volume V, driving off the moisture in a drying oven and determining the dry weight W_s. The unit dry weight $\gamma_{dry} = \dfrac{W_s}{V}$.

The void ratio of a soil in place can be determined from the total dry weight W_s of known original volume V and the specific gravity of the solids G_s. The volume of solids, $V_s = \dfrac{W_s}{G_s \gamma_0}$. Volume of voids, $V_v = V - V_s$. Void ratio, $e = \dfrac{V_v}{V_s}$.

The most difficult part of this problem is to determine the original volume of soil removed as the sample. There are three methods in general use for this purpose.

a. Undisturbed Sample

In this case, the sample is taken as nearly as possible in its original state in a sharpened cylinder of known volume.

The cylinder is carefully pressed a short distance into the soil and the soil excavated around the outside to the bottom of the cylinder. The cylinder is pressed a further short distance into the soil and the soil excavated to the bottom of the cylinder on the outside. This is repeated until the soil protrudes above the top of the cylinder. The excess soil is struck off with a steel straight edge and the top of the cylinder covered with a smooth flat plate, preferably in the form of a cap which remains in place when the cylinder is turned upside down. The cylinder can then be lifted out by inserting a trowel or shovel beneath the cylinder.   The covered cylinder is turned upside down and struck off to a smooth plane surface.   In taking samples of clean cohesionless materials, extreme care must be exercised that the soil in the cylinder is not disturbed by vibration or erratic movements of the cylinder.

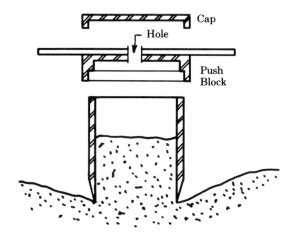

Figure 5.16a
Apparatus for Taking Undisturbed Sample

The soil in the cylinder can then be emptied into a clean container and sealed with an air tight cover to prevent the evaporation of moisture.   The soil will be dried out and weighed in the laboratory to determine W_s.

Samples can sometimes be taken from the bottom of a dry bore hole by pressing in a cylinder of thin wall tubing on the end of a drive pipe or rod.   Care must be exercised to see that the sample is not disturbed.

b. Sand Cone

Another commonly used method for finding the volume of the sample removed for determining the void ratio of soils in place at the surface or bottom of a test pit is by means of a so-called sand cone.

The apparatus consists of a steel plate with a hole approximately 4 inches in diameter in the center, a hollow cone which fits over this hole and with a fruit jar connection above, and a 1 gallon fruit jar. Between the cone and the fruit jar connection there is a valve which allows sand to run from the fruit jar into the cone below.

Before using the sand cone apparatus, it must be calibrated to determine the unit weight of the sand and the weight of the sand which runs out of the jar in excess of that required to fill the cavity from which the earth sample was removed. This is necessary in order to be able to compute the volume of the excavated earth from the weight of the sand used to fill the cavity. The apparatus is described in detail and instructions for its calibration and use are given in Chapter XII.

c. Rubber Balloon

Still another method of determining the volume of the sample removed is by the use of a rubber balloon which is expanded to fill the pit.

The apparatus consists of a density plate and a graduated glass or lucite cylinder enclosed in an airtight aluminum case with an opening in the bottom sealed by a latex balloon which may extend up into the cylinder or out through the bottom. The cylinder is partially filled with water. A double acting rubber bulb is attached to the bottom of the cylinder with which either a pressure or a vacuum can be applied to the inside of the cylinder for forcing the balloon out the bottom of the case through the hole in the density plate and into the excavated pit, or for pulling the balloon up into the cylinder. _The volume of the hole may be read directly on the Graduated cylinder._

Drawings and instructions for use of the apparatus are given in the chapter on Laboratory Testing.

REFERENCES

Grim, Ralph E., *Applied Clay Mineralogy,* New York: McGraw-Hill Book Company, Inc., 1962.

Lambe, T. William, "The Structure of Compacted Clay," *Journal Soil Mechanics and Foundations Division*, Proc. Am. Soc. C. E., Vol. 84, No. SM2, (May, 1958).

Lambe, T. William, "The Engineering Behavior of Compacted Clay," *Journal Soil Mechanics and Foundations Division*, Proc. Am. Soc. C. E., Vol. 84, No. SM2, (May, 1958).

Mitchell, James K., "Fundamental Aspects of Thixotropy in Soils," *Journal Soil Mechanics and Foundations Division*, Proc. Am. Soc. C. E., Vol. 86, No. SM3, (June, 1960).

Proctor, R. R., "Fundamental Principles of Soil Compaction," *Engineering News Record*, August 31, 1933.
(Also three articles in September issues.)

Seed, H. B., and C. K. Chan, "Structure and Strength of Compacted Clay," *Journal Soil Mechanics and Foundations Division*, Proc. Am. Soc. C. E., Vol. 85, No. SM2, (April, 1959).

Seed, H. B., C. K. Chan, and C. E. Lee, "Resilience Characteristics of Subgrade Soils and Their Relation to Fatigue Failures in Asphalt Pavements," Paper presented at International Conference on Structural Design of Asphalt Pavements, University of Michigan, August 20–24, 1962.

Trollope, D. H., and C. K. Chan, "Soil Structure and Step Strain Phenomena," *Journal Soil Mechanics and Foundations Division*, Proc. Am. Soc. C. E., Vol. 85, No. SM2, (April, 1960).

Wilson, S. D., "Small Soil Compaction Apparatus Duplicates Field Results Closely," *Engineering News Record*, November 2, 1950.

Standard Test, "Moisture Density Relations in Soils, ASTM Designation: D 698–58T and D 1557–58," *ASTM Standards, Part* 4, Philadelphia: American Society for Testing Materials, 1961.

Standard Test, "Density of Soil in Place by the Sand Cone Method, ASTM Designation: D 1556–58T," *ASTM Standards, Part* 4, Philadelphia: American Society for Testing Materials, 1961.

PROBLEMS

5.1 Describe the principal types of natural soil structure and explain the circumstances surrounding their formation. It is suggested that simple sketches be used for illustration.

5.2 Define (*a*) optimum moisture content and (*b*) maximum dry density. Explain clearly why these quantities are likely to have different values in the field than those obtained in the laboratory.

5.3 Discuss the effects of increased compactive effort on the optimum moisture content and maximum dry density.

5.4 Construct a rough graph of the zero air voids curve (γ_{dry} vs w) for a soil having a specific gravity of 2.70. Show derivation of the basic relationship among w, G_s, γ_{dry}, and S, and compute the value of γ_{dry} for $w = 10$ per cent.

5.5 For a given compactive effort the density of the soil is dependent upon the molding water content. Discuss the effect of variations in the molding water content upon the engineering properties of compacted soil. Explain how these properties may be changed during the service life of a compacted earthen dam.

5.6 The optimum moisture content for compaction of a certain soil of low plasticity is 12 per cent for a given compactive effort. If the same soil is compacted at a moisture content of 15 per cent, using the same compactive effort, describe the effect which the higher moisture content will have on the engineering properties of the compacted soil.

5.7 Specifications call for a fill to be compacted to 90 per cent of Standard Proctor Density. In the standard test the optimum moisture content was found to be 14 per cent. Estimate the moisture content which should be used in order to make the most efficient use of compaction equipment in compacting the fill. Justify your answer.

5.8 For one trial during a compaction test of a sandy silt a dry density of 110 lb/cu ft was obtained at an average water content of 10 per cent. What was the approximate degree of saturation of the compacted specimen?

5.9 State what a field density test is, describe one method of performing the test, and give the purpose of such tests.

5.10 An undisturbed sample of cohesionless soil is obtained in a sampling tube having an inside area of 30 cm² and a length of 25 cm. The soil and tube weigh 2076 g immediately after sampling. The weight of the tube is 900 g. If the average moisture content of the soil is found to be 7 per cent, estimate the natural void ratio of the soil.

5.11 An undisturbed specimen of lean clay had an average moisture content of 30 per cent. A block of the clay weighed 20 pounds in air and 9.5 pounds submerged in water (allowance having been made for a thin coat of wax on the block). Based on a reasonable estimate of G_s, determine the average natural void ratio of the clay.

5.12 During a field density test 3.82 lb of sandy clay soil were removed from a hole having a volume of 0.031 cu ft. The oven-dried weight of the soil was 3.31 lb.
(a) Determine the moisture content and dry density of the soil.
(b) Estimate a suitable value for G_s and determine the degree of saturation of the soil.

5.13 A sample of moist, fine sand was taken from an embankment by pressing a sharpened cylinder, having a volume of 0.033 cu ft, into the soil to exactly fill the cylinder. In the laboratory, the dry weight of the sand was found to be 3 lb. When placed in its loosest possible state, the dry sand occupied a volume of 0.036 cu ft. The volume was reduced to 0.028 cu ft by thorough rodding and vibration. Determine the relative density of the sand in the embankment.

5.14 (a) The following calibration data for a sand cone apparatus have been obtained in the laboratory:
Initial Wt jar + sand = 9182 g.   Final Wt jar + sand = 5012 g.
Vol of calibration cylinder = 970 cc.
Wt of sand in calibration cylinder only = 1528 g.
(b) The following data were obtained during a field test:
Initial Wt jar + sand = 8765 g.   Final Wt jar + sand = 4425 g.
Wt of soil taken from fill = 1783 g.
Dry wt of soil taken from fill = 1546 g.
(c) If G_s is estimated as 2.72, determine the water content, dry unit weight, and void ratio of the soil in the fill.

5.15 The unit weight of a sand used as subgrade was found from a field density test to be 118 lbs/cu ft at a moisture content of 10.1 per cent. In the laboratory, the specific gravity was found to be 2.66, and the loosest and densest void ratios were 0.621 and 0.433 respectively. What is the relative density of the subgrade?

5.16 The following data are obtained from a series of Proctor compaction tests using a cylindrical mold $\frac{1}{30}$ cu ft in size and weighing 4.52 lb. Soil is a moderately plastic, gray, silty loess.

Test Number	1	2	3	4	5	6
Water Content (dry weight basis)	13.7	15.8	17.7	20.0	24.7	28.6
Weight of mold and wet soil (Pounds)	7.86	8.09	8.22	8.49	8.53	8.46

(a) Plot the data in the form of a conventional Proctor curve with a scale for unit dry weight on the left and void ratio on the right. (Assume $G_s = 2.68$.) Also plot the zero air-voids curve.

(b) Determine the maximum Proctor density and optimum moisture content.

(c) Determine the degree of saturation of the soil at optimum moisture content.

CHAPTER VI

Classification of Soils

6.01 GENERAL

In addition to the general classifications of cohesionless and cohesive soils, which have already been discussed, several special classifications have been devised and are in use. None of these special classifications is completely satisfactory. The most useful are the Textural Classification and the Unified Classification. Size classification is also in common use but it is included in the more general Textural and Unified Classifications.

6.02 SIZE CLASSIFICATION

The grain size classification system has already been discussed under "Grain Size Distribution."

6.03 TEXTURAL CLASSIFICATION

Textural classifications are size classifications which take into account the gradation or distribution of sizes making up a soil.
  Probably the best known of these textural classifications is the tri-

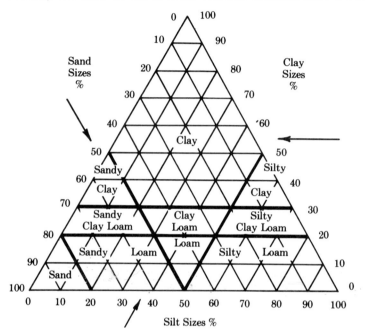

Figure 6.03a
U.S. Bureau of Soils Classification

angular classification of the U.S. Bureau of Soils. The classification is based upon the percentages of sand, silt, and clay sizes making up the soil. These percentage limits and designations of soil types are shown in the triangular representation of the classification. Sand sizes 2–0.05 mm, silt sizes 0.05–0.005 mm, clay sizes smaller than 0.005 mm.

Variation of this triangular classification, made by the Lower Mississippi Division of the U.S. Engineer Department, does not use the agricultural term "loam," using instead the terms sand, silty sand, clay sand, clay silt, sandy silt, silty clay, and clay.

Other textural classifications are based upon the entire grain size distribution curve and its relation to a set of master curves.

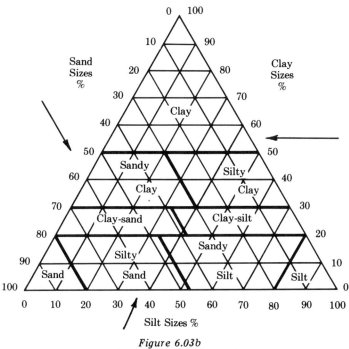

Figure 6.03b
U.S. Engineer Department

Any classification system based upon grain size alone is of relatively little importance to the engineer because of the slight dependence of the physical properties upon the grain size or the grain size distribution.

6.04   USE CLASSIFICATION

In order to avoid descriptions and in order to link physical properties of soils into the classification, several attempts have been made to classify soils with regard to their fitness for certain purposes. Although these use classifications have been in use for a considerable time, they have not been entirely satisfactory and have largely been abandoned in favor of the Unified Classification.

a. Bureau of Public Roads Classification

The U.S. Bureau of Public Roads, then Public Roads Administration, about 1929 classified soils in groups from A-1 to A-8, inclusive.

Soils fall into one of these classes depending upon grading of coarse and fine materials, plasticity limits, stability of binder, shrinkage, capillarity, and other properties which determine their fitness for use as a road surface or a base under a thin bituminous wearing surface. The PR system is deficient for classifying subgrades and is too complicated for practical use. Several modifications have been tried and others proposed, all contributing to the general confusion regarding classifications.

b. Texas Highway Department Classification

The Texas Highway Department has divided soils for use in highway construction into ten groups numbered 1 to 10. It is open to the same objections for general use as the PR classification. It has now been superceded by the Unified Classification.

c. Civil Aeronautics Administration Classification

The Civil Aeronautics Administration (CAA) has classified soils into ten groups, E-1 to E-10 inclusive, based upon grain size distribution, plasticity, frost heave, shrinkage, and the California Bearing Ratio (CBR). This classification has also been superceded by the Unified Classification.

6.05   UNIFIED CLASSIFICATION

The Unified Classification is a simple classification that owes many of its features to the Airfield Classification system that was developed by A. Casagrande and taught by him at Harvard University in army classes on "Control of Soils in Military Construction" during the period from 1942 to 1944. The system is based upon both grain size and plasticity properties of the soil and is, therefore, applicable to any use. The system is so simple that fairly accurate classifications of soils can be made by sight and feel as soils are examined in the field. The Unified Classification System was adopted jointly by the Corps of Engineers; U.S. Army and the U.S. Bureau of Reclamation in 1952. It has now been adopted by most of the agencies of the U.S. Government and state highway departments.

Soil types are designated by the following symbols:
Coarse grained soils are divided into two groups.

> G—Gravel and gravelly soils.
> S—Sand and sandy soils.

These two groups are further subdivided into four groups depending upon grading and inclusion of other materials.

> W—Well graded, clean.
> C—Well graded with excellent clay binder.
> P—Poorly graded, fairly clean.
> F—Containing fine materials not covered in other groups.

These symbols used in combination designate the type of coarse grained soils; i.e., GW means well graded, clean gravel; SC means well graded sand with excellent binder.

Fine grained soils are divided into four types:

> M—Inorganic silts and very fine sands.
> C—Inorganic clays.
> O—Organic silts and clays.
> Pt—Peat.

Because the liquid limit is a good index of compressibility, these three types of fine grained soils are grouped according to liquid limit:

> L—Liquid limit below 50 indicating low to medium compressibility.   Low plasticity.
> H—Liquid limit above 50 indicating high compressibility.   High plasticity.

Combinations of these symbols indicate soil classes, such as: ML means an inorganic silt with low to medium compressibility; CH means a highly compressible clay.

A weakness of these classification systems is that they deal with remolded soils and tell little or nothing of the structure of the soil in place.   This weakness can be alleviated to some extent by describing the soil according to origin.

6.06   SOIL TYPES ACCORDING TO ORIGIN

The structure and some of the physical properties of soil deposits are determined to a considerable extent by their origin and method of deposition.   Both geologists and engineers are familiar with the physical properties characteristic of certain types of soil.

Some terms indicating the origin and method of deposition are:

Residual....... A soil that was formed by weathering of the parent rock and still occupies the position of the rock from which it was formed. Residual soils are not as common as transported soils.

Transported.... Any soil that has been tranported from its place of origin by wind, water, ice or other agency and has been redeposited. Transported soils are classified according to the transporting agency and method of deposition as:

Alluvial......... Soils that have been deposited from suspension in running water.

Lacustrine....... Soils that have been deposited from suspension in quiet fresh water lakes.

Marine. .... Soils that have been deposited from suspension in salt water.

Aeolian ...... Soils that have been transported by wind.   Usually fine sands deposited in dunes, or silt which forms loess.

Glacial......... Deposits that have been transported by ice.

Some soils that have been formed by these methods of transportation and deposition are:

Loess................. A loose deposit of windblown silt that has been weakly cemented with calcium carbonate and montmorillonite.   Loess was formed in arid and semi-arid regions where the dust deposited by wind was cemented by a small amount of calcium carbonate left by the evaporation of seepage water or by montmorillonite.   Loess stands in nearly vertical banks.

Modified Loess.......... Loess that has been made denser by the collapse of the loess structure from immersion in water or decomposition.

Adobe................ Windblown clay similar to loess, except with a different structure that was formed by deposition in shallow water.

Caliche................ A soil (not calcareous siltstone) whose grains are rather strongly cemented with a fairly large amount of calcium carbonate. Caliche, like loess, was probably deposited by wind in a semiarid climate and later cemented by the calcium carbonate left from the evaporation of capillary water.   It contains much more calcium carbonate than loess.

Tuff.................. Small grained slightly cemented volcanic ash that has been transported by wind or water.

Bentonite.............. Chemically weathered volcanic ash or scoria.

Glacial Till............. Typically, a mixture of boulders, gravel, sand, silt, and clay as deposited by glaciers and not transported or segregated by water.

Glacial Outwash......... Gravel, sand, and silt formed by glacial action and transported by water from melting ice at the lower end of a glacier.

Varved Clay............ Alternate thin layers of silt and clay deposited in fresh water glacial lakes by outwash from glaciers. The silt is deposited in warm weather of summer during heavy runoff and clay is deposited during cool weather of fall and spring during small runoff. Generally, one band of silt and clay is deposited each year.

Marl.................. Very fine grained calcium carbonate soil of marine origin.

Gumbo................ Sticky, plastic, dark colored clay.

Peat.................. A highly organic soil consisting almost entirely of vegetable matter in varying stages of decomposition, brown to black in color and possessing a strong organic odor.   Peat is in general highly compressible.

Hardpan.............. A layer of extremely hard cohesive soil that can hardly be drilled with ordinary earth boring tools.

6.07   DESCRIPTION OF SOILS

Although complete description of soils cannot be considered as a classification system, description is necessary as a complement to any classification system.   Soils are described using generally accepted terminology based upon visual and manual inspections.   The properties usually considered in the description are grain size, grain shape, gradation, compactness, consistency, density, structure, plasticity, and dry strength.

Some of the terms and definitions used in describing soils are as follows:

Grain Size	*Size*	*Sieve No.*
Gravel		
Coarse..........................	100–20.0 mm $4''-\frac{3}{4}''$	
Medium.......................	20.0–6.0 mm $\frac{3}{4}''-\frac{1}{4}''$	State Size
Fine...........................	6.0–2.0 mm $\frac{1}{4}''-\frac{5}{32}''$	

Sand

Coarse	2.0–0.6 mm	No. 8–28
Medium	0.6–0.2 mm	No. 28–65
Fine	0.2–0.06 mm	No. 65–200

Silt

Coarse	0.06 –0.02 mm	No. 200
Medium	0.02 –0.006 mm	Dust when dry
Fine	0.006–0.002 mm	Not sticky when wet

Grain Shape

Angular.............. Corners and edges sharp and unworn.

Subangular.......... Corners worn off.   Angles not worn off.

Subrounded.......... Corners and angles worn off.   Flat surfaces remain.

Rounded............. Worn to almost spherical shape.

Gradation

Uniform.............. All grains approximately the same size.

Poorly graded........ Two or more sizes predominate.
Intervening sizes sparse.

Well graded.......... All sizes present from coarsest to finest.

Compactness or Density

Loose................ Large volume of voids.   Volume of cohesionless soils can be reduced by vibration.

Medium.............. Intermediate between dense and loose.

Dense................ Firm, compact.

Consistency

Very soft............. Core (height twice diameter) sags under own weight. Easily penetrated several inches by fist. $c < 0.125$ tons ft^{-2}.

Soft................. Can be pinched in two between thumb and forefinger. Easily penetrated several inches by thumb. $c = 0.125$–0.25 tons ft^{-2}.

Medium stiff......... Can be imprinted easily with fingers.   Can be penetrated several inches by thumb with moderate pressure. $c = 0.25$–0.5 tons ft^{-2}.

Stiff................. Can be imprinted with considerable pressure from fingers.   Readily indented by thumb but penetrated only with great effort. $c = 0.5$–1 tons ft^{-2}.

Hard................ Cannot be imprinted by fingers.   Indented with difficulty by thumbnail. $c = $ over 1 ton ft^{-2}.

Other terms for describing consistency.

Brittle.............. Ruptures with little deformation.

Friable.............. Crumbles or pulverizes easily.

Elastic.............. Returns to original length after small deformation.

Spongy.............. Is very porous, loose, and elastic.

Sticky.............. Adheres or sticks to tools and hands.

Degree of Plasticity
See Chapter on Plasticity

Cohesion near Plastic Limit
Weak

Firm

Tough

Very tough

Dry Strength
See Chapter on Plasticity

Structure
Slickensided......... Cut by fracture planes which are slick and glossy in appearance and constitute planes of weakness.

Fissured.............. Containing shrinkage cracks, frequently filled with fine sand or silt, usually more or less vertical.

Friable.............. Having a structure, usually dry, that can be powdered or broken easily with the hands.

Crumbly............. Pertaining to cohesive soils which break into small blocks and crumbs on drying; very active and usually organic; typical of clays identified as "buckshot."

Marly............... Containing particles and pieces of calcium carbonate.

Varved.............. Composed of thin laminae of varying color, usually grading from sand or silt at the bottom to clay at the top of each layer.

Interbedded......... Composed of alternate layers of different soil textures.

6.08   GUIDE FOR IDENTIFICATION AND CLASSIFICATION

To aid in the identification and classification of soils in the Unified System, A. A. Wagner of the Bureau of Reclamation devised a tabular guide similar to the one which is reproduced on a following page.

A. Casagrande devised a Plasticity Chart which is useful for identifying and classifying soils. This chart bases the classification upon the relationship between liquid limit w_L and the plastic index I_P. A line

Table 6.08a

Guide for Identification and Classification of Soils

Suggested by A. A. Wagner Guide Table

Field		Identification			Symbol	Typical Names
Coarse Grained Soils						
Gravels — More than 50% larger than ¼ in. — Clean Gravel		Wide range in grain size. Well graded.			GW	Well graded gravels, gravel-sand mixtures with few or no fines.
		Predominately one size. Poorly graded.			GP	Poorly graded gravels, gravel-sand mixtures with few or no fines.
Gravel with fines		Containing non-plastic fines.			GM	Silty gravels, poorly graded gravel-sand-silt mixtures.
		Containing plastic fines.			GC	Clayey gravels, poorly graded gravel-sand-clay mixtures.
Sands — More than 50% less than ¼ in. — Clean Sand		Wide range in grain size. Well graded.			SW	Well graded sands, gravelly sands with few or no fines.
		Predominately one size. Poorly graded.			SP	Poorly graded sands, gravelly sands with few or no fines.
Sand with fines		Containing non-plastic fines.			SM	Silty sands, poorly graded sand-silt mixtures.
		Containing plastic fines.			SC	Clayey sands, poorly graded sand-clay mixtures.
		Dry Strength	**Reaction to shaking**	**Toughness**		
Fine Grained Soils						
Silts & Clays — Liquid Limit less than 50%		None to slight	Quick to slow	None	ML	Inorganic silts and very fine sands, rock flour, silty or clayey fine sands.
		Medium to slight	None to slow	Medium	CL	Inorganic clays of low plasticity, gravelly clays, sandy clays, silty clays.
		Slight to medium	Slow	Slight	OL	Organic clays and organic silt-clay mixtures.
Silts & Clays — Liquid Limit greater than 50%		Slight to medium	Slow to none	Slight to medium	MH	Inorganic silts, micaceous fine sandy or silty soils, elastic silts.
		High to very high	None	High	CH	Inorganic clays of high plasticity. Fat clays.
		Medium to high	None to very slow	Slight to medium	OH	Organic clays of medium to high plasticity.
Highly organic soils		Identified by color, odor, spongy feel or fibrous texture			Pt	Peat or other highly organic soils.

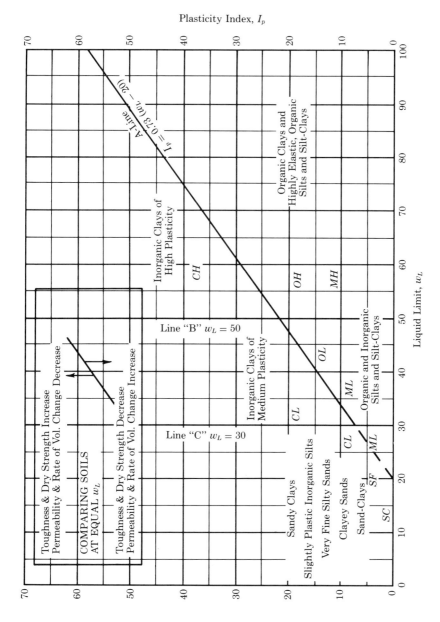

CHART 6.08a

PLASTICITY CHART

After A. Casagrande

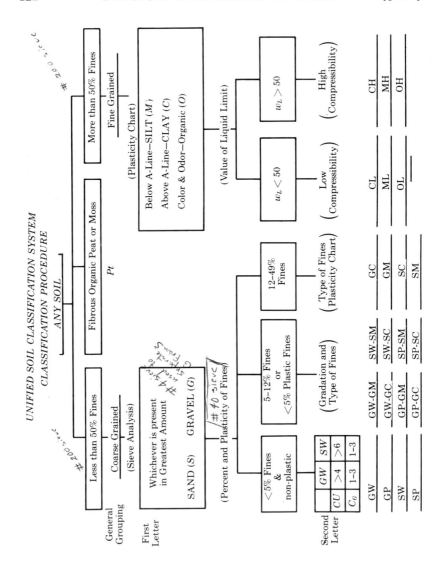

CHART 6.08b

From USAR Lesson Plan

Soils Engineering and Geology

U. S. Army Engineer School

Table 6.08b
Soil Classification for Airfield Projects

1		2	3	4	
Major Divisions		Soil Groups & Typical Names	Group Symbols	General Identification	
				Dry Strength	Other Pertinent Exams
COARSE GRAINED SOILS	Gravel and Gravelly Soils	Well Graded Gravel & Gravel-Sand Mixtures, Little or No Fines	GW	None	Gradation, Grain Shape
		Well Graded Gravel-Sand-Clay Mixtures, Excellent Binder	GC	Med. to High	Gradation, Grain Shape, Binder Exam. Wet & Dry
		Poorly Graded Gravel & Gravel-Sand Mixtures, Little or No Fines	GP	None	Gradation, Grain Shape
		Gravel with Fines, Very Silty Gravel, Clayey Gravel, Poorly Graded Gravel-Sand-Clay Mixtures	GF	Very Slight to High	Gradation, Grain Shape, Binder Exam. Wet & Dry
	Sands and Sandy Soils	Well Graded Sands & Gravelly Sands, Little or No Fines	SW	None	Gradation, Grain Shape
		Well Graded Sand-Clay Mixtures, Excellent Binder	SC	Med. to High	Gradation, Grain Shape, Binder Exam. Wet & Dry
		Poorly Graded Sands, Little or No Fines	SP	None	Gradation, Grain Shape
		Sand with Fines, Very Silty Sands, Clayey Sands, Poorly Graded Sand-Clay Mixtures	SF	Very Slight to High	Gradation, Grain Shape, Binder Exam. Wet & Dry
FINE GRAINED SOILS — Little or No Coarse Matter.	Fine Grained Soils Having Low to Medium Compressibility	Silts (Inorganic) & Very Fine Sands, Rock Flour, Silty or Clayey Fine Sands with Slight Plasticity	ML	Very Slight to Medium	Examination Wet (Shaking Test & Plasticity)
		Clays (Inorganic) of Low to Medium Plasticity, Sandy Clays, Silty Clays, Lean Clays	CL	Medium to High	Examination in Plastic Range
		Organic Silts & Organic Silt-Clays of Low Plasticity	OL	Slight to Medium	Examination in Plastic Range, Odor
	Fine Grained Soils Having High Compress.	Micaceous or Diatomaceous Fine Sandy & Silty Soils, Elastic Silts	MH	Very Slight to Medium	Examination (Wet Shaking Test & Plas.)
		Clays (Inorganic) of High Plasticity, Fat Clays	CH	High	Examination in Plastic Range
		Organic Clays of High to Medium Plasticity	OH	High	Examination in Plastic Range, Odor
Fibrous Soils V. High Comp.		Peat and Other Highly Organic Swamp Soils	Pt	Readily Identified	

Table 6.08b (Continued)
Soil Classification for Airfield Projects

3	5	6	7	8	
Group Symbol	Observations and Tests Relating to Material in Place	Principal Classification Tests (On Disturbed Samples)	Value as Foundation when not Subject to Frost Action	Value as Wearing Surface for Stage or Emergency Construction	
				With Dust Palliative	With Bituminous Surface Treatment
GW	Dry Unit Weight or Void Ratio,	Mechanical Analysis	Excellent	Fair to Poor	Excellent
GC	Degree of Compaction,	Mech. Analysis, Liquid & Plastic Limits on Binder	Excellent	Excellent	Excellent
GP	Cementation, Durability of Grains,	Mechanical Analysis	Good to Excellent	Poor	Poor to Fair
GF	Stratification & Drainage Characteristics,	Mechanical Analysis, Liquid & Plastic Lim. on Binder if Applicable	Good to Excellent	Poor to Good	Fair to Good
SW	Ground Water Condition,	Mechanical Analysis	Excellent to Good	Poor	Good
SC	Traffic Tests,	Mech. Analysis, Liquid & Plastic Limits on Binder	Excellent to Good	Excellent	Excellent
SP	Large Scale Load Tests or	Mechanical Analysis	Fair to Good	Poor	Poor
SF	California Bearing Tests.	Mechanical Analysis, Liquid & Plastic Lim. on Binder if Applicable	Fair to Good	Poor to Good	Poor to Good
ML	Dry Unit Weight, Water Content & Void Ratio,	Mechanical Analysis, Liquid & Plastic Lim. on Binder if Applicable	Fair to Poor		Poor
CL	Consistency—Undisturbed & Remolded,	Liquid & Plas. Limits	Fair to Poor		Poor
OL	Drainage & Ground Water Conditions,	Liq. & Plas. Limits before & after Drying	Poor	Very Poor	Very Poor
MH	Stratification, Root Holes, Fissures, etc.	Mechanical Analysis, L. & P. Lim. if Appl.	Poor		Very Poor
CH	Large Scale Load Tests, Traffic Tests,	Liquid & Plastic Lim.	Poor to Very Poor		Very Poor
OH	California Bearing Test, Compres. Test.	Liq. & Plas. Limits before & after Drying	Very Poor		Useless
Pt	Consistency, Texture, and Natural Water Content		Extremely Poor		Useless

Table 6.08b (Continued)
Soil Classification for Airfield Projects

3 Group Symbol	9 Potential Frost Action	10 Shrinkage, Expansion, Elasticity	11 Drainage Characteristics	12 Compaction Characteristics and Equipment	13* Solids at Opt. Compaction lb. per cu. ft. & Void Ratio	14 Cal. Bearing Ratio for Compacted & Soaked Specimen	15 Comparable Groups in Public Roads Classification
GW	None to Slight	Almost None	Excellent	Excellent, Tractor	>125 e < 0.35	>50	A-3
GC	Medium	Very Slight	Practically Impervious	Excellent, Tamp. Roller	>130 e < 0.30	>40	A-1
GP	None to V. Slight	Almost None	Excellent	Good, Tractor	>115 e < 0.45	25–60	A-3
GF	Slight to Medium	Almost None to Essential	Fair to Practically Impervious	Good, Close Control Rubber Tired Roller, Tractor	>120 e < 0.40	>20	A-2
SW	None to V. Slight	Almost None	Excellent	Excellent, Tractor	>120 e < 0.40	20–60	A-3
SC	Medium	Very Slight	Practically Impervious	Excellent, Tamp. Roller	>125 e < 0.35	20–60	A-1
SP	None to V. Slight	Almost None	Excellent	Good, Tractor	>100 e < 0.70	10–30	A-3
SF	Slight to High	Almost None to Medium	Fair to Practically Impervious	Good, Close Control Rubber Tired Roller, Tractor	>105 e < 0.60	8–30	A-2
ML	Medium to Very High	Slight to Medium	Fair to Poor	Good to Poor, Control Essential Rubber Tired Roller	>100 e < 0.70	6–25	A-4 A-6 A-7
CL	Medium to High	Medium	Practically Impervious	Fair to Good, Tamp. Roller	>100 e < 0.70	4–15	A-4 A-6, A-7
OL	Medium to High	Medium to High	Poor	Fair to Poor, Tamp. Roller	>90 e < 0.90	3–8	A-4 A-7
MH	Medium to V. High	High	Fair to Poor	Poor to Very Poor	>100 e < 0.70	<7	A-5
CH	Medium	High	Practically Impervious	Fair to Poor, Tamp. Roller	>90 e < 0.90	<6	A-6 A-7
OH	Medium	High	Practically Impervious	Poor to Very Poor	<100 e > 0.70	<4	A-7 A-8
Pt	Slight	Very High	Fair to Poor	Compaction not Practical			A-8

*These weights apply only to soils having specific gravities from 2.65 to 2.75.

called the A-line drawn diagonally across the chart divides the chart into areas representing different soil types.   Generally, the areas above the line represent inorganic soils and those below represent organic soils.   The w_L and I_P for a soil plotted as a point on this chart indicates the classification to which the soil belongs.

Several organizations and departments of government have devised charts and tables as guides for identification, classification, and use of soils.   A few of these charts and tables are reproduced on pages 120 to 125.

REFERENCES

Casagrande, A., "Classification and Identification of Soils," *Trans. Am. Soc. C. E.*, Vol. 113, (1948).

Wagner, A. A., "The Use of the Unified Soil Classification System by the Bureau of Reclamation," *Proc. 4th International Conference on Soil Mechanics and Foundations*, August, 1957.

Corps of Engineers, *The Unified Soil Classification System*, Vicksburg, Mississippi: U. S. Army Engineer Waterways Experiment Station, Technical Memorandum No. 3-357, Vol. 1, March, 1953.   Revised April 1960.

PROBLEMS

6.1   (*a*) What is the ultimate purpose of a soil classification system?
(*b*) Why are existing classification systems of relatively little benefit to foundation engineers?

6.2   Describe at least three different bases upon which soil classification systems have been devised.

6.3   Keeping in mind (as an aid to your judgment) the location of various soil types on the plasticity chart, state typical values for the liquid limits and plastic limits of the following soils:
(*a*) organic silt
(*b*) highly compressible inorganic clay
(*c*) very silty inorganic clay
(*d*) highly organic clay.

6.4   Give the Unified Soil Classification symbol for the following soils.
(*a*) Sticky, red clay for which $w_L = 67$, $I_P = 45$.
(*b*) A soil for which 60 per cent passes the No. 10 sieve and 10 per cent passes the No. 200 sieve.   That fraction which passes the No. 40 sieve has a $w_L = 30$ and a $I_P = 6$.

6.5   Using the Unified Soil Classification guide sheet as an aid, determine the classification symbols for the following soils.
(*a*) Light gray, fine-grained, cohesive soil, having a w_L of 35 and a w_P of 25.

(b) Tan, coarse-grained soil for which 40 per cent passes the No. 4 sieve, 20 per cent passes the No. 40 sieve, and 4 per cent passes the No. 200 sieve. The fraction passing the No. 40 sieve has a I_P of 10.

(c) A sandy soil, all of which passes the No. 4 sieve. 43 per cent of the soil passes the No. 200 sieve. That fraction which passes the No. 40 sieve has a w_L of 27 and a w_P of 23.

6.6 Using the Unified Soil Classification guide sheet as an aid, determine the classification symbols for the following soils.

(a) A dark gray, sticky soil having a strong odor of hydrogen sulfide. *from Marshy Land* $w_L = 80$, $w_P = 50$.

(b) Dark brown, very fibrous, spongy soil, having a $w_{nat} = 420\%$. *← Peat organic*

(c) A cohesionless soil for which 60 per cent passes the No. 4 sieve and 10 per cent passes the No. 200 sieve. The fraction passing the No. 40 sieve is non-plastic.

(d) Tan colored loess: $w_L = 35$, $w_P = 32$.

(e) A light gray soil having a very high dry strength, and which becomes very slick and sticky when wet.

CHAPTER VII

Hydraulics Applied to Soil Mechanics

7.01 VISCOSITY

All molecules of matter are vibrating at a velocity which is sensed as temperature. In a solid body the molecules are close enough together that they maintain a definite relative position relationship. The molecular bond at this close range is so strong that this relationship is maintained against the force of gravity. Shearing force causes distortion of this arrangement without destroying the relative positions. Once the force is released, the molecular bond causes the molecules to return to their original positions (elastic rebound).

As the mass absorbs energy, the velocity of the molecules (temperature) increases, forcing the molecules farther apart and increasing the volume (thermal expansion). When the velocity has become great enough to force the molecules far enough apart, the bond between molecules becomes so small that the molecules can no longer maintain their relative rigid positions against the force of gravity. In this state the mass is liquid.

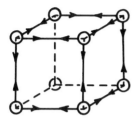

Figure 7.01a
Simple   Molecular
Arrangement of
Solid Crystal

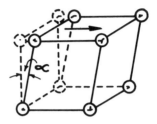

Figure 7.01b
Molecular Ar-
rangement Dis-
torted by Shear

Although in the liquid state, the molecules do not maintain their relative positions, a bond between molecules continues to exist so that the material in the liquid state possesses considerable resistance to being pulled apart (tensile strength) and some resistance to having their positions changed relative to each other (shear strength). The force required to move the molecules relative to each other is dependent upon the velocity with which the molecules are moved.

As discussed earlier, the molecules of a liquid are attracted to the surface of some materials so that at the surface the molecules of the liquid become immobile. Therefore, if a surface of such a material is moved through the liquid, the resistance to movement is produced by the resistance between liquid molecules and not between the molecules of the other material and the molecules of the liquid.

Newton discovered that, if two plates that will be wet by a liquid are separated by a sheet of the liquid, the shear resistance offered by the sheet of liquid is proportional to the area A; inversely proportional to the thickness of the sheet of liquid d; and directly proportional to the velocity with which the two surfaces are moved relative to each other v.

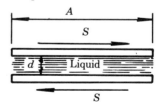

Figure 7.01c
Plates Separated by
Sheet of Liquid

$\dfrac{S}{S_1} = \dfrac{A d_1 v}{A_1 d v_1}$ which may be written $\dfrac{S}{A} = \dfrac{S_1 d_1 v}{A_1 v_1 d} = s_1 \dfrac{d_1 v}{v_1 d}.$ If s_1 be unit shear in g cm^{-2} for a velocity v_1 of 1 cm sec^{-1} and a distance d_1 of 1 cm, then $s_1(d_1/v_1)$ becomes a constant expressing a property of the liquid under a given set of conditions of temperature, etc. This constant is called absolute viscosity, η, and has the dimensions g sec cm^{-2}.

The shear resistance of a liquid can then be expressed as $s = \eta \dfrac{v}{d}.$

$\dfrac{v}{d}$ may be referred to as the average velocity gradient. The instantaneous velocity gradient is dv/dd.

Often in relationships involving viscosity, the term η/γ_w occurs. Because both viscosity and unit weight of water are affected by temperature, it is sometimes convenient to use this ratio as a single property. In this context the ratio of absolute viscosity to the unit weight of water is designated as the kinematic viscosity ν having the dimensions cm sec. Sometimes, kinematic viscosity is defined as the absolute viscosity η (g sec cm^{-2}) divided by the mass density ρ (g sec^2 cm^{-4}), in which case the dimensions of kinematic viscosity are cm^2 sec^{-1}. If kinematic viscosity is defined as the viscosity in poises (dyne sec cm^{-2}) divided by the mass density ρ, the dimensions of kinematic viscosity are the same as used in this context, cm sec.

7.02   SETTLEMENT OF SPHERE IN SUSPENSION— STOKES' LAW

The relationship expressing the rate of settlement of a sphere out of suspension in a fluid was first derived rationally by the English physicist George Gabriel Stokes and published in the *Philosophical Magazine* in 1846 and in *Transactions* of the Cambridge Philosophical Society in 1849. The rational development of this relationship, known as Stokes' Law, involves the use of mathematics beyond the scope of most students beginning a study of Soil Mechanics.

Stokes' Law has been verified experimentally and can be derived by the use of experimental data by equating the settlement force F_s to the force F_r resisting the movement of the body through the fluid. The force resisting the settlement of the body through the fluid is often called the drag force.

The settlement force is the weight of the body in air minus the weight of the displaced fluid. For a sphere this settlement force is

$$F_s = \tfrac{4}{3}\pi R^3(\gamma_s - \gamma_f). \quad \text{(Eq. 7.02a)}$$

The force resisting settlement consists of three parts; force producing acceleration of the particle mass, force producing acceleration of the displaced fluid mass, and shearing force due to viscosity of the fluid as the body moves through the fluid. After the settling body has attained a uniform velocity with the settling and resisting forces F_s and F_r balanced, the force producing acceleration is zero. If the velocity of settlement is slow enough that the flow of the fluid around the body is laminar at every point, the pressure of the fluid below the body resisting movement is balanced by the pressure above the body as the fluid flows back to its original relative position. Under these conditions, the resistance to settlement is offered by the viscous shear strength of the fluid.

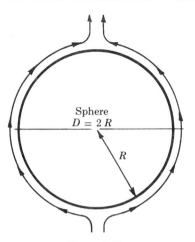

Figure 7.02a
Sphere Settling Out of Suspension

If it is assumed that the resistance to settlement is proportional to some power of each of the following entities; the mass density of the fluid $\left(\rho_f = \dfrac{\gamma_f}{g}\right)$, the projected area A of the body normal to the direction of movement, the velocity of settlement v, and the viscosity of the fluid η, the drag resistance can be written as

$$F_r = k\rho_f{}^a A^b v^c \eta^d. \quad \text{(Eq. 7.02b)}$$

Dimensionally this relationship is

$$g^1 = k\left(\frac{g\ cm^{-3}}{cm\ sec^{-2}}\right)^a (cm^2)^b (cm\ sec^{-1})^c (g\ sec\ cm^{-2})^d.$$

Equating exponents for dimensional correctness

for g $1 = a + d$
for sec $0 = 2a - c + d$
for cm $0 = -4a + 2b + c - 2d.$

Solving for the values of coefficients a, b, and c in terms of d

$$a = 1 - d$$
$$c = 2 - 2d + d = 2 - d$$
$$b = 2 - 2d - 1 + \frac{d}{2} + d = 1 - \frac{d}{2}.$$

Inserting these values of the coefficients equation 7.02b becomes

$$F_r = k\rho^{(1-d)} A^{\left(1-\frac{d}{2}\right)} v^{(2-d)} \eta^d$$
$$F_r = k \frac{\rho}{\rho d} \frac{A}{A^{d/2}} \frac{v^2}{v^d} \eta^d = k\rho A v^2 \left(\frac{\eta}{\rho A^{\frac{1}{2}} v}\right)^d.$$

Dimensionally, the square root of A is a distance D allowing the equation to be written as

$$F_r = k\rho A v^2 \left(\frac{\eta}{\rho D v}\right)^d. \qquad \text{(Eq. 7.02c)}$$

Reynold's number N_R is the dimensionless ratio $Dv\rho/\eta$ which is the reciprocal of the term $\eta/\rho Dv$ in the equation above. Equation 7.02c can be written

$$F_r = \rho A v^2 \left(\frac{k}{N_R{}^d}\right) = \frac{\rho A v^2}{2}\left(\frac{2k}{N_R{}^d}\right).$$

The term $2k/N_R{}^d$ is often called the drag coefficient and designated C_D,

$$F_r = C_D \frac{\rho A v^2}{2}. \qquad \text{(Eq. 7.02d)}$$

It has been determined experimentally that only viscous flow exists around a settling sphere at values of N_R less than 1 and that, in this range of laminar flow, the value of C_D for a sphere is $24/N_R$. Inserting this value of C_D in equation 7.02d

$$F_r = \frac{\rho A v^2}{2}\left(\frac{24\eta}{Dv\rho}\right) = \frac{\rho\pi R^2 v^2}{2}\frac{24\eta}{2Rv\rho} = 6\pi Rv\eta.$$

Equating this value of F_r to the value of F_s in equation 7.02a

$$\tfrac{4}{3}\pi R^3(\gamma_s - \gamma_f) = 6\pi Rv\eta.$$

Solving for v

$$v = \frac{2}{9}\frac{(\gamma_s - \gamma_f)}{\eta} R^2 \text{ or } \frac{2}{9}\frac{(\gamma_s - \gamma_f)}{\eta}\left(\frac{D}{2}\right)^2,$$

which is Stokes' Law.

The derivation of Stokes' Law given above is not the same as presented by Stokes. It is not presented here as proof of Stokes' Law but as a simple means of showing the conditions under which Stokes' Law is valid.

Obviously, Stokes' Law applies only to a single sphere settling out of suspension. The law does not apply to a body of heavy enough unit weight, or of a large enough size, or in a fluid of low enough viscosity to produce a velocity that will cause turbulent flow of the fluid around the sphere.

7.03   TYPES OF FLOW

The flow of water through a passageway may be laminar or turbulent. In laminar flow a droplet of water moves directly in a straight line from one position to another with the least possible loss of energy. In turbulent flow a droplet takes a devious path and expends more energy in moving the same distance along its line of flow than in laminar flow. The two types of flow may be illustrated by allowing water to flow from a reservoir through a transparent tube provided with an adjustable outlet for controlling the head loss through the tube. If a glass tube with a fine nozzle be inserted into the reservoir with the nozzle pointing into the entrance of the tube, a dye introduced through the nozzle will show the path of flow as the water passes through the tube.

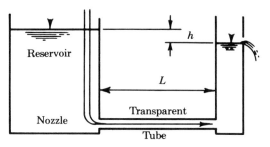

Figure 7.03a
Laminar Flow

When the velocity through the tube is made very low by reducing the head, the dye will pass through the tube in a straight line indicating laminar or streamline flow. When the velocity is increased sufficiently, the dye will take a devious path as it passes through the tube. A little experimentation will show that from a

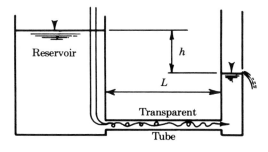

Figure 7.03b
Turbulent Flow

condition of laminar flow, as the velocity is gradually increased by increasing the head loss through the tube, the flow will remain laminar up to a certain velocity when it will change to turbulent.   Then, after turbulent flow has been established and the velocity is gradually reduced, the flow will again become laminar at some lower velocity. But, the velocity at which the change from laminar to turbulent flow

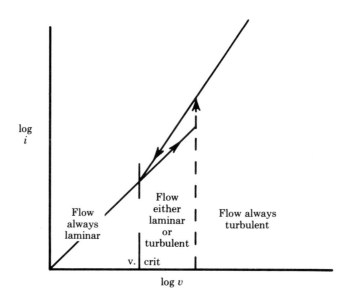

Figure 7.03c
Relation for Laminar and Turbulent Flow Between Hydraulic Gradient and Velocity

takes place is considerably higher' than the velocity at which the change takes place in passing from turbulent to laminar.   Once the flow has been established it tends to remain in that state until forced to change.   Below a certain velocity under a given set of conditions the flow will always be laminar and above a certain velocity the flow will always be turbulent.   The velocity below which the flow is always laminar is called the critical velocity v_{crit}.

The velocity through the tube is a function of the loss in head per unit of length of the tube, h/L, which is the meaning of hydraulic gradient i.   It will be shown later that the velocity during laminar flow is proportional to the first power of the hydraulic gradient.   The velocity during turbulent flow is proportional to approximately $\frac{4}{7}$ power of the hydraulic gradient.   If v be plotted against i to logarithmic scales, the line indicating the relationship for laminar flow will be a straight 45° line and that indicating turbulent flow will be a straight line at a steeper angle than 45°.

Reynolds determined experimentally that the critical velocity of water as it flows through a round tube is dependent slightly upon the temperature and is inversely proportional to the diameter of the tube.

The loss in energy within the mass of water due to turbulent flow as it moves to a lower potential is illustrated by the fact that a loaded boat displacing a large volume of water will float down a river in which the flow is turbulent at a faster rate than small pieces of debris or foam on the surface and which follows the devious path of the turbulent water.

7.04   FLOW THROUGH CAPILLARY TUBE—
POISEUILLE'S LAW

Early in the 19th century in considering the flow of blood in capillaries, Poiseuille first worked out the relationship governing the laminar flow of liquids through tubes. Poiseuille's work was based on the earlier work of Newton which was discussed under "Viscosity."

During laminar flow any particle of water flows in a direct line and there are no sudden jumps in velocity. At the surface of the tube of material which will be wet by water, the molecular attraction holds a very thin layer of water immobile so that at the surface the velocity is zero. The velocity increases gradually to a maximum at the center.

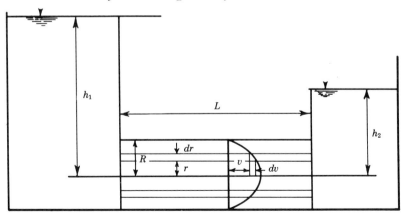

Figure 7.04a
Laminar Flow Through Tube

Let Figure 7.04a represent the conditions for laminar flow of water through a tube of radius R and length L with an average entrance head h_1 and a discharge head h_2.   If v_r is the velocity at distance r from the center, the velocity gradient at r is $-\dfrac{dv}{dr}$ and the unit shear at distance from the center is $s_r = -\eta \dfrac{dv}{dr}$.

The forces acting on a cylinder of water of radius r taken from the center of the tube are shown in Figure 7.04b.

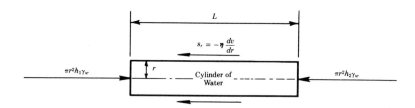

Figure 7.04b
Free Body for Cylinder of Water from Center of Tube

Equating forces acting on the cylinder

$$s_r L 2\pi r = (h_1 - h_2)\pi r^2 \gamma_w.$$

$$2\pi r L \eta \left(-\frac{dv}{dr}\right) = (h_1 - h_2)\pi r^2 \gamma_w$$

$$dv = -\frac{h_1 - h_2}{L}\frac{\gamma_w}{2\eta} r\, dr$$

$$v_r = -\frac{h_1 - h_2}{L}\frac{\gamma_w}{4\eta} r^2 + C.$$

At $r = R$, $v = 0$, so

$$C = \frac{h_1 - h_2}{L}\frac{\gamma_w}{4\eta} R^2$$

and

$$v_r = \frac{h_1 - h_2}{L}\frac{\gamma_w}{4\eta} (R^2 - r^2).$$

Now consider the quantity of water flowing in the thin cylindrical sheet dr thick as shown in Figure 7.04a.

$$dq = v_r 2\pi r\, dr = \frac{h_1 - h_2}{L}\frac{\gamma_w}{4\eta} (R^2 - r^2)2\pi r\, dr.$$

The total quantity of water flowing in the tube per unit of time is

$$q = \frac{h_1 - h_2}{L} \frac{\gamma_w}{4\eta} 2\pi \int_0^R (R^2 - r^2) r \, dr$$

$$q = \frac{h_1 - h_2}{L} \frac{\gamma_w}{8\eta} \pi R^4.$$

Since $\dfrac{h_1 - h_2}{L}$ = the hydraulic gradient i,

$$q = \frac{\gamma_w \pi R^4}{8\eta} i.$$

If a = area of the tube, the average velocity $= \dfrac{q}{a} = \dfrac{q}{\pi R^2} = v_{av}$

$$v_{av} = \frac{\gamma_w R^2}{8\eta} i \text{ or } \frac{\gamma_w}{8\eta\pi} ia$$

which is a statement of Poiseuille's Law and indicates that the velocity during laminar flow is proportional to the first power of the hydraulic gradient.

7.05 SURFACE TENSION

The molecules on the surface of a liquid are attracted by other molecules on the surface and inside the body of the liquid.   Because there is no pull from outside, the surface molecules are pulled toward the inside of the liquid mass tending to reduce the surface to a minimum. The resultant surface tension is constant in all directions and independent of the surface area.   When the surface is increased more molecules come to the surface so that the pull between molecules is the same along a unit of length regardless of the extent of the surface.

The existence of surface tension may be illustrated by floating a loop of fine thread on the surface of water containing dissolved soap.   The loop may be moved about into any shape while it lies on the surface. A loop of wire larger than the thread may be passed carefully under the thread and an unbroken soap film with the thread loop lifted free. The loop of thread retains the same shape as that in which it was formed so long as the film is unbroken over the entire area of the wire loop; but, when the film inside the thread is broken, the thread loop is pulled into a perfect circle by the tension of the remaining film between the thread and the wire loop.   A further illustration is that of a greased steel needle which can be floated on the surface of water, but which sinks when pushed beneath the surface.

Surface tension exists also on the surface of solids, but compared to the shearing resistance of the material is too small to be significant.

When a piece of material that will be wet with water is inserted

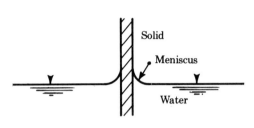

Figure 7.05a
Meniscus Around Rod

vertically into a vessel of water so that it extends below and above the surface, the water molecules, due to the attraction between the molecules of water and the material, climb the solid surface forming a curved meniscus adjacent to the solid. The water in the meniscus is supported above the horizontal surface by the tensile strength due to the attraction of the surface molecules for each other and for the tube wall.

The surface tension for water, designated T_s, is approximately constant for ordinary temperatures and is equal to 75 dynes per cm or 0.076 g cm^{-1}.

7.06   RELATION BETWEEN DIAMETER OF TUBE AND CAPILLARY RISE

When the lower end of a hollow tube of material which will be wet by water is inserted vertically below the surface of water, the water climbs the inside of the tube with a concave meniscus surface. The water in the tube hangs in tension from the surface film as from an inverted dome supported from the side of the tube around the edge of the meniscus.   If the tube is inserted into a liquid which will not wet the tube, a convex meniscus will be formed and the surface of the liquid inside the tube will be depressed as shown in Figure 7.06a.

The vertical component of the surface tension in the tube depends upon the angle of incidence α between the meniscus and the tube surface as shown in the enlargement of A above and is equal to $T_s \cos \alpha$.

The relation between the height of rise h_1 and the diameter of the tube d may be shown by reference to Figure 7.06b.

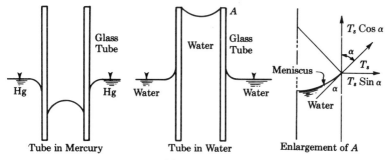

Figure 7.06a
Negative and Positive Menisci in Capillary Tubes

The column of water in the tube weighs

$$\frac{\pi d^2}{4}\, h_1 \gamma_w$$

and the vertical component of the reaction of the meniscus against the inside circumference of the glass tube is

$$\pi d T_s \cos \alpha.$$

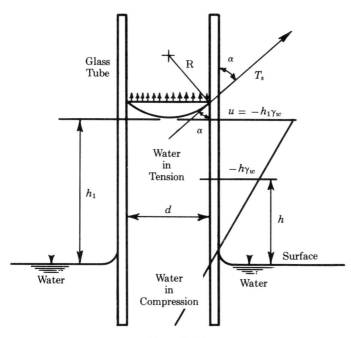

Figure 7.06b
Capillary Tube in Water

When the water has stopped moving in the tube, these two forces are in equilibrium and may be equated.

$$\frac{\pi d^2}{4} h_1 \gamma_w = \pi d T_s \cos \alpha$$

Solving for h_1,

$$h_1 = \frac{4 T_s \cos \alpha}{\gamma_w d}.$$

If the glass tube is clean and wet, the meniscus will be approximately semispherical when fully developed, in which case the angle α is zero and the maximum height of capillary rise h_c will be reached and is equal to

$$h_c = \frac{4 T_s \cos \alpha}{\gamma_w d} = \frac{4 T_s}{\gamma_w d}.$$

Since $T_s = 0.076$ g cm^{-1} and $\gamma_w = 1$ g cm^{-3} (Approx.) when d is in cm

$$h_c = \frac{4 \times 0.076}{1 \times d} = \frac{0.304}{d} \text{ cm.} \quad \text{Good Estimate}$$

For practical purposes

$$h_c = \frac{0.3}{d} \text{ cm.}$$

Checking dimensions for h_1 or h_c

$$h = \frac{\text{g cm}^{-1}}{\text{g cm}^{-3} \text{ cm}} = \text{cm.}$$

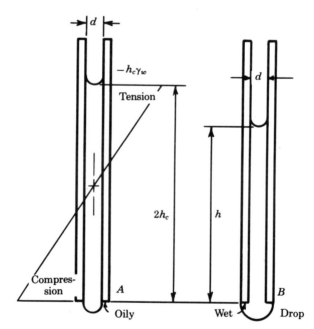

Figure 7.06c
Water in Vertical Open End Capillary Tubes

If a long glass tube of capillary size is filled with water and lifted free and held vertical as shown in Figure 7.06c, a complete meniscus of diameter d will be formed at the bottom, if the bottom of the tube is oily, and also at the top when the inside surface of the tube is clean and wet. Such a condition is shown at A. The bottom meniscus holds the water in the bottom half of the tube in compression and the top meniscus holds the water in the top half in tension. The total length of the column of water held in the tube is

$$2h_c = \frac{2 \times 0.3}{d} \text{ cm.}$$

If the bottom of the tube is wet, a drop as shown at B in Figure 7.06c will form on the bottom end of the tube. The length of the column of water in this case will depend upon the size of the drop as well as the diameter of the tube and may be more than h_c but never greater than $2h_c$.

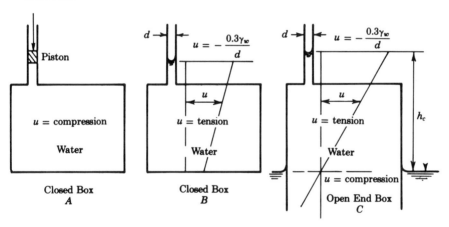

Figure 7.06d
Pressure Distribution in Vessel with Capillary Outlet

The stress in the water, u, whether tension or compression, is distributed equally in all directions throughout the water. If the water in the closed box shown at A in Figure 7.06d is compressed by a load on the piston, the compressive stress in the water in excess of the hydrostatic pressure is equal to $u = +p$ and is constant in all directions throughout the entire mass of water. If the tube is of capillary size small enough that h_c is greater than the depth of the box and the water extends up into the tube as shown at B in Figure 7.06d, the water will attempt to climb the walls of the tube, forming a meniscus subjecting the water in the box to tension equally in all directions. If the box with the open end capillary tube as in C has no bottom and is immersed so that water is up in the tube, the box may be lifted as shown in C and the water will be held in the box by the meniscus in the capillary tube and will climb the tube to a height of h_c, if h_c is greater than the depth of the box, the same as though the tube were of capillary size the full length. If the box is large enough, tons of water may be supported in tension from the meniscus in the capillary tube.

7.07 RELATION BETWEEN RADIUS OF MENISCUS AND STRESS IN WATER

The water hanging from the meniscus is stressed in tension. The stress in the water is equal to the unit weight of water times the distance from the surface, being tension above the water surface outside the tube and compression below.

$$u = h\gamma_w$$

From Figure 7.07a, if R is the radius of the meniscus,

$$\frac{d}{2} = R\cos\alpha$$

or

$$d = 2R\cos\alpha.$$

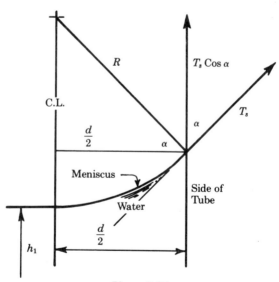

Figure 7.07a
Relation Between Radius and Stress

Placing this value of d in the equation for h_1 in terms of T_s and d,

$$h_1 = \frac{4T_s \cos \alpha}{\gamma_w 2R \cos \alpha}$$
$$= \frac{4T_s}{\gamma_w 2R}$$
$$u_1 = \gamma_w h_1 = \frac{2T_s}{R}.$$

Since T_s is constant, the maximum stress in the water is inversely proportional to the radius of the meniscus, R.

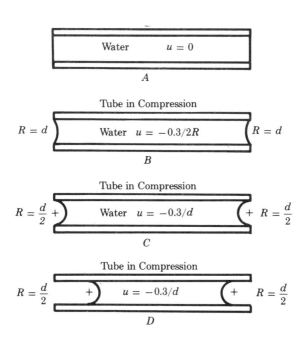

Figure 7.07b
Evaporation of Water in Capillary Tube

If a capillary tube of constant diameter is completely filled with water and laid in a horizontal position as shown in Figure 7.07b, the surface of the water is plane as shown at A and the surface tension acts radially producing no stress in the water. As evaporation takes place a meniscus is formed with the radius becoming smaller as evaporation progresses. Likewise the tension in the water increases, being at any time equal to $u = 2T_s/R$.

Since the stress in the water is transmitted throughout its volume, the radii will be equal at both ends of the tube regardless of where the evaporation takes place. After enough water has been evaporated, the menisci will become fully developed at both ends as shown at C with the radii equal to $d/2$. Since the radii of the menisci can be no further reduced, there can be no further increase in the tensile stress in the water. Further evaporation causes recession of the menisci from both ends with no further increase in the tension in the water. At every stage the menisci react against the walls of the tube stressing the walls between the menisci in compression.

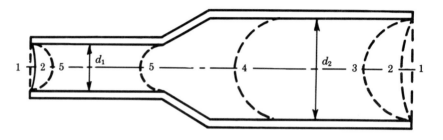

Figure 7.07c
Capillary Tube with Varying Diameter

If a horizontal capillary tube having a diameter d_1 at one end and a larger diameter d_2 at the other end as shown in Figure 7.07c is filled with water, there will be no menisci and no tension in the water in the beginning. After some evaporation, menisci of radius R_2 will be formed at both ends of the tube and a resulting tension in water of $-2T_s/R_2$. Further evaporation reduces the radius of the meniscus at each end with no recession until the meniscus at the large end is fully developed with a radius of $d_2/2$. The tension in the water increases with decreasing R. Still further evaporation cannot reduce the radius so there is a recession of the meniscus at the large end with no further increase in tension in the water until the meniscus reaches a smaller diameter of the tube. Then the meniscus at the small end cannot become fully developed and therefore does not recede until the meniscus at the large end has receded to the diameter d_1 and has become fully developed as shown at 5 in Figure 7.07c. At this stage the radius can become no smaller, the tension in the water will have become a maximum after which the two menisci will recede together with no further increase in the tension in the water and compression in the walls of the tube.

7.08   EFFECT OF TENSION UPON AIR BUBBLES

Inside a capillary tube the tension in the water may be much greater than one atmosphere regardless of whether air is present or not.   On the other hand in larger bodies of water the tension cannot be greater than one atmosphere if entrapped air is present in the water.   This

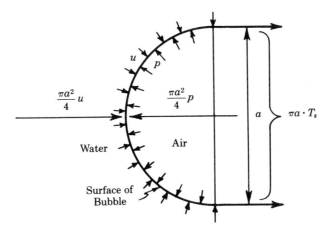

Figure 7.08a
Forces Acting on Bubble in Water

may be seen after consideration of the effect of tension in the water on an air bubble.   Call the stress in excess of the gravitational pressure in the water u and the pressure inside the bubble p.   The forces acting on the bubble are shown in Figure 7.08a.   Writing $\Sigma H = 0$,

$$\frac{\pi a^2}{4} (p - u) = \pi a T_s.$$

Solving for diameter of bubble, a

$$a = \frac{4 T_s}{p - u}$$
$$= \frac{4 \times 0.076}{p - u}$$
$$= \frac{0.3}{p - u} \text{ cm.}$$

Thus, if the pressure in the water is 1 atmosphere (zero gage or $u = 1000$ g cm^{-2} absolute) the pressure inside a bubble of 0.1 cm diameter would be 1003 g cm^{-2} absolute or 3 g cm^{-2} above atmosphere.

For the following purpose u and p are referred to atmospheric pressure; i.e., at 1 atmosphere u and p equal zero gage pressure.

If the tension in the water is 1 atmosphere (-1000 g cm^{-2}), the diameter of a bubble in which the pressure is 1 atmosphere ($p = 0$) is

$$a = \frac{0.3}{0 + 1000} = 0.0003 \text{ cm or 3 microns.}$$

Thus, for 1 atmosphere tension in the water (vacuum) the pressure will be greater than 1 atmosphere ($p > 0$ gage) in all bubbles smaller than 3 microns in diameter and less than 1 atmosphere ($p < 0$ gage) in all bubbles larger than 3 microns in diameter.   Therefore, in a tube of less than 3 microns diameter a bubble will fill the tube and part the column under a tension in the water of 1 atmosphere ($u = -1000$ g cm^{-2}) and a pressure in the bubble of 1 atmosphere ($p = 0$) as is the case in an open end tube.

Consider a bubble inside a capillary tube of diameter d filled with water and with fully developed menisci at both ends as shown in Figure 7.08b.

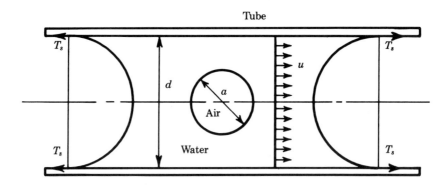

Figure 7.08b
Air Bubble Inside Capillary Tube

Equating the total stress in the water to the total pull of the meniscus at the end of the tube

$$\frac{\pi d^2}{4} u = -\pi d T_s \qquad \text{(minus because } u \text{ is tension)}$$

Solving for u

$$u = -\frac{4T_s}{d}.$$

Putting this value of u in the equation for the diameter of the bubble

$$a = \frac{4T_s d}{pd + 4T_s} = \frac{0.3d}{pd + 0.3} \text{ cm.}$$

Thus, we see that when the pressure in the bubble is 1 atmosphere ($p = 0$), which is the case for an open end tube as shown in Figure 7.08b, the diameter of the bubble equals the diameter of the tube, the bubble completely fills the tube and the water column is parted. The tension in the water will be greater or less than 1 atmosphere depending upon the diameter of the tube. If the tube is smaller than 0.0003 cm diameter, the tension in the water will be greater than 1 atmosphere even though the column may be parted by a bubble of air. The pressures on the water at the ends of the tube and in the bubble are equal, making the tension in the water dependent only on the diameter of the tube. Conversely, the pressure inside an air bubble whose diameter is equal to the diameter of the tube ($a = d$) must be equal to the pressure at the ends of the tube; i.e., p must be zero in the relationship $a = \dfrac{0.3d}{pd + 0.3} = \dfrac{0.3a}{pa + 0.3}$ or $\dfrac{0.3}{pa + 0.3} = 1.$

It will be observed from this analysis that in moving water through pipes or spaces larger than 0.0003 cm diameter continuity cannot be maintained in the water column at near vacuum pressures if there is any air present in the water. If, however, there were no air in the water, a continuous column could be maintained under a tensile stress greater than 1 atmosphere; i.e., less than absolute zero.

PROBLEMS

7.1 A liquid having an absolute viscosity of η g-sec/cm² flows under a hydraulic gradient i between two parallel plates D cm apart. Using Poiseuille's derivation as a guide, derive expressions for
(a) the rate of flow per unit of width, and
(b) the mean velocity of flow.

7.2 A single capillary tube is formed by joining together two tubes having inside diameters d_1 and d_2. The tube is filled with water and placed in a horizontal position. When menisci are formed by evaporation,
(a) at which end will the meniscus first become fully developed? Explain.

(b) Assuming the meniscus to be fully developed at one end, derive an expression for the contact angle at the other end in terms of the two diameters.

7.3 The glass vessel shown in the sketch is filled with water. A fully developed meniscus has formed in a small hole of diameter $d_1 = 0.01$ cm in the upper wall of the vessel. Another hole of diameter d_2 exists in the lower wall.

(a) What is the greatest value which d_2 may have?

(b) If $d_2 = d_1$ find the contact angle α in the lower hole.

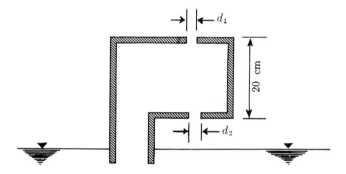

CHAPTER VIII

Permeability

8.01 GENERAL

The properties of soil materials discussed in the preceding chapters are of interest to the engineer only indirectly. The engineer is primarily interested in permeability, time-stress-deformation, and strength characteristics which he may use directly in the design of engineering structures. The properties studied up to this point are of interest mainly in that they are indicative of these needed properties. From this point on, this text will be concerned with the direct measurement of permeability, time-stress-deformation relationships, and strength of soil materials.

Permeability, or ease with which water or other fluid flows through materials, is of interest to the engineer in varying degrees in the design of structures of any material. The durability of concrete is dependent

149

upon permeability. But, in the solution of a great many problems involving earth materials, the permeability may be of primary importance.

Some of the engineering problems in which permeability is of primary importance are as follows:

The rate of settlement of a layer of saturated compressible soil subjected to load is dependent upon the permeability. The sustained rate of flow from wells is dependent upon the permeability of the aquifer. The permeability of the soil will determine the spacing of wells and amount of water to be pumped in lowering ground water to facilitate excavation for construction purposes. It also will affect the amount of loss from reservoirs through, under, and around dams. Permeability is a factor in the need for and design of filters to prevent piping or uplift, and to control the hydraulic stability of slopes of earth materials.

8.02 FLOW OF WATER THROUGH SOIL—DARCY'S LAW

After a great deal of experimentation, Darcy in 1856 established the empirical relationship for flow of water through fine grained soils which is known as Darcy's Law, $Q = kiAt$, in which Q is the total quantity of water flowing through a soil having a total area A, in time t, under a hydraulic gradient i. k is a coefficient expressing the permeability of the soil and is called the coefficient of permeability. The quantity of water flowing through the soil in a unit of time $= \frac{Q}{t} = kiA$. Since q is a quantity per unit of time, $cm^3\ sec^{-1}$; and A is an area, cm^2; then ki must be a velocity cm sec^{-1}. Also, because A is the total cross sectional area of the soil through which the water flows, the velocity $\frac{q}{A} = ki$ is not the true velocity, but is the velocity at which the quantity of water would flow through a tube of the same area as the total area of the soil.

Applying Darcy's Law to the conditions shown in Figure 8.02a

$$q = k \frac{h_1 - h_2}{L} A = kiA = v_d A.$$

It is readily apparent that the velocity v_d is that through the tube after the water emerges from the sample. This velocity as determined by Darcy's Law is called the discharge velocity.

Darcy's Law indicates that the velocity of water through soil is proportional to the first power of the hydraulic gradient. Poiseuille's Law, which was developed in Chapter VII, indicates that in a capillary tube for laminar flow the velocity is proportional to the first power of the hydraulic gradient. By comparison one would conclude that Darcy's Law is valid only for those soils and under those conditions for which the flow is laminar and in which the pore spaces are capillary tubes.

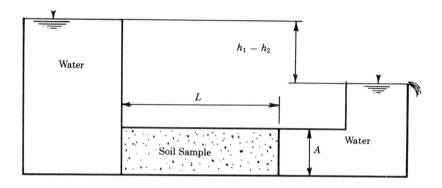

Figure 8.02a
Flow of Fluid Through Soil

Tests made subsequent to Darcy's statement of his law have shown that Darcy's Law is not valid for very coarse grained soils, which leads to the conclusion that the flow through these soils is not laminar or that the tubes are not capillaries. Under these conditions the quantity of flow could not be determined for any hydraulic gradient by the application of Darcy's Law when the quantity is known for one value of the hydraulic gradient. Fortunately, Darcy's Law does apply to all soils, except the very coarse grained ones, such as very coarse sand.

8.03 COEFFICIENT OF PERMEABILITY, k

From a statement of Darcy's Law, the discharge velocity, $v_d = ki$. Because the hydraulic gradient i is a dimensionless quantity, the dimensions of the coefficient of permeability k are the same as those of velocity. In Soil Mechanics the dimensions are usually cm sec^{-1} and will be so used throughout these notes unless otherwise stated.

The coefficient of permeability might be defined as the discharge velocity through soil under a hydraulic gradient of unity.

The value of k varies from 1 cm sec^{-1} for clean gravels down to 10^{-9} cm sec^{-1} for homogeneous clays. For purposes of comparison, the values of k are usually stated in terms of 10^{-1}, 10^{-4}, and 10^{-9}. If the coefficient of permeability of a medium sand were 0.0036 cm sec^{-1}, it would be stated as 36×10^{-4} cm sec^{-1}.

8.04   SEEPAGE VELOCITY, v_s

As has already been shown, the discharge velocity, v_d, as determined from the application of Darcy's Law is the velocity at which the same quantity of water as flows through the soil would flow through an area equal to the total area of the soil. Because the water does not flow through the solid particles of the soil but only through the void spaces, the velocity of the water parallel to the direction of flow must be somewhat greater than the discharge velocity. This velocity along the direction of flow is called the seepage velocity and may be designated v_s. The seepage velocity is the velocity at which the line of wetting will progress in the direction of flow through dry soil, and may be observed under proper conditions.

It is obvious that the seepage velocity is not the true velocity of the water through the soil pores. The water as it passes through the soil cannot go directly through but must detour around solid particles following the open spaces between these particles. The true velocity through the pores is variable as the water passes through large void spaces and through much smaller pores connecting the larger spaces. The discharge and seepage velocities represent average velocities through a large number of pores, and for any homogeneous soil are constant and measurable.

8.05   RELATION BETWEEN DISCHARGE AND SEEPAGE VELOCITIES

Because of the definitions of discharge and seepage velocities, the relationship between them may be shown by equating the quantity flowing through the entire area to the same quantity flowing through the area of voids only.

$$q = v_d A = v_s A_v$$

If the total area be considered as $1 + e$, the area of voids is e, and the

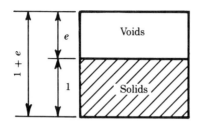

Figure 8.05a
Idealized Section Through Soil

area of the solids is 1, as shown in Figure 8.05a.   Then,

$$q = v_d(1 + e) = v_s e, \text{ and } v_d = v_s \frac{e}{1 + e}. \qquad v_d = n \, v_s$$

Similar relationships in terms of porosity, n, may be used.

8.06   SEEPAGE OF WATER THROUGH SOILS

The stability of earth structures and of natural deposits is dependent not only upon the static properties of the soil but also upon the forces produced by water as it flows or seeps through the pores of the soil. As an aid to his judgment in the design of earth structures or the stabilization of earth deposits, the engineer should be able to estimate through analyses the magnitude of seepage forces and pressures and the quantities of water flowing through the soil.

a. Basic Conditions of Flow

Darcy's Law for smooth, one directional flow through a length of soil of uniform cross section has been discussed in the preceding sections.   In general, the flow of water through a natural deposit of soil is not uniform over the entire area perpendicular to the flow and the flow does not occur in one direction only.

If piezometer tubes which measure the head of water at different points were placed in the ground through which water is flowing, the top of the water in the tubes would lie on a smooth curved surface. A plane tangent to this phreatic surface at any point indicates the gradient in any direction at that point.   The maximum slope of the plane indicates the steepest gradient and the direction of maximum flow.

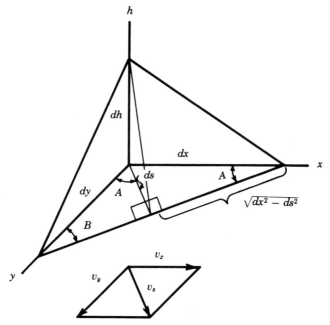

Figure 8.06a
Element of Tangent Plane and Velocity Vectors on Phreatic Surface

The hydraulic gradients causing flow in all directions are illustrated in Figure 8.06a.

From the geometry of the upper portion of Figure 8.06a, $\dfrac{ds}{dy} = \dfrac{\sqrt{dx^2 - ds^2}}{dx}$ and $\dfrac{ds^2}{dy^2} = \dfrac{dx^2 - ds^2}{dx^2} = 1 - \dfrac{ds^2}{dx^2}$, from which, $\dfrac{1}{ds^2} = \dfrac{1}{dy^2} + \dfrac{1}{dx^2}$.   After multiplying the above equation by dh^2, it becomes

$$\frac{dh^2}{ds^2} = \frac{dh^2}{dy^2} + \frac{dh^2}{dx^2}. \qquad \text{(Eq. 8.06a)}$$

Darcy's Law applied to flow in s, x, and y directions give the relationships as follows:

$$v_s = k_s i_s = k_s \frac{dh}{ds}, \; v_x = k_x i_x = k_x \frac{dh}{dx}, \text{ and } v_y = k_y \frac{dh}{dy}.$$

The maximum velocity v_s and its components v_x and v_y must conform to the relationship shown by the vector diagram, which is

$$v_s{}^2 = v_x{}^2 + v_y{}^2.$$

Applying Darcy's Law to the vector relationship,

$$k_s{}^2 \frac{dh^2}{ds^2} = k_y{}^2 \frac{dh^2}{dy^2} + k_x{}^2 \frac{dh^2}{dx^2}.$$

If the k values are all the same, they cancel out of the equation which makes the equation become

$$\frac{dh^2}{ds^2} = \frac{dh^2}{dy^2} + \frac{dh^2}{dx^2}. \qquad \text{(Eq. 8.06}b\text{)}$$

Equations 8.06a and 8.06b are identical, indicating that Darcy's Law satisfies both the geometric and vectorial relationships if the coefficients of permeability are the same in all directions (isotropic material). Therefore, Darcy's Law is valid for flow in all directions in isotropic materials and

$$v_s = k \frac{dh}{ds}, \; v_y = k \frac{dh}{dy}, \text{ and } v_x = k \frac{dh}{dx}.$$

It is also of interest to know if the velocity of flow in all directions is proportional to any but the first power of the hydraulic gradient.

The general case for any power of the hydraulic gradient is

$$v_s = k \left(\frac{dh}{ds}\right)^m, \; v_y = k \left(\frac{dh}{dy}\right)^m, \text{ and } v_x = k \left(\frac{dh}{dx}\right)^m.$$

Substituting these general values in the relationship which must be true for velocities, $v_s{}^2 = v_y{}^2 + v_x{}^2$, the relationship becomes

$$k^2 \left(\frac{dh}{ds}\right)^{2m} = k^2 \left(\frac{dh}{dy}\right)^{2m} + k^2 \left(\frac{dh}{dx}\right)^{2m}. \qquad \text{(Eq. 8.06}c\text{)}$$

But, to satisfy the requirements of geometry, the relationship must be $\frac{dh^2}{ds^2} = \frac{dh^2}{dy^2} + \frac{dh^2}{dx^2}$. This condition can be met by Equation 8.06c only when m is equal to unity. Therefore, the velocity of flow in different directions is proportional to the first power of the hydraulic gradient only; i.e., for laminar flow.

It is also desirable to know if the developed relationships are true for flow in different directions in soils having different coefficients of permeabilities in different directions (anisotropic materials).

Consider an elemental block of the soil through which water is flowing as shown in Figure 8.06b.   If there is steady flow through the block so that there is no volume change produced in the block during the flow, the quantity of water flowing into the block in an interval of

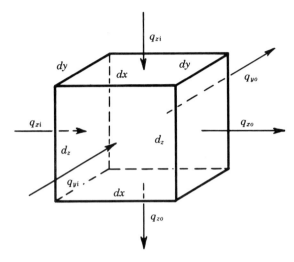

Figure 8.06b
Elemental Block Through Which Water Is Flowing

time is equal to the amount flowing out.   For steady flow the velocity is a function of the head h and the coordinates, x, y, and z; i.e., $v = f(h)(x,y,z)$.

If the hydraulic gradient at a face of the block in a given direction z is $i_z = \dfrac{\partial h}{\partial z}$, the hydraulic gradient at the opposite face dx away is $i_z + \dfrac{\partial i}{\partial z} dz = \dfrac{\partial h}{\partial z} + \dfrac{\partial^2 h}{\partial z^2} dz$.   The quantity of water flowing in a z direction into face $dx\, dy$ in a unit of time is

$$q_{zi} = k_z \frac{\partial h}{\partial z} dx\, dy$$

and the quantity of water flowing out the opposite face is

$$q_{zo} = k_z \left(\frac{\partial h}{\partial z} + \frac{\partial^2 h}{\partial z^2} dz \right) dx\, dy = k_z \frac{\partial h}{\partial z} dx\, dy + k_z \frac{\partial^2 h}{\partial z^2} dx\, dy\, dz.$$

The difference in the quantities of water flowing in and out of the $dx\, dy$ faces is

$$q_{zo} - q_{zi} = k_z \frac{\partial^2 h}{\partial z^2} dz\, dy\, dx.$$

Similarly, the difference in the quantities flowing in and out of the block in an x direction is

$$q_{xo} - q_{xi} = k_x \frac{\partial^2 h}{\partial x^2} dx\, dy\, dz,$$

and the difference in quantities flowing in and out of the block in a y direction is

$$q_{yo} - q_{yi} = k_y \frac{\partial^2 h}{\partial y^2} dy\, dx\, dz.$$

Because the volume of the block does not change during steady flow, the differences in the quantity of water flowing into the block must be equal to the quantity flowing out, or

$$k_z \frac{\partial^2 h}{\partial z^2} dz\, dx\, dy + k_x \frac{\partial^2 h}{\partial x^2} dx\, dy\, dz + k_y \frac{\partial^2 h}{\partial y^2} dy\, dx\, dz = 0$$

or,
$$k_z \frac{\partial^2 h}{\partial z^2} + k_x \frac{\partial^2 h}{\partial x^2} + k_y \frac{\partial^2 h}{\partial y^2} = 0.$$

If the material through which the water flows is isotropic, the k values are all equal and the equation becomes

$$\frac{\partial^2 h}{\partial z^2} + \frac{\partial^2 h}{\partial x^2} + \frac{\partial^2 h}{\partial y^2} = 0. \qquad \text{(Eq. 8.06d)}$$

This equation represents three families of curves meeting at 90°.

If the water is assumed to be flowing in two directions only as might be the case for water flowing through and under earth dams, under impervious cutoff walls, etc., the relationship becomes

$$\frac{\partial^2 h}{\partial x^2} + \frac{\partial^2 h}{\partial y^2} = 0.$$

This equation represents two families of curves meeting at 90°. One family of these curves might represent the flow lines in a flow net and the other family the equipotential lines in the flow net. In the simple case shown in Figure 8.06c which is a sheet pile wall driven half way into a pervious layer of soil, one-half of a family of ellipses and of hyperbolas meeting at 90°, except immediately adjacent to the cut off wall,

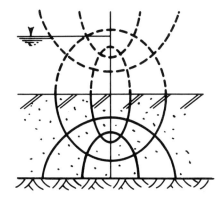

Figure 8.06c
Two Families of Curves Meeting at 90°

form the true flow net for the flow of water under the cut off wall due
to a head of water on one side.

For anisotropic soils in which k_x and k_y are not equal the relationship
for two directional flow is

$$k_x \frac{\partial^2 h}{\partial x^2} + k_y \frac{\partial^2 h}{\partial y^2} = 0.$$

This relationship represents two families of curves which do not meet
at 90°.

b. Flow Lines

A droplet of water, in flowing from one position A to another position
B under a hydraulic gradient, follows a certain path through the soil.
The path of the droplet through the soil is referred to as the flow line.

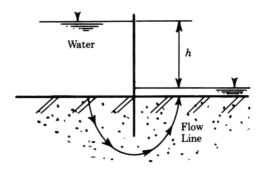

Figure 8.06d
Flow Line

Except for simple symmetrical cases, the shape of the flow lines cannot
be readily determined mathematically.   But, enough is known about
the flow lines in isotropic materials and under certain conditions in
anisotropic materials that a flow net of flow lines and equipotential
lines can be drawn experimentally.

Water is forced to flow through soil against frictional resistance by
a difference in head.   The head is a function of the coordinates x and
y.   The hydraulic gradients in the x and y directions are $i_x = \frac{\partial h}{\partial x}$ and
$i_y = \frac{\partial h}{\partial y}.$

At a point along the flow line the water flows with a velocity v_s.   The
components of v_s are $v_x = k_x \frac{\partial h}{\partial x}$ and $v_y = k_y \frac{\partial h}{\partial y}.$ The slope of the flow

line at this point is

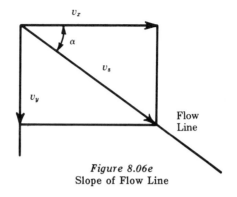

$$\frac{v_y}{v_x} = \frac{k_y \dfrac{\partial h}{\partial y}}{k_x \dfrac{\partial h}{\partial x}}.$$

For isotropic soils, the slope of the flow line is

$$\tan \alpha = \frac{\dfrac{\partial h}{\partial y}}{\dfrac{\partial h}{\partial x}} = \text{Slope of Flow Line.}$$

Figure 8.06e
Slope of Flow Line

c. Equipotential Lines

If a piezometer is placed at a point A in the soil through which water is flowing, the water will rise to some height h above a reference plane. This height to which the water rises in the piezometer is a measure of the potential head at point A. If another piezometer is placed at point B where the water rises to the same elevation as in the piezometer at point A, the two points are of equal potential. A line drawn through all the points in the soil having the same potential is an equipotential line.

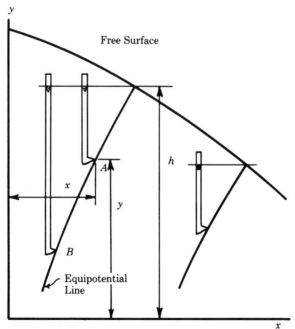

Figure 8.06f
Equipotential Line

It should be observed that equipotential lines do not join points of equal pressure. The pore water pressure at any point is equal to the head of water in the tube times the unit weight of water, which is not equal to the pressure at another point on the equipotential line.

The potential head h is a function of x and y; i.e., $h = f(x,y)$. For equipotential lines, h is constant. Therefore, $dh = 0$ and the total differential is

$$dh = \frac{\partial h}{\partial x}\, dx + \frac{\partial h}{\partial y}\, dy = 0.$$

The slope of the equipotential line at x, y is

$$\frac{dy}{dx} = -\frac{\dfrac{\partial h}{\partial x}}{\dfrac{\partial h}{\partial y}},$$

which is the negative reciprocal of the slope of the flow line at point x, y in an isotropic material. Therefore, flow lines and equipotential lines in isotropic materials meet at 90°.

d. Flow Nets

A family of flow lines and equipotential lines is called a flow net. Forchheimer devised a method of choosing flow lines and equipotential

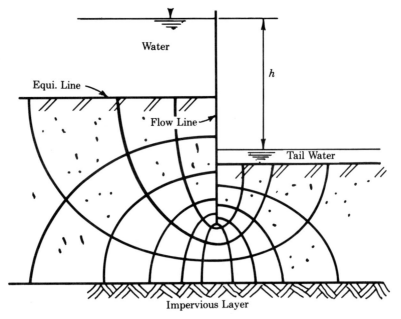

Figure 8.06g
Flow Net

lines in such a way that the quantity of water q_1 flowing between any two adjacent flow lines (flow channel) is equal to the quantity flowing between any other two flow lines. In such a flow net the total quantity of water flowing per unit of time through the soil q_T is equal to the quantity flowing through one flow channel q_1 times the number of flow channels in the net N_f or $q_T = N_f q_1$. Forchheimer also chose equipotential lines such that the drop in head between any two adjacent equipotential lines is equal to the drop between any other two adjacent equipotential lines. In such a flow net, the total drop in head h is equal to the number

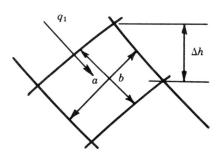

Figure 8.06h
One Rectangle of Flow Net

of equipotential intervals N_e times Δh, in which case $\Delta h = \dfrac{h}{N_e}$.

Consider one rectangle of a flow net for isotropic material having an area between flow lines equal to a for a unit thickness and a distance between equipotential lines equal to b as shown in Figure 8.06h.

Applying Darcy's Law to the flow through this channel,

$$q_1 = kia = k\,\frac{\Delta h}{b}\,a.$$

Therefore, if a flow net is drawn consisting of rectangles in which the ratio of breadth to length is the same $\left(\dfrac{a}{b} = \text{constant}\right)$, the drop in head, Δh, between any two adjacent equipotential lines is h/N_e and the flow through each flow channel is q_T/N_f.

Obviously, the simplest ratio of sides for use in constructing a flow net is unity. If, then, a flow net made up of "squares" is drawn to scale on a section through the structure or deposit through which the water is flowing, the net will be a true one. With such a flow net pore pressures, uplift pressures, quantity of water flowing, and seepage forces at any point in the flow net can be determined.

When the flow net consists of squares, the total quantity of water flowing in a unit of time is

$$q_T = q_1 N_f$$

in which

$$q_1 = k\,\Delta h \cdot 1 = k\,\frac{h}{N_e}\cdot 1$$

so that

$$q_T = kh\,\frac{N_f}{N_e}. \tag{Eq. 8.06e}$$

Equation 8.06e is a statement of Darcy's Law in another form for the special case in which the flow net consists of squares. The ratio N_f/N_e is often called the shape factor.

Flow nets drawn by eye fulfilling the conditions listed above are quite accurate for isotropic soils and can be drawn rather quickly by a series of successive trial sketches.

In a flow net an impervious surface is a flow line and a surface of permeable material into which the water enters or from which it flows is an equipotential line, except for the special case of the free surface. The free surface, or phreatic surface, is a flow line having the special property that vertical distances between intersections with equipotential lines are constant and equal to Δh.

e. Free Surface—Seepage Line

As water flows through soil, except where it is confined by an impervious surface, there exists a surface on which the pressure in the pore water is zero. This surface of zero pressure is called the free surface or the phreatic surface. The soil above the free surface may be saturated with capillary water in tension. Below the free surface the water is in compression.

Along the free surface the change in potential head, Δh, is the actual vertical distance from one point on the free surface to another. Free surfaces exist either above or below the surface of the soil. In this text the line of intersection between a vertical plane and the free surface below the surface of the soil is called the seepage line. The seepage line is a flow line and marks the boundary of the flow net. Equipotential lines must meet the seepage line at 90° and they must be spaced so that the vertical distances between intersections are all the same.

Before a flow net for water flowing through soil can be drawn, the seepage line must be defined. In simple cases where water stands on each side of an impervious barrier, the free surface is a level plane.

The shape of the seepage line can be determined approximately by means of the assumptions suggested by Dupuit that the hydraulic gradient is equal to $\dfrac{dy}{dx} = \tan \theta$, and that the hydraulic gradient is constant in a vertical plane from the free surface to an impervious layer or a surface in which there is no transverse flow. Neither of these assumptions is accurate in a general sense. The true hydraulic gradient is $i_s = \dfrac{dy}{ds} = \sin \theta$. But, for values of θ less than 30°, the error in using $\tan \theta$ instead of $\sin \theta$ is small.

Using the Dupuit assumptions, the shape of the seepage line can be determined as follows:

Assume that one point on the free surface and seepage line is known at $x = d$, $y = h$, and that the free surface meets the impervious layer or tail water at point 0.

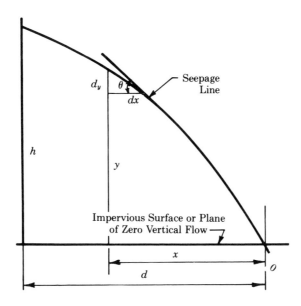

Figure 8.06i
Seepage Line

The same quantity of water in a thickness of unity and in a unit of time flows through any vertical plane between the free surface and the impervious layer as through any other. The quantity of water flowing through a vertical plane at x distance from 0 having an area equal to y is

$$q = kiy = ky \frac{dy}{dx}.$$
$$q\,dx = ky\,dy.$$
$$\int q\,dx = \int ky\,dy = qx = k\frac{y^2}{2} + C.$$

At $x = d$, $y = h$ the equation above becomes $qd = k\dfrac{h^2}{2} + C$ from which C is found to be $C = qd - k\dfrac{h^2}{2}.$ Placing this value of C in the equation above, the equation for the seepage line becomes

$$q(x - d) = \frac{k}{2}(y^2 - h^2).$$ (Eq. 8.06f)

The above equation for the seepage line represents a parabola.

The free surface for flow in the soil must come to the surface of the slope at some distance above an impermeable layer or a tail water surface. The seepage line meets the slope surface in such a manner

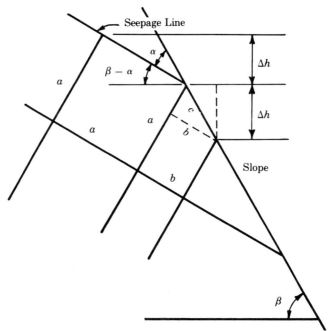

Figure 8.06j
Intersection of Seepage Line with Slope

that the water leaves the soil either tangent to the surface or at some angle α. Since the surface represented by the seepage line and the surface of the slope from which water is emerging is a free surface, the equipotential lines must meet the seepage line and the slope at a constant Δh. If the flow net consists of squares above the exit of the seepage line, it must consist of squares below. From Figure 8.06j,

$$\frac{\Delta h}{a} = \sin(\beta - \alpha)$$

$$\frac{\Delta h}{c} = \sin\beta$$

$$\frac{b}{c} = \cos\alpha.$$

Then $\dfrac{\Delta h}{b} = \dfrac{\sin \beta}{\cos \alpha}.$ When $a = b$ (squares), since Δh is constant,

$$\frac{\Delta h}{a} = \frac{\Delta h}{b} = \sin (\beta - \alpha) = \frac{\sin \beta}{\cos \alpha}.$$

From which, $\sin \beta = \cos \alpha \sin (\beta - \alpha).$
The only possible solution for this relationship is for $\alpha = 0$ and $\cos \alpha = 1.$ Therefore, the seepage line is tangent to the discharge face of the slope when β is less than $90°.$

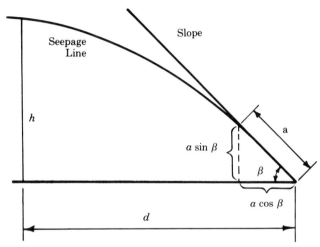

Figure 8.06k
Tangent Point Seepage Line and Slope

Before the seepage line can be drawn as the boundary of the flow net, it is necessary to determine the point of tangency of the seepage line and the slope. This can be determined approximately by the following reasoning based on the Dupuit assumptions:

At the discharge point, the vertical area through which the water flows is $a \sin \beta$ and the hydraulic gradient is the slope of the discharge face, $\dfrac{dy}{dx} = \tan \beta.$

The quantity of water flowing through this area per unit of time is

$$q = k \frac{dy}{dx} A = k \tan \beta \, a \sin \beta.$$

Putting this value of q in the equation for the seepage line (Eq. 8.06f),

$$ka \tan \beta \sin \beta (x - d) = \frac{k}{2} (y^2 - h^2)$$

or, $$a \tan \beta \sin \beta (x - d) = \tfrac{1}{2}(y^2 - h^2).$$

To determine a, put its coordinates $x = a \cos \beta$, $y = a \sin \beta$ in the equation and solve for a.

$$a \tan \beta \sin \beta (d - a \cos \beta) = \tfrac{1}{2}(h^2 - a^2 \sin^2 \beta)$$

$$a \frac{\sin^2 \beta}{\cos \beta} (d + a \cos \beta) = \tfrac{1}{2}(h^2 - a^2 \sin^2 \beta)$$

$$ad - a^2 \cos \beta = \frac{h^2 \cos \beta}{2 \sin^2 \beta} - \frac{a^2 \cos \beta}{2}$$

$$a^2 \cos \beta - 2ad + \frac{h^2 \cos \beta}{\sin^2 \beta} = 0$$

$$a = \frac{2d \pm \sqrt{4d^2 - 4 \dfrac{h^2 \cos^2 \beta}{\sin^2 \beta}}}{2 \cos \beta}$$

$$a = \frac{d}{\cos \beta} - \sqrt{\frac{d^2}{\cos^2 \beta} - \frac{h^2}{\sin^2 \beta}} \qquad \text{(Eq. 8.06}g\text{)}$$

Schaffernach devised a simple graphical solution for Equation 8.06g as follows: (See Figure 8.06l).

a. Extend vertically from $x = d$ at 2 to the slope surface at 1.   Distance along slope $0-1 = \dfrac{d}{\cos \beta}$.

b. With $0-1$ as a diameter draw the semicircle $0-1$.

c. Extend horizontally from point $3(d,h)$ to intersection with slope at 4.

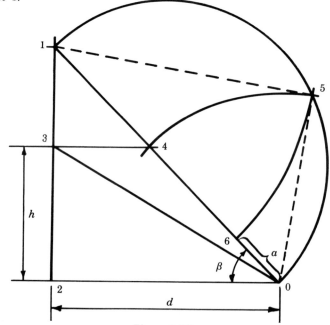

Figure 8.06l
Schaffernach Solution for Exit of Seepage Line

d. With 0 as a center, strike arc 4–5.

e. With point 1 as a center, strike arc 5–6.

f. Point 6 is where seepage line joins the discharge slope and distance 0–6 = a.

Proof of Schaffernach solution.

Distance
$$0\text{–}1 = \frac{d}{\cos \beta}$$

$$0\text{–}4 = \frac{h}{\sin \beta} = 0\text{–}5$$

$$1\text{–}5 = \overline{0\text{–}1}^2 - \overline{0\text{–}5}^2 = \sqrt{\frac{d^2}{\cos^2 \beta} - \frac{h^2}{\sin^2 \beta}} = 1\text{–}6$$

$$0\text{–}6 = (0\text{–}1) - (1\text{–}6) = a = \frac{d}{\cos \beta} - \sqrt{\frac{d^2}{\cos^2 \beta} - \frac{h^2}{\sin^2 \beta}}$$

Except at the seepage line, the water is emerging from the slope under a positive head. The slope surface covered with the flow is a free surface but not an equipotential line. Flow lines may intersect this portion of the slope at any angle between 0° and 90°. If the toe of the slope is under tail water, that portion of the slope covered with tail water is an equipotential line and the flow lines are perpendicular to the slope.

On the upstream face of an earth dam, which is an equipotential line, the water enters the embankment at 90° to the face. But immediately the flow lines turn to form the squares of the flow net, the free surface following the parabolic seepage line.

A. Casagrande determined experimentally that a parabola drawn from the upstream surface of water above the dam at a point $m/3$ from the intersection of the water surface with the upstream face to the exit

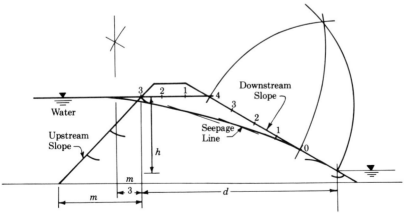

Figure 8.06m
Section Through Dam of Homogeneous, Isotropic Material

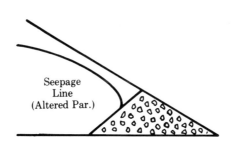

Figure 8.06n
Rock Toe Filter

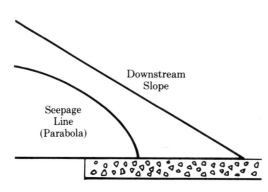

Figure 8.06o
Blanket Filter

point at distance a up from the toe closely followed the seepage line, except for a small portion at the upstream slope. A short curve perpendicular to the slope of the parabola corrects the seepage line in that portion.

A graphical method of laying out the seepage line for a dam of homogeneous, isotropic material without a downstream filter is shown in Figure 8.06m. Such a dam which allows the seepage line to come to the downstream surface is hydraulically unstable. In order for the dam to be stable, the seepage line must remain well below the downstream surface. The seepage line can be kept below the downstream face by means of properly designed filters at or near the toe of the dam.

The design of filters, the flow of water through layers of materials with different coefficients of permeability, and the flow of water through anisotropic materials is beyond the scope of this text. For further study of this subject the reader is referred to the references listed at the end of this chapter.

f. Equipressure Lines

In Figure 8.06p is shown a portion of a flow net. In the figure the following symbols are used:

p = Pore pressure at point in flow net.

h = Potential head from reference plane to free surface.

z = Head from reference plane to point.

Subscripts 1 and 2 make symbols apply to points 1 and 2.

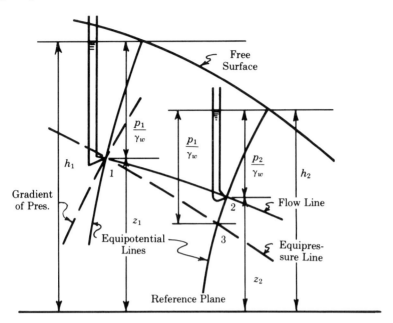

Figure 8.06p
Partial Flow Net Showing Equipressure Line

If a tube were inserted to any point on the equipotential line through point 1, the water would rise to the height of the intersection of the equipotential line with the free surface. This height of water in the tube represents the pressure head at point 1; i.e., $p_1 = (h_1 - z_1)\gamma_w$. The head of water in the tube at point 1 is p_1/γ_w. The head of water at point 2 on the next equipotential line on the same flow line is p_2/γ_w, which is not the same as at point 1. If the head at point 1, p_1/γ_w, is laid off vertically from the intersection of the equipotential line through 2 with the free surface to the equipotential line through 2, point 3 having the same pressure as point 1 will be determined. Then, an equipressure line for p_1 can be drawn by laying off the distance p_1/γ_w from the intersection of each equipotential line with the free surface to the same equipotential line inside the flow net.

Lines drawn perpendicular to these equipressure lines are called pressure gradient lines. Nets of these equipressure and pressure gradient lines are sometimes drawn instead of the more common flow nets of flow lines and equipotential lines.

g. Seepage Forces

A drop of water as it flows through the soil is acted upon by the weight of the drop and by a force parallel to the pressure gradient line.

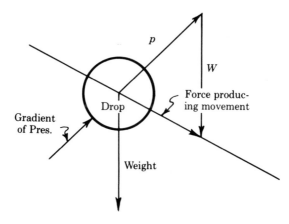

Figure 8.06q
Equilibrium Forces on a Drop of Water Flowing Through Soil

The resultant of these forces produces movement resisted by friction in the direction of the flow line along which the drop moves.

A volume of water q_1 moving in a flow channel through a drop in potential head from one equipotential line to the next, Δh, does an amount of work J equal to the weight of the water times the change in potential head Δh,

$$J = q_1 \gamma_w \, \Delta h.$$

This quantity of water in dropping through the change in potential head from one equipotential line to the next Δh travels a distance b equal to the distance between equipotential lines along the flow line. A unit volume of water in traveling a unit distance along the flow line does work equal to

$$j = \frac{q_1 \gamma_w \, \Delta h}{q_1 b} = \frac{\gamma_w \, \Delta h}{b}.$$

But $\dfrac{\Delta h}{b} = i_s =$ hydraulic gradient along the flow line. Then

$$j = \frac{\gamma_w \, \Delta h}{b} = i_s \gamma_w.$$

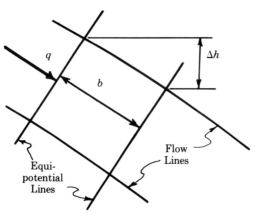

Figure 8.06r
Square of Flow Net

This value of j is equal to the force exerted by a unit volume of water dropping through the head lost per unit of distance traveled. This force equal to j is the force exerted by the moving water and has no relation to the pore water pressure nor to the direction of flow.

The engineer is interested in the seepage force that would cause piping in sand at the bottom of an excavation or at the toe of a dam or levee.

The submerged weight in water of the solids in a unit volume of soil is

$$\gamma_{\text{sub}} = \frac{G_s - 1}{1 + e} \gamma_w.$$

Equating this submerged weight of soil solids in a unit cube of soil, to the force exerted by a unit volume of water flowing the unit distance through the cube with its head loss between entrance and exit,

$$\frac{G_s - 1}{1 + e} \gamma_w = i\gamma_w.$$

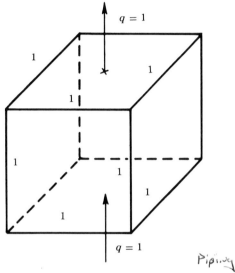

Figure 8.06s
Unit Cube of Soil with Unit Volume of Water Flowing Through

The hydraulic gradient required to hold grains of sand in suspension and cause piping to start is computed for sand having a specific gravity of solids equal to 2.65 and of different densities and shown in the following table:

Density	Void Ratio	Relationship	Hydraulic Gradient to Start Piping
Loose	1.0	$\frac{1.65}{2}$	0.825
Medium	0.65	$\frac{1.65}{1.65}$	1.00
Dense	0.40	$\frac{1.65}{1.4}$	1.18

This analysis indicates that the critical hydraulic gradient that will cause piping of sand by water moving vertically through the sand is approximately unity.

The effect of the force of moving water can be illustrated with a device consisting of a container for the sand through which water can be allowed to flow downward or forced upward at will. When the

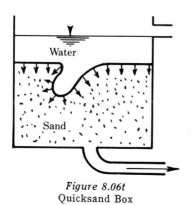

Figure 8.06t
Quicksand Box

water flows downward through the sand, slopes in the submerged sand are stable even when undercut. When the water moves upward through the sand, the slopes collapse. If the hydraulic gradient is made greater than 1, the sand grains are lifted in suspension and may be washed out the top of the vessel.

This knowledge of seepage forces forms the basis for the design of filters.

h. Example of Seepage Problem

Assume a low concrete dam as shown in Figure 8.06u on cohesionless soil having a coefficient of permeability of 20×10^{-4} cm sec^{-1}.

1. Draw flow net.
2. Determine the quantity of water flowing under the dam per foot of width per day.
3. Determine the uplift pressure on the bottom of the dam 5 ft upstream from the toe.
4. Determine the seepage force of the flowing water at the toe of the dam.   Is there danger of piping at this point?

Solution:

1. See flow net in Figure 8.06u.
2. $N_f = 4$, $N_e = 17$

$$Q = \frac{20 \times 10^{-4} \times 60 \times 60 \times 24}{2.54 \times 12} \times 10 \times \frac{4}{17}$$

$$= 44.5 \text{ cu ft per day per ft.}$$

3. $(13 - \frac{12}{17} \times 10) \times 62.4 = (13 - 7.06) \times 62.4 = 371 \text{ lb ft}^{-2}$.
4. $\Delta h = \frac{10}{17} = 0.59$. L of square at toe $= 2$ ft.

$$i \text{ at toe} = \frac{0.59}{2} = 0.29.$$

Piping will not occur at the toe, but surface must be protected against erosion.

8.07   DETERMINATION OF PERMEABILITY OF SOILS

Tests for the determination of the coefficient of permeability of soils may be divided into two general classes depending upon whether the

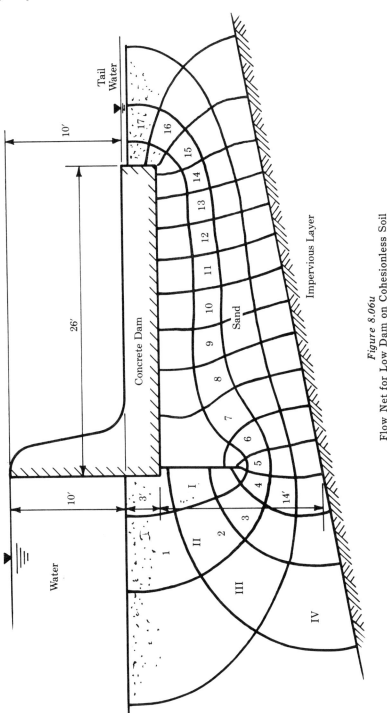

Figure 8.06u
Flow Net for Low Dam on Cohesionless Soil
Scale $\frac{1}{8}'' = 1' - 0''$

value of the coefficient is determined directly by measuring the flow of water through the soil or indirectly by computation from the results of other test data or properties of the soil.

Methods in common use for determining the coefficient of permeability of soils may be classified as follows:

Direct Methods
    a. Constant Head Permeability Test
    b. Falling Head Permeability Test
    c. Pumping Test on Soil in Place
Indirect Methods
    a. Consolidation Test Data
    b. Horizontal Capillarity Test
    c. Computation from Grain Size by Hazen's Formula

All the tests listed, except the pumping test, are made in the laboratory or in the field on small samples of the soil removed from their original positions. The pumping test is made in the field on the soil in place as it exists.

In making permeability tests on undisturbed samples attempts are sometimes made to use large samples, assuming that a test on a large sample is more representative of the composite soil than several tests made on small samples. A large number of tests on small samples about 2 inches in diameter by a few inches long can be made with the effort required for a single test on a sample 1 cu ft in volume. The small specimens are more easily handled, making them less likely to be damaged than the large sample. Thus, the small samples may yield results which more nearly represent the soil in place. Although the large sample contains more soil, it is not much more representative of a thick layer of nonhomogeneous soil than is a small sample. Several small samples scattered throughout the layer will be more representative of the layer than one large sample and they can be taken and handled more easily.

8.08 CONSTANT HEAD PERMEABILITY TEST

The coefficient of permeability of a soil sample is determined by the constant head permeability test by measurement of the quantity of water that flows through the sample of cross sectional area A, and length L in a measured time t while maintaining a constant head of water h on the sample. By application of Darcy's Law the coefficient

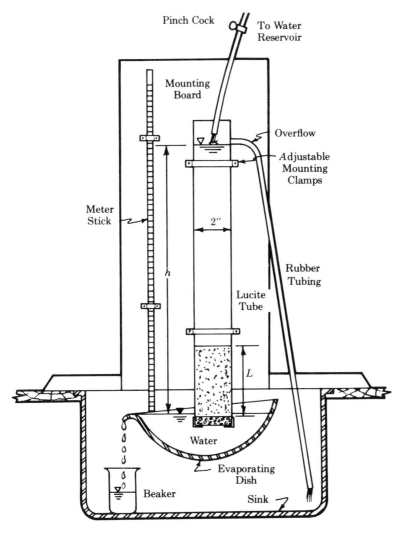

Figure 8.08a
Constant Head Permeameter

of permeability can be determined as $k = \dfrac{QL}{hAt}$. If all measurements

are made as g, cm, and sec, the dimensions of k are cm sec^{-1}.

A simple constant head permeameter can be made with a 2 in. diameter lucite tube 30 to 36 inches long provided with an outlet near the top. The outlet can be made by cementing into the lucite tube a short piece of glass tubing over which will fit a short length of rubber tubing. For convenience the lucite tube can be clamped to a board provided

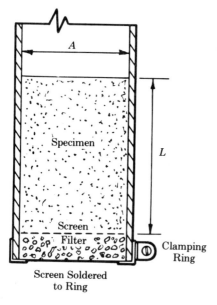

Figure 8.08b
Detail of Sample

with legs to hold the perme-ameter upright over a sink. The sample is prepared for testing by clamping a disk of copper wire mesh over the bottom of the lucite tube, placing about an inch of coarse sand above the mesh, placing a copper screen disk above the filter, and placing the sample above the copper screen.

This type of permeameter is suited to cohesionless soils of high enough permeability that a measurable quantity of water will flow through the specimen in a few min-utes under a low hydraulic gradient.

Detailed instructions for performing a constant head permeability test in the permeameter described above are given in Section 8.08L in the chapter on Labora-tory Testing.

8.09 FALLING HEAD PERMEABILITY TEST

For those fine grained soils for which the constant head permeability test is not suitable the coefficient of permeability can be determined with a falling head permeability test in which the small quantity of water flowing through the specimen is measured in a standpipe of smaller diameter than the specimen. The ratio of the area of the standpipe to the area of the specimen can be adjusted to suit the permeability of the material being tested.

Because the head of water during a test in the falling head perme-ability test is not constant, Darcy's Law cannot be used directly in computing the coefficient of permeability from data obtained in a fall-ing head test. The relationship expressing the coefficient of permeability in terms of data from the falling head test can be determined as follows:

Let A = Area of Specimen

a = Area of Standpipe

h = Head at given Time t

h_1 = Head at beginning of Test

h_2 = Head at end of Test

h_c = Capillary Rise in Standpipe

L = Length of Specimen

t = Elapsed Time during which head falls from h_1 to h_2.

In the falling head test, the water is allowed to flow through the specimen until as much air as possible has been removed from the soil. When ready to start the test the water is allowed to fall in the stand-pipe to some head h_1 at which time a stop watch is started. When the head has fallen to some head h_2, the watch is stopped to determine the time t.

As the head decreases, the velocity of the water through the specimen decreases. At some head h between h_1 and h_2 the velocity $v = k\dfrac{h}{L}$. During an interval of time dt the head will fall a distance dh and the quantity of water dq flowing through the specimen can be expressed by Darcy's Law as

$$dq = k\,\frac{h}{L}\,A\,dt.$$

This quantity of water as measured in the standpipe is equal to $a\,dh$, so

$$dq = k\,\frac{h}{L}\,A\,dt = -a\,dh \qquad (h \text{ decreases with time}).$$

Separating variables

$$dt = -\frac{La}{kA}\frac{dh}{h}.$$

Integrating from h_1 to h_2

$$t = -\frac{La}{kA}(\ln h_2 - \ln h_1) = \frac{La}{kA}\ln\frac{h_1}{h_2} = \frac{La}{kA}2.303\log\frac{h_1}{h_2}.$$

Solving for k

$$k = \frac{La}{tA}2.303\log\frac{h_1}{h_2}.$$

$$k = \frac{La}{tA}\,2.303\,\log\left(\frac{h_o - h_c}{h_c - h_c}\right)$$

A simple falling head permeameter can be made with a piece of 2 in. diameter lucite tubing about 8 inches long, a rubber stopper with a length of glass tubing in the center hole to serve as a standpipe, a copper screen arranged to clamp over the bottom of the lucite tube and two copper screen disks.   This arrangement can be fastened to a board and set up for running a test as shown in Figure 8.09b.   The permeameter described for the constant head test may also be used to run a falling head test for soils of such permeability that the water falls too rapidly in the small standpipe for accurate measurement.

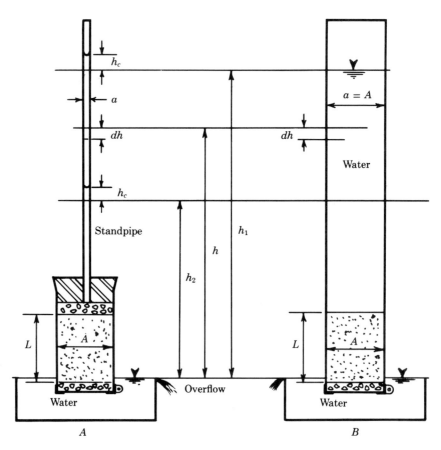

Figure 8.09a
Falling Head Permeability Test

a. Cohesionless Material

Detailed instructions for performing a permeability test in the permeameter illustrated in Figure 8.09*b* are given in Section 8.09*L* in the chapter on Laboratory Testing.

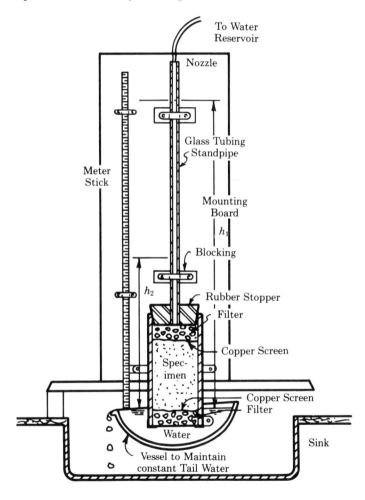

Figure 8.09b
Falling Head Permeameter

In the range of permeability from 10^{-3} to 10^{-6} cm sec^{-1} for which the falling head test is suited, a state of instability exists because of the tendency of soils having coefficients in this range to collect and hold air entrapped in the pore spaces. If successive tests are made on such soils using ordinary tap water or distilled water containing dissolved air, the coefficient of permeability will be found to decrease with time during which water flows through the specimen. On the other hand, if deaired water is used for the test, the permeability may increase with time. In one case, the soil collects and traps air from the water, and, in the other case, the deaired water dissolves the entrapped air and removes it from the pores of the soil.

b. Undisturbed Samples

In this category are included samples in their undisturbed natural state or in an undisturbed state as compacted. Because only soils possessing some cohesion can be formed into an undisturbed sample, this discussion applies only to those soils having sufficient cohesion to allow them to be formed into a stable specimen. In general, these soils are of such low permeability that only tests made in a falling head permeameter with a standpipe of small area are practical. The principal difficulty in performing permeability tests on such undisturbed samples is in preventing water from flowing outside the specimen in the joint between the specimen and the wall of the container.

A successful method suggested by A. Casagrande consists of sealing the specimen in the permeameter with a small amount of wax at the bottom and with bentonite gel between the specimen and the side of the permeameter. A sand cover to protect the top of the specimen against erosion is provided above the specimen. Care must be exercised to prevent sealing the surface of the specimen with the molten paraffin or wax.

The arrangement for making a falling head permeability test on a specimen of undisturbed soil is shown in Figure 8.09c.

Detailed instructions for making constant head and falling head permeability tests are given in Section 8.09L in the chapter on Laboratory Testing.

c. Low Permeability Soils

The coefficient of permeability of soils in the range from 10^{-6} to 10^{-9} cm sec^{-1} can also be determined in a falling head permeameter of special design used in connection with the consolidation test. In this device the area of the specimen A is made large, the length L is small, and the head h is made large. Because only a small quantity of water flows through the specimen, the quantity of water is measured in a burette. In order to avoid the use of a high standpipe, the high head is produced with air pressure. A diagram of the device is shown in Figure 8.09d.

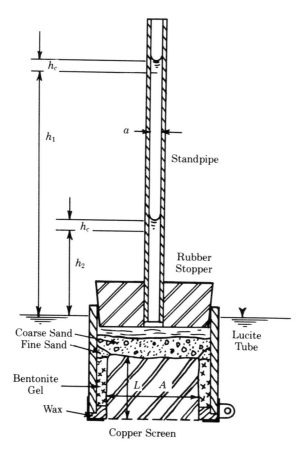

Figure 8.09c
Permeability Test for Undisturbed Sample

The apparatus for applying the head of water may be mounted on a board as shown in the figure. This portion of the apparatus consists of an arrangement of glass tubing with a connection for applying a high head from a compressed air reservoir. The head applied by air pressure is measured by a mercury column with a scale marked Hg scale and reading in cm of water. This head applied by the air h_a is read on the Hg scale as the difference in elevations of the mercury columns in the manometer. A water reservoir is provided so that water can be run through the specimen to provide saturation before starting a test and to set the head of water in the burette. A two-way stop is provided so that water can be run directly from the reservoir to the

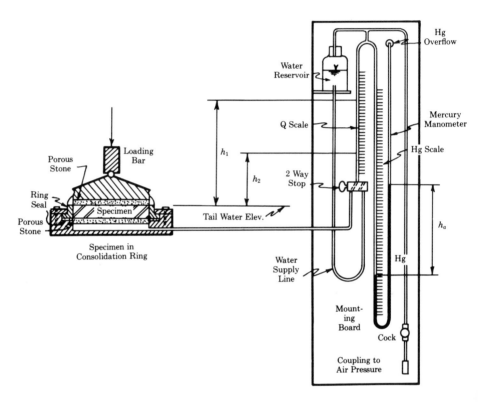

Figure 8.09d
Apparatus for Permeability Test in Conjunction with Consolidation Test

specimen or directly from the burette only. During a test the stop is turned so that all the water going through the specimen is registered on the Q scale which is calibrated in cc of water.

The specimen is cut to fit snugly into a ring with a flange so that it may be bolted into the base of the consolidation device using a gasket to prevent leakage outside the ring. In fitting the specimen to the ring care must be taken to provide a snug fit to prevent leakage between specimen and ring. Porous stones are placed above and below the specimen to allow water to flow freely through the specimen.

Water under the applied head is admitted to the bottom of the specimen through the lower porous stone and forced upward through the
specimen. The water comes to a free surface at the top of the specimen. Load is applied to the specimen through the upper porous
stone.

Before starting a permeability test with this arrangement the load
applied to the specimen should remain long enough to produce complete consolidation under this load. The head acting on the specimen
at the beginning of the test is $h_1 + h_a$ measured in cm of water and the
head at the end of the test is $h_2 + h_a$. The total pressure applied to
the bottom of the specimen by the head of water should be considerably
less than the load applied to the top of the specimen.

Because h_1 and h_2 are only a small portion of the total applied head,
the test may be considered as a constant head test without appreciable
error by using for the head the average of h_1 and h_2 plus h_a; i.e.,
$\dfrac{h_1 + h_2}{2} + h_a.$

8.10 PERMEABILITY OF SOIL IN PLACE

The average coefficient of permeability of a fairly large mass of soil
lying immediately below the water table can be determined by a pumping test. This test is made by sinking a well from which water is
pumped and two or more observation wells at different distances from

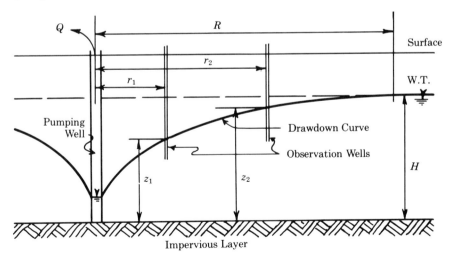

Figure 8.10a
Drawdown Around Well During Pumping

the main well, and measuring the drawdown in the observation wells during pumping from the main well.

A relationship between the average coefficient of permeability k and drawdown measurements were derived by Dupuit about 1863.

The following symbols are used in the derivation and statement of the relationship between k and drawdown.

H = Head from impervious layer (reference plane) to water table before pumping.

q = Quantity of water pumped per unit of time to maintain steady state flow.

R = Radius of influence (Distance from well to zero draw-down during steady flow.)

r = Distance from well to observation well.

z = Head from impervious reference plane to seepage line after drawdown at r distance from well.

i = Hydraulic gradient.

A = Area of cylinder of radius r and height z.

From Darcy's Law, $q = kiA$.

The area through which the water flows on its way to the well at r distance from the well is the surface of a cylinder z high and of radius r.

$$A = 2\pi rz.$$

At the surface of the seepage line at r distance from the well the hydraulic gradient is

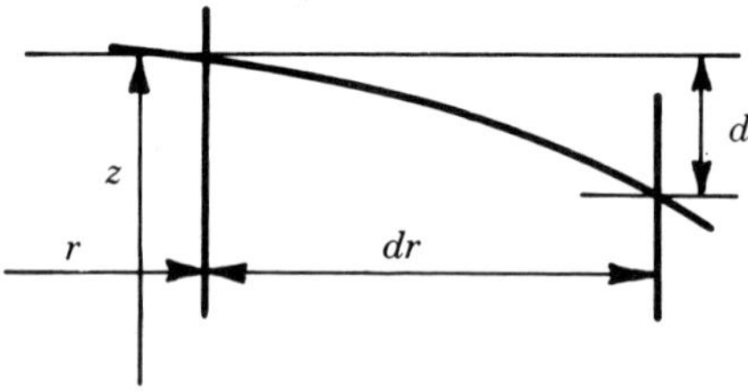

Figure 8.10b
Phreatic Surface

$$i = \frac{dz}{\sqrt{dr^2 - dz^2}} = \sin \alpha.$$

To simplify the mathematical solution, Dupuit assumed $i = \frac{dz}{dr}$ $= \tan \alpha$, and also assumed i to be constant along any vertical line extending from the phreatic surface to the impermeable base.

Although these assumptions may limit to some extent the practical application of the results, a solution to the problem may be simply obtained as follows:

$$q = k2\pi rz \frac{dz}{dr} \text{ or } z \, dz = \frac{q}{2\pi k} \frac{dr}{r}.$$

Integrating, $\dfrac{z^2}{2} = \dfrac{q}{2\pi k} \ln r + C.$

When $r = R$, $z = H$.

Then $\dfrac{H^2}{2} = \dfrac{q}{2\pi k} \ln R + C$ from which $C = \dfrac{H^2}{2} - \dfrac{q}{2\pi k} \ln R$.

The general relationship becomes

$$\frac{z^2}{2} = \frac{q}{2\pi k} \ln r + \frac{H^2}{2} - \frac{q}{2\pi k} \ln R \text{ or } z^2 - H^2 = \frac{q}{\pi k} \ln \frac{r}{R}.$$

Because H is greater than z and R is greater than r, for convenience the relationship may be written

$$H^2 - z^2 = \frac{q}{\pi k} \ln \frac{R}{r}.$$

Solving for k

$$k = \frac{q}{\pi(H^2 - z^2)} \ln \frac{R}{r} = \frac{q}{\pi(H^2 - z^2)} \, 2.303 \log \frac{R}{r}.$$

Because in a pumping test R is usually not known, data from two or more observation wells can be used to determine k independently of H and R.

In the general relationship, $\dfrac{z^2}{2} = \dfrac{q}{2\pi k} \ln r + C$. Inserting the values of z_1, r_1 for observation well No. 1 and z_2, r_2 for well No. 2,

$$\frac{z_1^2}{2} = \frac{q}{2\pi k} \ln r_1 + C$$

and

$$\frac{z_2^2}{2} = \frac{q}{2\pi k} \ln r_2 + C.$$

Eliminating C

$$z_2^2 - z_1^2 = \frac{q}{\pi k} (\ln r_2 - \ln r_1)$$

$$k = \frac{q}{\pi(z_2^2 - z_1^2)} \ln \frac{r_2}{r_1} = \frac{q}{\pi(z_2^2 - z_1^2)} \, 2.303 \log \frac{r_2}{r_1}.$$

For locations where the slope of the drawdown curve, α, is less than about 30°, the tan α is not much different from sin α and the equipotential lines are nearly parallel thus making the relationship developed by using the Dupuit assumption accurate enough for practical purposes. Near the well where the slope is quite steep, the relationship developed above is not applicable.

8.11 HORIZONTAL CAPILLARITY TEST

When the bottom surface of a vertical column of dry soil comes in contact with water, menisci are formed in the pores between the grains, causing the water to be pulled upward through the soil. The tension in the water pulling it upward through the soil is $u = -\dfrac{4T_s \cos \alpha}{d}$. This upward pull is resisted by the weight of water and the resistance to flow through the length of soil h between the upper line of wetting and the free water surface. Both the motive force and the resistance to flow are dependent upon the effective pore diameter of the soil. Because the permeability of the soil is dependent upon the pore diameter, the rate of capillary rise is a measure of the permeability of the soil.

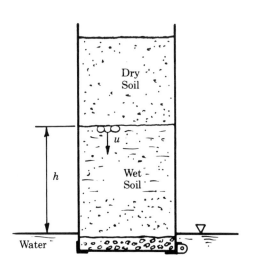

Terzaghi devised a simple field test for permeability by preparing a set of master curves for comparison with the rate of rise in the line of wetting. Each curve was prepared by plotting the height of wetting h against the elapsed time t in a test on a soil of known permeability. The nature of these curves is illustrated in Figure 8.11b. A field determination of the coefficient of permeability could then be made by comparing rise h against the elapsed time t required for the line of wetting to reach this height. The coefficient of permeability of soils for which points found in a capillary rise test fall between two curves can be estimated by interpolation.

A rational development of the relationship between rate of capillary rise and permeability is complicated and difficult and does not agree well with experimental results.

Figure 8.11a
Capillary Rise

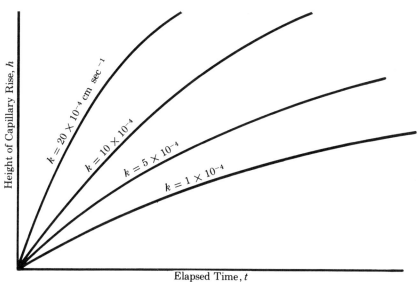

Figure 8.11b
Master Curves for Capillary Rise Test

A simpler and more accurate relationship can be developed for a sample of soil exposed to seepage forces in a horizontal position as shown in Figure 8.11c.

In this position the variable resistance of the weight of water below the surface of wetting is eliminated so that the resistance is due entirely to the resistance to flow through the soil. If it be assumed that the menisci at the line of wetting are fully developed, which appears to be true for soils of permeability in the range from 10^{-2} to 10^{-5} cm sec^{-1}, the hydraulic gradient i is equal to the head for maximum capillary rise h_c divided by the distance to the line of wetting x. Therefore, at any elapsed time t, $i = \dfrac{h_c}{x}$. At any time t, in an increment of time dt, x will increase by an increment dx.

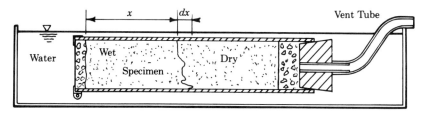

Figure 8.11c
Horizontal Capillarity Test

The seepage velocity v_s at any elapsed time t is dx/dt.

From Darcy's Law the discharge velocity $v_d = ki = k\dfrac{h_c}{x}$.

Because $v_d = v_s \dfrac{e}{1+e}$, we may write

$$\frac{e}{1+e}\frac{dx}{dt} = k\frac{h_c}{x} \qquad \text{or} \qquad x\,dx = \frac{1+e}{e}kh_c\,dt.$$

Integrating

$$\frac{x^2}{2} = \frac{1+e}{e}kh_ct + C.$$

When $t = 0$, $x = 0$, therefore $C = 0$.

The relationship is $x^2 = 2kh_c\dfrac{1+e}{e}t$.

For a given soil specimen e, k, and h_c are constant, making it possible to write the relationship between distance to the line of wetting and elapsed time as

$$x^2 = 2k\frac{1+e}{e}h_ct = mt.$$

The dimensions of m are cm^2 sec^{-1}.

In order to determine the value of the coefficient of permeability k it is necessary to express e and h_c as functions of k. The unknown factor in this relationship is h_c.

From capillarity,

$$h_c = \frac{4T_s \cos \alpha}{\gamma_w d}.$$

Substituting this value of h_c in the expression for m results in the following expression

$$m = 2k\frac{1+e}{e}\frac{4T_s \cos \alpha}{\gamma_w d}$$

from which an expression for d can be obtained.

$$d = \frac{8k\dfrac{1+e}{e}T_s \cos \alpha}{\gamma_w m}$$

The effective diameter of the tubes cannot be determined from this relationship.

Another expression for d can be obtained from Poiseuille's Law

$$v_m = \frac{\gamma_w}{8\eta\pi}i_m a.$$

In this case i_m is the hydraulic gradient and v_m is the velocity along the zig zag path of the water as it flows between and around the soil particles.

The relation between the horizontal projection of the distance a droplet of water travels as measured by v_s and the actual distance the water travels through the voids as measured by v_m can be expressed as shown in Figure 8.11e as $L_x = L_m \cos \beta$ or as $v_s = v_m \cos \beta$.

Then $\quad i_m = \dfrac{h_c}{L_m} = \dfrac{h_c \cos \beta}{L_x}$.

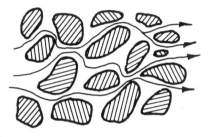

We may then write

$$v_m = \frac{v_s}{\cos \beta} = \frac{k h_c}{L_x \cos \beta} \frac{1 + e}{e}.$$

Inserting these values of v_m and i_m in Poiseuille's Law,

$$\frac{k h_c}{L_x \cos \beta} \frac{1 + e}{e} = \frac{\gamma_w h_c \cos \beta}{8 \eta \pi L_x} a.$$

Figure 8.11d
Path of Water Through Soil

Since $a = \dfrac{\pi d^2}{4}$ the above may be written as

$$\frac{k h_c}{\cos \beta} \frac{1 + e}{e} = \frac{\gamma_w h_c \cos \beta}{8 \eta \pi} \frac{\pi d^2}{4}.$$

Solving for d^2

$$d^2 = \frac{k(1 + e)32\eta}{e \gamma_w \cos^2 \beta}.$$

Equating this value of d from Poiseuille's Law to d from capillarity

$$\frac{64 k^2 \left(\dfrac{1 + e}{e} \right)^2 T_s^2 \cos^2 \alpha}{\gamma_w^2 m^2} = \frac{k \left(\dfrac{1 + e}{e} \right) 32 \eta}{\gamma_w \cos^2 \beta}.$$

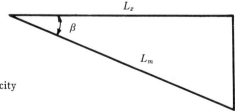

Figure 8.11e
Relation Between Seepage Velocity and Actual Velocity

Solving for k,

$$k = \frac{m^2}{\dfrac{2T_s^2 \cos^2 \alpha \cos^2 \beta}{\gamma_w \eta}\left(\dfrac{1 + e}{e}\right)}.$$

For any given soil $T_s \cos \alpha$ is constant, and for the same soil structure $\cos \beta$ is approximately constant. For most soils e varies from $e = 1$ to $e = 0.4$. Using these values of e, the range of $\dfrac{1 + e}{e}$ varies from 2 to 3.5.

If $\dfrac{2T_s^2 \cos^2 \alpha \cos^2 \beta}{\gamma_w \eta} \dfrac{1 + e}{e}$ be assumed constant for a given soil and designated as z,

$$k = \frac{m^2}{z}.$$

The dimensions of z are cm^3 sec^{-1}.

The value of z can be determined in the laboratory by running a permeability test to determine k and a horizontal capillarity test on the same soil to determine m. Then z can be determined from the relationship $z = \dfrac{m^2}{k}$.

If desired, the values of m can be computed on the basis of x in cm and t in minutes. In this case the value of z will be different than if t is in seconds. The dimensions of z can also be made such that the value found will be $k \times 10^{-4}$. In any case, the units used for the determination of m and z must be used for determining k from a horizontal capillarity test.

In performing the horizontal capillarity test, the specimen in the lucite tube as shown in Figure 8.11c is immersed in water. A stop watch is started at the instant of immersing. As the line of wetting progresses, a series of distances it has traveled x are marked on the tube and the time for each x distance noted. The x distances and the corresponding times t are recorded. The values of x^2 are computed and plotted as the ordinate and corresponding values of t plotted as the abscissa as shown in Figure 8.11f. The slope of the line best fitting these points represents the value of m.

The value of z for the soil can be found by determining the coefficient of permeability in a falling head permeability test and computing z from the relationship $z = \dfrac{m^2}{k}$.

Because the permeability is dependent upon the void ratio, if the value of z is to be correct, the void ratio of the horizontal capillarity test specimen and the void ratio of the falling head test specimen must be the same.   It will usually be necessary, therefore, to correct the results of one of the tests to the void ratio of the other from the relationship

$$\frac{k_1}{k_2} = \frac{e_1{}^2}{e_2{}^2}.$$

Sometimes during a horizontal capillarity test on loose fine grained cohesionless soil the fairly high capillary force may pull the specimen

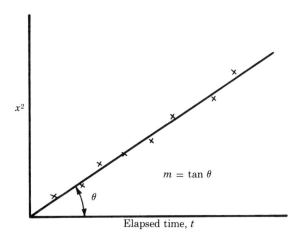

Figure 8.11f
Seepage Rate in Capillarity Test

in two.   Such a parting of the sample is indicated by a change in slope of the $x^2 - t$ plot and a parallel shift in the $\log x^2 - \log t$ plot.

There is some advantage in plotting the values of x^2 and t to logarithmic scales because of the fact that all second degree parabolas are represented as straight 45° lines when plotted to logarithmic scales. When plotted in this manner all the $x^2 - t$ points should fall on a straight 45° line.   The intersection of this 45° line representing the parabola $x^2 = mt$ with the line for $t = 1$ indicates the value of m.

When plotted to such logarithmic scales one value of x and its corresponding value of t is sufficient to determine the value of m, although several points are usually plotted and a 45° line best fitting these points drawn.   An advantage of the logarithmic plot is that parting of the sample at the advance line of wetting shows up as a shift in the line on the logarithmic plot as shown at B and B' on Figure 8.11g; whereas, such parting of the sample produces a non-linear relationship when plotted to arithmetic scales. Often the line of separation of the sample during test can be seen in the sample. The location of the break in the sample will usually be found to correspond to the position of shift in the line plotted to logarithmic scales. Another advantage of the logarithmic plot is that several test results can be plotted on the same diagram with the same degree of accuracy and convenience.

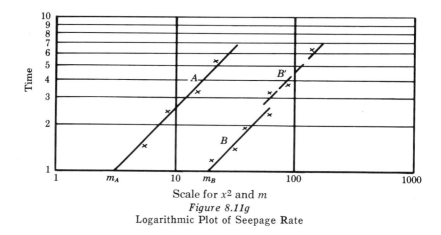

Scale for x^2 and m

Figure 8.11g

Logarithmic Plot of Seepage Rate

The horizontal capillarity test is useful as a field test for comparison purposes. It is also useful for laboratory testing of soils in the unstable permeability range from 10^{-3} to 10^{-6} cm sec^{-1}. In this range great care must be exercised in the determination of the permeability by direct measurement. Although greater accuracy can be obtained by direct measurement when sufficient care is taken, approximate values can be determined with little experience by the horizontal capillarity test. After a careful determination of the coefficient of permeability has been made, a large number of determinations can be made easily with horizontal capillarity tests on the soil.

8.12 PERMEABILITY FROM CONSOLIDATION TEST

The rate at which water is squeezed out of saturated soil during consolidation under load is dependent upon the permeability of the soil. The relationship between permeability and rate of consolidation is developed in the chapter on consolidation as $T = \dfrac{k(1 + e)t}{\gamma_w a_v H^2}$ in which

T is a dimensionless time factor, t is elapsed time from the start of consolidation under an increment of load, H is the maximum distance through which the water must travel to a free surface, and a_v is the slope of the void ratio-pressure curve. For a consolidation specimen drained both sides, H is one half the thickness of the specimen. The slope of the void ratio-pressure curve $\Delta e / \Delta p$ is called the coefficient of compressibility and is denoted as a_v. T is related to the average degree of consolidation of the loaded specimen. For 50 per cent consolidation, T is approximately 0.2. If the time for 50 per cent consolidation and the slope of the e-p curve are determined from a consolidation test, and if e for the soil is known, the coefficient of permeability can be determined from the relationship

$$k = \frac{0.2\gamma_w a_v H^2}{(1 + e)t_{50}}.$$

The unit weight of water can be determined from the temperature or for practical purposes may be taken as unity when using the cgs system.

The time for 50 per cent consolidation can be found by converting a time-dial reading curve for a load increment of the consolidation test into a time-consolidation curve and reading the time for 50 per cent consolidation from this curve. The construction of a time-consolidation curve from a time-dial reading curve is described in detail in the chapter on consolidation.

The change in void ratio during consolidation under a load increment may be used to determine the slope of the e-p curve.

Example. Assume a consolidation test specimen 2.54 cm thick between porous stones subjected to a load increment under which the time-dial reading curve obtained was as shown in Figure 8.12a. Assume the void ratio after complete consolidation under a unit pressure $p_1 = 1.473$ kg cm^{-2} to be $e_1 = 0.5847$ and after complete consolidation under an added increment to produce a unit pressure $p_2 = 2.946$ kg cm^{-2} to be $e_2 = 0.4986$. Under these conditions the average value of e is 0.5416 and the value of

$$a_v = \frac{\Delta e}{\Delta p} = \frac{0.5847 - 0.4986}{2946 - 1473} = \frac{0.0861}{1473} = 0.0000584 \text{ cm}^2 \text{ g}^{-1}.$$

Assume $t_{50} = 12$ min. or 720 sec.

The coefficient of permeability during consolidation under the load increment considered is

$$k = \frac{0.2(1)(0.584 \times 10^{-4})1.27^2}{(1.5416)(720)} = 169 \times 10^{-9} \text{ cm sec}^{-1}.$$

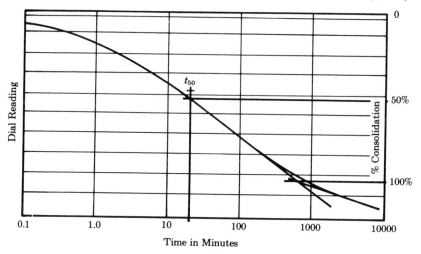

Figure 8.12a
Time-Consolidation Curve

The coefficient of permeability of clays having coefficients of permeability in the range from 10^{-7} to 10^{-9} cm sec^{-1} can be more easily and accurately determined indirectly from a consolidation test than from a direct determination. A direct determination of the permeability of such soils would be made with the apparatus shown in Figure 8.09d in conjunction with a consolidation test.

8.13 PERMEABILITY FROM GRAIN SIZE DISTRIBUTION

Darcy's Law applies only to soils in which the flow is laminar. It has been known for a long time that Darcy's Law does not apply to coarse grained soils. Work done at the Pennsylvania State University indicated laminar flow in all soils of grain size equal to or smaller than diameters of about 0.5 mm, but showed that the flow may be turbulent for soils coarser than 0.5 mm diameter.

While studying the characteristics of filter sands, Allan Hazen in the 1890's defined a grain size D_e, called effective grain size, as that diameter of grains in a soil of one grain size which would have the same coefficient of permeability as the soil as it exists with its varying grain sizes. He found that the coefficient of permeability for coarse grained soils could be estimated from the relationship $k = CD_e^2$, in which case C is a constant which for coarse sands and gravels is approximately equal to 100 cm^{-1} sec^{-1}. D_e is in cm and k is cm sec^{-1}. Hazen found that for practical purposes the effective grain size can be taken as the 10 per cent size.

For clean coarse sands and gravels the coefficient of permeability may be reliably computed from Hazen's formula,

$$k = 100D_{10}^2.$$

8.14 EFFECT OF VOID RATIO ON PERMEABILITY

From Poiseuille's Law the velocity of water flowing through a single capillary tube is

$$v_m = \frac{\gamma_w}{8\eta\pi} i_m a.$$

The quantity of water flowing in the tube in a unit of time is

$$q_1 = v_m a = \frac{\gamma_w}{8\eta\pi} i_m a^2.$$

In an area of soil equal to $(1 + e)$ the total area through which the water flows may be considered as some number of tubes N times the area of the average tube a in which case the total quantity of water flowing through the soil is

$$q = Nq_1 = \frac{\gamma_w}{8\eta\pi} i_m a^2 N.$$

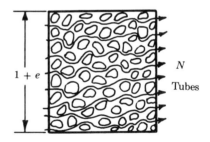

Figure 8.14a
Idealized Section

It is conceivable that the total area through which the water flows is some proportional part of the total voids, which in the case of an area $(1 + e)$ is e. Reasoning so, $Na = ce$ or

$$a = \frac{ce}{N} \quad (c \text{ is a constant}).$$

Inserting this value of a in Poiseuille's Law

$$q = \frac{\gamma_w}{8\eta\pi} i_m \frac{c^2 e^2}{N}.$$

From Darcy's Law the water flowing through a total area of soil $(1 + e)$ is

$$q = ki_x(1 + e).$$

In this relationship ki_x is the discharge velocity.

By reference to Figure 8.14b,

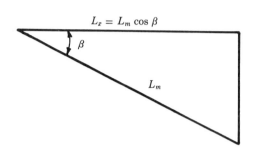

$$i_x = \frac{h}{L_x} \text{ and } i_m = \frac{h}{L_m}.$$

Then from Darcy's Law

$$q = k\frac{h}{L_x}(1 + e)$$

and from Poiseuille's Law

Figure 8.14b
Relation Horizontal Distance to Length of Tube

$$q = \frac{\gamma_w}{8\eta\pi} \frac{h}{L_m} \frac{c^2 e^2}{N}.$$

Equating these equal values of q

$$k\frac{h}{L_x}(1 + e) = k\frac{h}{L_m \cos \beta}(1 + e) = \frac{\gamma_w}{8\eta\pi} \frac{h}{L_m} \frac{c^2 e^2}{N}.$$

From which

$$k = \frac{\gamma_w \cos \beta}{8\eta\pi} \frac{c^2}{N} \frac{e^2}{(1 + e)}.$$

If the structure of the soil is not destroyed, the number of tubes N does not change, and c and β are comparatively little affected by a change in void ratio. γ_w and η are changed by temperature only. It may, therefore, be reasoned that the coefficient of permeability is proportional to the ratio $e^2/1 + e$.

Other rationalizations show the coefficient of permeability as being proportional to a function of e; i.e., $k = k_1 F(e)$ with limiting values of $F(e)$ of $F_1(e) = \frac{2e^2}{1 + e}$ and $F_2(e) = \frac{2e^3}{1 + e}$. The geometric mean of these limiting values is $F_3(e) = \frac{2e^{2.5}}{1 + e}$ and the arithmetic mean is $F_4(e) = \frac{1}{2}\left[\frac{2e^2}{1 + e} + \frac{2e^3}{1 + e}\right] = e^2$. For any of these values of $F(e)$, k_1 is the coefficient of permeability for a void ratio of unity.

For the limited range in which soils commonly vary, the difference given by these relationships is quite small, making little practical difference which is used. Because $k = k_1 e^2$ is the simplest of these relationships and because it gives as reliable results as any, A. Casagrande has suggested it for general use.

So, for all but very fine grained soils either of the relationships between k and e listed above is satisfactory. But, for very fine grained soils, especially clays at about the plastic limit, test observations indicate that the permeability is reduced to zero at void ratios equal to about 0.1. Terzaghi derived a formula for the relationship between coefficient of permeability and void ratio which gives a value of $k = 0$ when $e = 0.1$. Schlicter also derived a similar formula. In a soil

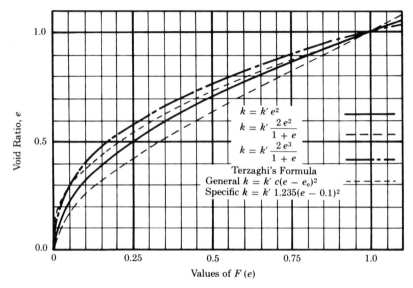

Figure 8.14c
Relation Between Void Ratio and Coefficient of Permeability

with a void ratio as low as 0.1, the diameter of pores is probably so small that the water is so viscous that it will not flow but is held to the walls of the tubes by molecular attraction. For these very fine grained soils with void ratios below 0.4, Terzaghi's formula is recommended.

A comparison of different relationships between k and e is shown in Figure 8.14c.

8.15 EFFECT OF TEMPERATURE ON PERMEABILITY

In the relationship $k = \dfrac{\gamma_w \cos \beta}{8\eta\pi} \dfrac{c^2}{N} \dfrac{e^2}{(1+e)}$, the only factors affected by temperature are the unit weight of water γ_w and viscosity η.

This relationship could be written for a particular soil specimen as $k = \dfrac{C}{\dfrac{\eta}{\gamma_w}}.$ Kinematic viscosity ν is defined as η/γ_w.

The coefficient of permeability is inversely proportional to the kinematic viscosity of the water. The relationship can be written

$$k_{20} = k_T \frac{\nu_T}{\nu_{20}}.$$

For practical purposes, the ratio of the absolute viscosities η_T/η_{20} may be used in place of the ratio of the kinematic viscosities.

8.16 WATER SUPPLY FOR PERMEABILITY TESTS

For those permeability tests for which deaired water is required, a supply of deaired water must be available. An apparatus for removing the dissolved air from water and a method of storing the deaired water is shown and described in Section 8.16L in the chapter on Laboratory Testing.

8.17 SUMMARY

The reader should be warned that the determination of the permeability of a natural deposit of soil is not as simple as the foregoing discussion of laboratory testing may imply.

If the deposit of soil covering the affected area consists of a few thick layers of uniform quality and thickness, the results of permeability tests on undisturbed samples from the different layers can give reliable information for estimating the flow of water through the deposit. Even for this simplest of cases an analysis based upon the results of laboratory tests can be used only as an aid to the judgment.

Generally, natural deposits of soil consist of thin and thick layers of soils of widely different permeabilities. Some of the layers may be fairly uniform in thickness, others may vary greatly in thickness and permeability over the affected area. A layer of sand discovered in a bore hole which indicates a drainage layer may be only a lens through which water cannot flow except as permitted by the surrounding soil. Laboratory tests on samples of the soils in such deposits are of little value in estimating the flow of water through the deposits.

Determination of the permeability of heterogenous natural deposits can be made with a greater degree of accuracy by testing of the soil in place than by laboratory testing of samples. Below the ground water table measurements of drawdown caused by pumping from wells are useful in estimating the average permeability of the deposit. Above the water table pumping water into a well and measuring the rise in observation wells can give useful information. Measurement of the amount of water lost at different elevations in a well into which water is pumped can give useful information also.

In the interpretation of these in place tests, a knowledge of the geologic nature and history of the deposit is necessary. The results of these in place tests must be considered only as aids to the judgment and must be used wisely. Wisdom in the use of such results can be developed only from experience. When properly evaluated, the experience of others can be as useful as one's own. One should read critically reports of the experience of other engineers as presented in papers to technical societies and in articles in technical publications. The engineer also has an obligation to report the results of those portions of his experience that contribute to the knowledge of others.

Permeability tests on laboratory samples are useful in the design of structures of artificially placed deposits of soil made under rigid control conditions, such as, dikes, levees, dams, and filters.

A. Casagrande has arranged in tabular form a brief summary of the relationship between permeability and other properties of soils and a comparison of different methods for determining the coefficient of permeability for different types of soils.

This chart is reproduced on the following page.

REFERENCES

Casagrande, A., "Seepage Through Dams," *Journal New England Water Works Association*, June, 1937, and in *Contributions to Soil Mechanics*, 1925 to 1940, Boston Society of Civil Engineers.

Hall, Howard P., "A Historical Review of Investigations of Seepage Toward Wells," *Journal Boston Society of Civil Engineers*, July, 1954.

Harr, Milton E., *Groundwater and Seepage*, New York: McGraw-Hill Book Company, Inc., 1962.

COEFFICIENT OF PERMEABILITY
k in. cm. per sec. (log scale)
After A. Casagrande

Coefficient of Permeability scale: 10^2 — 10^1 — 1.0 — 10^{-1} — 10^{-2} — 10^{-3} — 10^{-4} — 10^{-5} — 10^{-6} — 10^{-7} — 10^{-8} — 10^{-9}

Drainage Property: Good Drainage | Poor Drainage | Practically Impervious

Application Earth Dams and Dikes: Pervious Sections of Dams & Dikes | Impervious Sections of Earth Dams

Types of Soil:
- Clean Gravel
- Clean Sands, Clean Sand and Gravel Mixtures
- Very Fine Sands, Organic and Inorganic Silts, Mixtures of Sand, Silt and Clay, Glacial Till, Stratified Clay Deposits etc.
- "Impervious" Soils e.g., Homogeneous Clays below zone of Weathering.
- "Impervious Soils" which are modified by effects of Vegetation & Weathering

Direct Determination of Coefficient of Permeability.
- Direct Testing of Soil in Place (Wells) Reliable – Experience Required.
- Constant Head Permeameter Little Experience Required.
- Reliable Little Experience Required
- Falling Head Permeameter Range of Unstable Permeability – Much Experience Necessary.
- Fairly Reliable Experience Necessary

Indirect Determination of Coefficient of Permeability.
- Computation from Grain Size Distribution (Hazen's Formula) Cohesionless Sands & Gravels
- Horizontal Capillarity Test Little Experience Necessary Useful for Field Testing
- Computations from Consolidation Tests – Exp. Read.

PROBLEMS

8.1 Distinguish among the terms discharge velocity, seepage velocity, and true velocity as they apply to flow through soils. Starting with Darcy's Law, derive an expression for the seepage velocity v_s.

8.2 Define the "critical hydraulic gradient" for the upward flow of water through the soil, and derive an expression for it in terms of G_s and e.

8.3 Upon what factors does the permeability of a soil depend? Which factor has the greatest practical significance? Explain.

8.4 Define coefficient of permeability k as used in soil mechanics.

8.5 What simple relationships are used to express the variation of the coefficient of permeability with void ratio for (a) sands, and (b) clays? Why is it necessary to use different types of expressions for these two soils?

8.6 Derive an expression for the coefficient of permeability k based on Darcy's Law, and employing those physical data measured in a typical constant head permeability test in the laboratory.

8.7 Sketch a typical set-up for a *falling head* permeability test and derive an expression for the coefficient of permeability making use of physical quantities which may be measured using the apparatus of your sketch.

8.8 Explain how the coefficient of permeability of a clay may be calculated from the data obtained in a consolidation test.

8.9 Explain in some detail, but without deriving any formulas, how the coefficient of permeability of soil may be determined in place.

8.10 What difficulties in sampling may lead to important errors in the evaluation of the permeability of sand strata?

8.11 A sand specimen 35 sq cm in area and 20 cm long was tested in a constant head permeameter. Under a head of 50 cm the discharge was 105 cc in 5 min. The dry weight of the sand was 1105 g and $G_s = 2.67$. Compute
(a) the coefficient of permeability k of the sand,
(b) the discharge velocity, and
(c) the seepage velocity.

8.12 A soil specimen 10 cm in diameter and 5 cm thick was tested in a falling head permeameter. The head dropped from 45 cm to 30 cm in 4 min and 32 sec. The area of the standpipe was 0.5 cm². Considering the effect of capillary rise in the standpipe, compute the coefficient of permeability of this soil in units of cm/sec.

8.13   A falling head permeability test on a specimen of fine sand, 16 cm² in area and 9 cm long, gave a coefficient of permeability of 7×10^{-4} cm/sec. The specific gravity of the sand was 2.68 and its dry weight 210 g. The test temperature was 26° C. Compute the coefficient of permeability of the sand for a void ratio of 0.70 and a temperature of 20° C.

8.14   A horizontal capillarity test on a representative sample of fine sand from a borrow area gave the following results:

Time (min)	0.5	1.5	4.0
Advancement of Line of Wetting (cm)	3.3	5.8	9.6

(The coefficient of permeability was determined directly to be 14×10^{-4} cm/sec.)
Compute the value of the calibration constant z based on units of cm²/min for m and units of 10^{-4} cm/sec for k.

8.15   A laboratory specimen of clay (2 cm thick, drained top and bottom) required 5 minutes to reach 50 per cent consolidation when the pressure was increased from 0.75 to 1.50 kg/sq cm. The total change in void ratio under this increment of load was 0.15. Determine the coefficient of permeability of the clay. Assume initial $e = 1.65$.

8.16   A specimen of fine sand has a coefficient of permeability of 3×10^{-4} cm/sec at a void ratio of 0.72. Estimate the permeability of this soil when its void ratio is 0.5.

8.17   The coefficients of permeability of a clay at void ratios of 1.55 and 1.25 are 58×10^{-9} cm/sec and 35×10^{-9} cm/sec, respectively. Determine the coefficient of permeability for a void ratio of 0.80.

8.18   Direct permeability tests on a particular clay produced the following data:

$$e_1 = 0.95 \qquad k_1 = 25 \times 10^{-9} \text{ cm/sec}$$
$$e_2 = 0.78 \qquad k_2 = 16 \times 10^{-9} \text{ cm/sec}$$

Estimate by some reasonable method the coefficient of permeability for this soil when the void ratio is 0.50.

8.19   The coefficient of permeability of a certain clay is 50×10^{-9} cm/sec when its porosity is 0.5. Make a reasonable determination of its coefficient of permeability for a porosity of 0.4.

8.20   A drainage pipe beneath a dam has become clogged with sand whose coefficient of permeability is found to be 85×10^{-4} cm/sec. For an average difference in headwater and tailwater elevation of 72 feet it has been observed that there is a flow of 6 cu ft/day through the pipe. The pipe is 320 ft long and has a cross-sectional area of 0.2 sq ft. What proportion of the pipe is filled with sand?

CHAPTER IX

Compressibility and Consolidation

9.01 GENERAL

All materials are deformed when subjected to stress. Within the elastic range solid materials are deformed by a relative distortion of the position of atoms in the molecular structure. Upon release of the stress the atoms spring back to their original positions. This type of deformation is illustrated in Figure 7.01b. When a solid crystalline material, such as steel, is deformed past the yield point, crystals slip along adjoining faces to new positions of stability. That portion of the deformation which is due to slipping of the crystals to new positions, is inelastic and is not recovered after release of stress.

Soils are composed of small solid particles not bonded or grown together, except by the small van der Waals forces and adsorbed and double layer water. A stress applied to these solid individual grains produces a deformation of the solid particles, but, in general, this elastic deformation of the solid particles is small, possibly negligible, when compared to the deformation caused by change in relative positions of the discrete particles and the resulting decrease in the volume of voids. This deformation and accompanying decrease in volume is referred to as consolidation.

Most of the deformation of structural materials occurs immediately upon application of the load and is not accompanied by volume change. Under sustained loads a further small deformation known as creep may occur after considerable time. In using these structural materials, the engineer is interested primarily in the initial deformation. Creep, in general, tends to relieve high stress concentrations and contributes to the safety of structures. On the other hand, the elastic deformation of the individual grains is of minor importance in soils used as structural material. The deformation which accompanies volume change is of primary importance in the use of soils. This deformation occurs at varying rates, depending to a large extent upon the permeability of the soil. In coarse grained soils, the volume change and deformation may occur fairly rapidly after application of the load. For fine grained soils the deformation and volume change occur at a much slower rate. Settlement of buildings on a thick layer of saturated clay may progress for several years. Sometimes, differential settlement sufficient to produce cracking of masonry walls may not occur until several years after erection of the building.

9.02   CONSOLIDATION OF COHESIONLESS SOILS

Cohesionless soils consist of bulky grains with a very small percentage of adsorbed water attached to their surfaces. They may occur in a single grained or in a honeycomb structure. Deformation of these materials is affected to a great extent by the structure or arrangement of the grains. Resistance to movement of the grains to positions of greater density is dependent upon friction between the grains produced by pressure.

When a static load is applied to a layer of sand which is confined against lateral deformation, a rapid vertical deformation occurs. This early rapid deformation is followed by a continuing deformation occurring at a gradually decreasing rate. If the sand is saturated, the initial rapid deformation can occur only as the incompressible water is squeezed out of the void spaces. The rate at which this deformation can take place depends upon the permeability of the soil and upon the distance the water must travel to reach a free surface. In coarse sand the volume change takes place almost immediately, while in fine sand considerable time may be necessary for the initial deformation to occur.

The first or primary consolidation which occurs upon the addition of load is caused by the displacement of the grains from their former stable positions against the frictional resistance under the applied load to positions of greater density and higher initial frictional resistance. Against this greater internal resistance, the rate of deformation gradually slows down until a density is reached in which the soil grains are nested together enough to prevent further deformation under this applied load. Obviously, the primary consolidation of loose sand is much greater than that of dense sand.

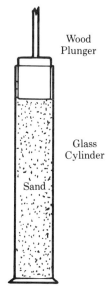

Figure 9.02a
Compressibility of
Sand—Demonstration

As has already been stated, the consolidation of clean sands produced by static loads is small (possibly insignificant) in comparison with that caused by vibration. The relative compressibility of sand can be illustrated by filling a 1000 cc glass cylinder to the calibration mark with clean dry sand. A hand should be placed over the open end and the cylinder slowly turned upside down to allow the sand to run to the opposite end of the cylinder. The cylinder should then be slowly turned upright allowing the sand to run to the bottom of the cylinder in a loose state. If a plunger that will fit loosely inside the cylinder is placed on the surface of the sand and pressure applied, the sand will be deformed very little. If, however, the plunger is removed and the cylinder of sand shaken and vibrated, the volume of the sand will be seen to decrease markedly. The volume of the loose sand may be decreased by as much as 20 per cent by vibration. The amount of volume decrease depends primarily upon the gradation of the sand.

Fine grained cohesionless soils having a honeycomb structure follow a similar pattern of deformation under light loads in which the deformation is relatively small. As the load is increased, the deformation is increased about the same as for loose sand until a critical load is reached which causes collapse of the arches forming the honeycomb structure. Loads above this critical one produce large deformations.

9.03 COMPRESSIBILITY AND CONSOLIDATION OF CLAY

Clay soils contain a significant quantity of minute, flat, scale-like particles. These scale-like particles are deposited from suspension in

Figure 9.03a
Loose Freshly Deposited Clay

water in a haphazard loose state as shown in Figure 9.03a. Attached to the surfaces of these particles is a layer of solid adsorbed water about 10 Å thick. Beyond the adsorbed water there may exist a layer of viscous double layer water of about 400 Å thickness on kaolinite to 200 Å thick on montmorillonite. Because the kaolinite particles may be 1000 Å thick, the double layer water makes up $\frac{800}{1820}$ or 44 per

cent of the total volume. Montmorillonite particles may be only 10 Å thick so that the double layer water could account for $\frac{400}{430}$ or 93 per cent of the total volume. The viscosity of this double layer water varies from a solid state near the surface of the particle to the viscosity of free water at the extremity of the double layer.

The clay particles are separated by the double layer water. Pressure applied to a layer of clay particles separated by the viscous water forces some of the viscous water from between the particles until they are separated by water of high enough viscosity to resist the applied pressure. The amount and viscosity of water between the particles depends upon the applied pressure and the period of time during which the pressure has acted. In freshly deposited, loose, flocculent clay the points of contact are separated by attached water whose viscosity is dependent upon the pressure. The void spaces between the grains are filled with free water protected from pressure by the soil structure.

When pressure is applied to a layer of the loose freshly deposited clay, the high stress at points of contact causes the particles to slip along contact faces to positions of greater density. As the volume begins to decrease, the free pore water is placed in compression and forced out. The reduction in void ratio and volume can occur only as the incompressible water is forced out of the pores.

The mineral particles of clay are themselves elastic. Like sheets of mica, they bend under flexural stresses and straighten back to their original shape when the stresses are released.

The compressibility of clays may be caused by these three factors: the double layer water may be forced from between the grains, the particles may be slipped to new positions of greater density, or the particles may be bent as elastic sheets.

Figure 9.03b
Consolidated Clay

Deformation of a clay having a dispersed structure in which the scale-like particles are oriented more nearly parallel to each other horizontally is produced primarily by forcing double layer water from between the grains. Artificial compaction of clays usually produces a more or less dispersed structure. Deformation of loose, flocculent clay is produced primarily by slipping of the grains to new positions and by elastic bending of the particles.

In any case, the volume change of the clay can take place only as free water is forced out of the void spaces between the grains. Because the permeability of clay is extremely small, time is an important factor in the consolidation of clays.

When a load is first applied to saturated clay, the free incompressible water is placed in compression so that initially the pore water carries all the pressure produced by the added load and the pressure between the grains is the same as before the load was added. But as free water leaves the void spaces, pressure decreases in the pore water. As the pressure between the grains increases with consolidation, pressure is applied to the viscous water between points of contact forcing some of it out from between the grains. This highly viscous asphalt-like water flows out very slowly, much slower than the free water leaves the void spaces.

Thus, the consolidation of saturated clay under load can be divided into two parts: primary consolidation during which free water is forced from the void spaces, and secondary compression during which some of the highly viscous water between the points of contact is forced out from between the grains. Secondary compression occurs, generally, at a much slower rate than primary consolidation and continues for a very long time after primary consolidation is complete.

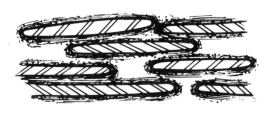

Figure 9.03c
Dispersed Clay

When the pressure that produced consolidation of a clay layer is removed, that portion of the deformation caused by elastic flexibility of the soil particles and that caused by squeezing out attached water is recovered as elastic rebound or swelling. The attached water that was forced out by pressure is drawn back to the particle surfaces by the electro-magnetic forces. This elastic recovery can occur only as enough water is drawn into the pores to equal the volume increase.

The elastic recovery or swelling of cohesionless soils after consolidation under pressure is extremely small.

9.04 THEORETICAL CONCEPT OF CONSOLIDATION

The theoretical concept of the consolidation process for saturated clays under load was stated and a mathematical statement for the progress of consolidation was developed by Terzaghi and Frölich during the decade between 1910 and 1920 while at the University of Vienna.

In the development of the mathematical statement of the consolidation process, the following simplifying assumptions are made.

1. Deformation of the soil is due entirely to change in volume.
2. The soil is homogeneous from top to bottom of the affected layer.
3. The soil is completely saturated.
4. Change in volume of the soil is due entirely to forcing free water from the voids.

5. Load is applied in one direction only and deformation occurs only in the direction of the load application; i.e., the soil is restrained against lateral deformation.

6. During consolidation the water flows only in the direction of the load application.

7. The boundary is a free surface offering no resistance to the flow of water from the soil.

8. The change in thickness of the layer during consolidation is insignificant compared to the thickness of the layer.

9. The coefficient of consolidation $c_v = \dfrac{k(1 + e)}{\gamma_w a_v}$ is constant during the consolidation process.

In evaluating the results of a consolidation test and in performing a settlement analysis, these simplifying assumptions should always be considered.

The concept of consolidation of a saturated clay under load can be simply illustrated by assuming a cylinder containing pistons having watertight and frictionless joints between pistons and cylinder as shown in Figure 9.04a. The pistons are separated by elastic springs dividing the cylinder into spaces which are filled with water. Each piston is pierced with a pin hole through which water can flow from one space to another.

Assume that a load p_1 has been applied to the top piston and water has been allowed to flow until the springs carry the entire load and the water is subjected to hydrostatic pressure only. Assume now that a load Δp is added to the top piston.

The springs cannot carry any of the added load until they are depressed. Because the water in the spaces is relatively incompressible, it carries all the added load, and the excess hydrostatic pressure u in the water in all the spaces is equal to Δp. Immediately water under this pressure begins to flow through the pin hole in the top piston. The volume decrease in the upper space allows the spring to be depressed and take part of the added load thereby reducing the pressure in the water in that space. After some water has flowed from the upper space and its pressure has been decreased, a pressure differential exists between the upper and middle spaces causing water to begin to flow from the middle space into the upper one. The reduction in volume of the middle space allows the spring in that space to carry part of the load and the pressure in the water to be reduced. Water then begins to flow from the lower chamber into the middle one giving to the spring in the lower space some of the added load.

At some time t after application of the load Δp to the top piston, the pressure in the water u and the pressure carried by the springs σ will be as shown in the right side of Figure 9.04a.   It is easy to see that the rate at which the pressure is changed from water to springs depends upon the size of the pin holes in the pistons.

If it be assumed that the clay particles act as tiny elastic springs with myriads of pore spaces filled with water, a layer of clay will consist of an infinite number of spaces and the relationship between pore

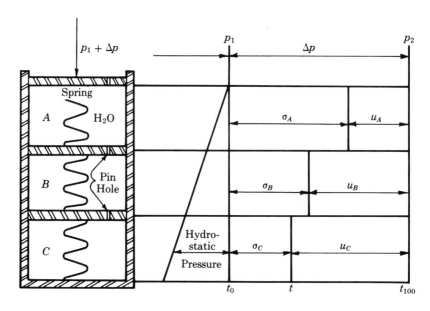

Figure 9.04a
Fundamental Concept of Consolidation

water pressure and pressure between the soil grains with depth becomes a smooth curve as shown in Figure 9.04b.

Consolidation of a clay layer occurs instantaneously at the surface and progresses downward with time to an impervious layer.   The curves drawn show the relationship among pore water pressure u, time t, and depth z.   The degree of consolidation at depth z after an elapsed time t after application of the load is the ratio of the pressure in the soil σ_z produced by the added load Δp to the total added load Δp; i.e., $U_z =$ degree of consolidation at depth $z = \dfrac{p_2 - p_1 - u_z}{p_2 - p_1}$.

If a layer of clay is drained both top and bottom, the pore water can flow in either direction.   For symmetrical conditions the water will flow upward from middepth the same as though an impervious layer existed at middepth.   Similarly, the water in the bottom half of the

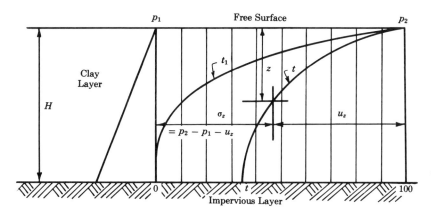

Figure 9.04b
Relation of u and σ with Depth and Time in Clay Layer

layer will flow downward.   Such a layer drained on both sides may be considered as having a thickness $2H$ and the same relationship will apply as for the layer H thick drained one side only; i.e., the thickness H in the consolidation relationship should be taken as one-half the total thickness of a layer that is drained on both sides.

A layer of sand or other material having a much larger coefficient of permeability than the clay will act as a drainage layer and the joint between the sand layer and the clay may be considered as a free surface.

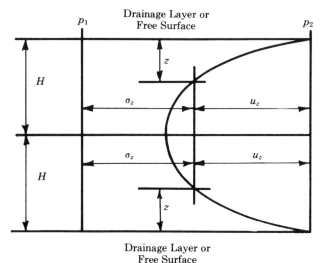

Figure 9.04c
Clay Layer Drained Two Sides

9.05 CONSOLIDATION EQUATION

From the preceding discussion, it is easy to see that the excess pressure in the pore water is a function of time t and depth z from the free surface; i.e., $u = F(z,t)$. The problem is to write an equation for this function so that the relationship between u and z can be determined.

If $u = F(z,t)$, the rate of change in u with depth z when time t is held constant is $\partial u/\partial z$ and the total change in u in a change in depth dz is $\dfrac{\partial u}{\partial z}\, dz$.

When t is changed and z is held constant, the change in u is $\dfrac{\partial u}{\partial t}\, dt$. At $z + dz$ and $t + dt$, the change in u is

$$du = \frac{\partial \dfrac{\partial u}{\partial t}\, dt}{\partial z}\, dz = \frac{\partial^2 u}{\partial t\, \partial z}\, dt\, dz,$$

or

$$du = \frac{\partial \dfrac{\partial u}{\partial z}\, dz}{\partial t}\, dt = \frac{\partial^2 u}{\partial z\, \partial t}\, dz\, dt.$$

This relationship is shown in Figure 9.05b which is an enlargement of the relationship shown in Figure 9.05a.

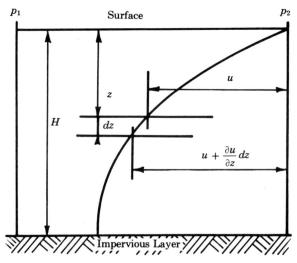

Figure 9.05a
Relationship Between Pore Water Pressure and Depth at Time t

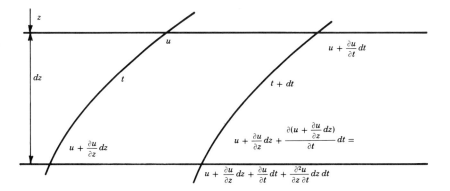

Figure 9.05b
**Relationship Between Change in Pore Water Pressure with Time and
with Depth**

Now consider an element of the soil layer at depth z, of area A, and
thickness dz, as shown in Figure 9.05c.

Assume that the flow of pore water due to pressure from the added
load is upward. Applying Darcy's Law, the flow of water into the

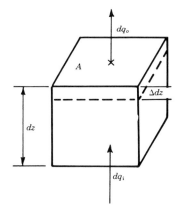

Figure 9.05c
Soil Element During Consolidation at Time t

element $dq_i = ki_i A \, dt$ and the flow out $dq_0 = ki_0 A \cdot dt$. The hydraulic
gradient or change in head per unit of distance at the entrance face

of the element is $i_i = \dfrac{\partial\left(u + \dfrac{\partial u}{\partial z}\,dz\right)}{\gamma_w \, \partial z} = \dfrac{\partial u}{\gamma_w \, \partial z} + \dfrac{\partial^2 u}{\gamma_w \, \partial z^2}\,dz$, and the

hydraulic gradient at the exit face is $i_0 = \dfrac{\partial u}{\gamma_w \, \partial z}$.

Then
$$dq_i = \frac{kA}{\gamma_w} \frac{\partial u}{\partial z} dt + \frac{kA}{\gamma_w} \frac{\partial^2 u}{\partial z^2} dz\, dt$$

and

$$dq_0 = \frac{kA}{\gamma_w} \frac{\partial u}{\partial z} dt.$$

The volume change of the saturated element in an element of time dt is

$$dq_0 - dq_i = - \frac{kA}{\gamma_w} \frac{\partial^2 u}{\partial z^2} dz\, dt.$$

Because the element is confined laterally, all the deformation occurs in a vertical direction so that the volume change in time dt is

$$A\, \Delta\, dz = - \frac{kA}{\gamma_w} \frac{\partial^2 u}{\partial z^2} dz\, dt, \text{ and } \Delta\, dz = - \frac{k}{\gamma_w} \frac{\partial^2 u}{\partial z^2} dz\, dt.$$

All this volume change occurs in the voids. The unit deformation ϵ in terms of void ratio as shown by Figure 9.05d is $\epsilon = \dfrac{de}{1 + e_1}.$ The deformation of the element dz thick is

$$\epsilon\, dz = \frac{de}{1 + e_1} dz = \Delta\, dz.$$

Then
$$\frac{de}{1 + e_1} dz = - \frac{k}{\gamma_w} \frac{\partial^2 u}{\partial z^2} dz\, dt$$

from which

$$de = - \frac{k(1 + e_1)}{\gamma_w} \frac{\partial^2 u}{\partial z^2} dt. \qquad \text{(Eq. 9.05a)}$$

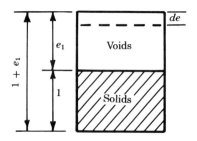

Figure 9.05d
Deformation and Void Ratio

The coefficient of compressibility a_v is defined as the slope of the curve expressing the void ratio–pressure relationship. For con-

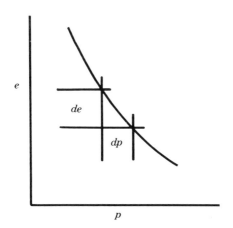

Figure 9.05e
Void Ratio–Pressure Curve

venience, a_v is taken as a positive number. Every point on the curve in Figure 9.05e represents the void ratio after complete primary consolidation under the corresponding pressure. Then $a_v = \dfrac{de}{dp}$ or $de = a_v \, dp$. Equation 9.05a can then be written

$$a_v \, dp = - \frac{k(1 + e_1)}{\gamma_w} \frac{\partial^2 u}{\partial z^2} \, dt.$$

But, as p increases, u decreases by the same amount or $dp = -du$.

So, $a_v \, du = \dfrac{k(1 + e_1)}{\gamma_w} \dfrac{\partial^2 u}{\partial z^2} dt,$ or $du = \dfrac{k(1 + e_1)}{\gamma_w a_v} \dfrac{\partial^2 u}{\partial z^2} dt.$

The total change du in time dt is the rate of change with respect to time $\partial u / \partial t$ times the total time dt or $du = \dfrac{\partial u}{\partial t} dt$. The above equation can then be written

$$\frac{\partial u}{\partial t} dt = \frac{k(1 + e)}{\gamma_w a_v} \frac{\partial^2 u}{\partial z^2} dt \quad \text{or} \quad \frac{\partial u}{\partial t} = \frac{k(1 + e)}{\gamma_w a_v} \frac{\partial^2 u}{\partial z^2}. \quad \text{(Eq. 9.05}b\text{)}$$

As consolidation progresses k and e are both decreased. a_v also decreases as consolidation progresses. If it be assumed that $k(1 + e)$ decreases at the same rate as a_v, the ratio $\dfrac{k(1 + e)}{\gamma_w a_v}$ is a constant and may be designated c_v and called the coefficient of consolidation. The consolidation equation can then be written

$$\frac{\partial u}{\partial t} = c_v \frac{\partial^2 u}{\partial z^2}. \quad \text{(Eq. 9.05}c\text{)}$$

This equation can be integrated between the limits p_1 and p_2 and 0 and H as

$$u = (p_2 - p_1) \sum_{N=0}^{N=\infty} \left\{ \frac{4}{(2N+1)\pi} \sin \left[\frac{(2N+1)\pi}{2} \frac{z}{H} \right] \right\} \xi^{-\left[\frac{(2N+1)^2 \pi^2}{4} c_v \frac{t}{H^2} \right]}.$$

(Eq. 9.05d)

In Equation 9.05d, N = all the whole numbers from 0 to ∞, and ξ = base of the natural logarithms.

The relationship between u and z for any time t can be determined by summing the values of u for $N = 0$, $N = 1$, $N = 2$, $N = 3$, etc. It will be found that as the whole numbers become larger, the contribution to the summation for u becomes smaller. For nearly all cases, the summation for $N = 0$, $N = 1$, $N = 2$, and $N = 3$ is sufficient for a high degree of accuracy.

9.06   DEGREE OF CONSOLIDATION

As already described, the degree of consolidation at depth z and time t for an increase of external pressure from p_1 to p_2 is represented by the ratio $U_z = \dfrac{p_2 - p_1 - u}{p_2 - p_1}$. Over a depth dz the degree of consolidation is

$$U_z = \frac{(p_2 - p_1 - u)\, dz}{(p_2 - p_1)\, dz}.$$

It follows that the average degree of consolidation for a layer of thickness H is equal to the ratio of the shaded area in Figure 9.06a to

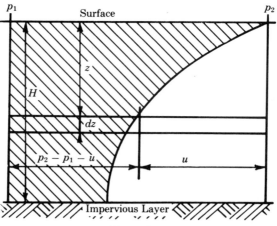

Figure 9.06a
Degree of Consolidation

the entire area $(p_2 - p_1)H$. Thus, the average degree of consolidation at time t is

$$U = \frac{\int_0^H (p_2 - p_1 - u) \, dz}{(p_2 - p_1)H}. \qquad \text{(Eq. 9.06}a\text{)}$$

$$U = \frac{\int_0^H (p_2 - p_1) \, dz - \int_0^H u \, dz}{(p_2 - p_1)H} = 1 - \frac{1}{(p_2 - p_1)H} \int_0^H u \, dz.$$
$$\text{(Eq. 9.06}b\text{)}$$

Evaluating $\int_0^H u \, dz$ after substituting for u its value given in Equation 9.05d,

$$\int_0^H u \, dz = \int_0^H (p_2 - p_1) \sum_{N=0}^{N=\infty} \xi^{-\left[\frac{(2N+1)^2\pi^2}{4} c_v \frac{t}{H^2}\right]}$$
$$\left[\frac{4}{(2N+1)\pi} \sin \frac{(2N+1)\pi}{2} \frac{z}{H} \, dz\right]$$
$$= (p_2 - p_1) \sum_{N=0}^{N=\infty} \xi^{-\left[\frac{(2N+1)^2\pi^2}{4} c_v \frac{t}{H^2}\right]}$$
$$\left[\frac{4}{(2N+1)\pi} \int_0^H \sin \frac{(2N+1)\pi}{2} \frac{z}{H} \, dz\right].$$

The statement under the integral sign in this last equation is of the form $\int \sin ax \, dx$, which is equal to $-\frac{1}{a} \cos ax$. Carrying out this integration,

$$\int_0^H u \, dz = (p_2 - p_1) \sum_{N=0}^{N=\infty} \xi^{-\left[\frac{(2N+1)^2\pi^2}{4} c_v \frac{t}{H^2}\right]}$$
$$\frac{4}{(2N+1)\pi} \int_0^H \left[-\frac{2H}{(2N+1)\pi} \cos \frac{(2N+1)\pi}{2H} z\right]$$
$$= (p_2 - p_1) \sum_{N=0}^{N=\infty} \xi^{-\left[\frac{(2N+1)^2\pi^2}{4} c_v \frac{t}{H^2}\right]}$$
$$\frac{8H}{(2N+1)^2\pi^2} \int_0^H \left[-\cos \frac{(2N+1)\pi}{2H} z\right].$$

Inserting the upper limit $z = H$,

$$-\cos \frac{(2N+1)\pi}{2H} z \text{ becomes } -\cos \frac{(2N+1)\pi}{2}.$$

For any value of N, the angle $\dfrac{(2N + 1)\pi}{2}$ is an odd number times $\pi/2$

making $-\cos \dfrac{(2N + 1)\pi}{2} = 0$.   Inserting the lower limit $z = 0$,

$$-\cos \frac{(2N + 1)\pi}{H} z = -\cos 0 = -1.$$

Therefore,

$$\int u \, dz = (p_2 - p_1) \sum_{N=0}^{N=\infty} \xi^{-\left[\frac{(2N+1)^2\pi^2}{4} c_v \frac{t}{H^2}\right]} \frac{8H}{(2N + 1)^2\pi^2} [-0 - (-1)]$$

$$= (p_2 - p_1)H \sum_{N=0}^{N=\infty} \frac{8}{(2N + 1)^2\pi^2} \xi^{-\left[\frac{(2N+1)^2\pi^2}{4} c_v \frac{t}{H^2}\right]}. \quad \text{(Eq. 9.06}c\text{)}$$

Inserting this value of $\int u \, dz$ in Equation 9.06b for the average degree of consolidation,

$$U = 1 - \frac{(p_2 - p_1)H}{(p_2 - p_1)H} \sum_{N=0}^{N=\infty} \frac{8}{(2N + 1)^2\pi^2} \xi^{-\left[\frac{(2N+1)^2\pi^2}{4} c_v \frac{t}{H^2}\right]}$$

and the per cent consolidation,

$$U\% = 100 \left\{ 1 - \sum_{N=0}^{N=\infty} \frac{8}{(2N + 1)^2\pi^2} \xi^{-\left[\frac{(2N+1)^2\pi^2}{4} c_v \frac{t}{H^2}\right]} \right\}. \quad \text{(Eq. 9.06}d\text{)}$$

Thus, the average degree of consolidation of a layer of soil H thick drained one side only can be evaluated for any time t by evaluating the infinite series of Equation 9.06d.   If the layer of soil is drained both sides, the value of H in the consolidation equation is one half the thickness of the layer.

9.07   TIME FACTOR, T

In the equation for degree of consolidation U, all the properties and dimensions of the soil are included in the relationship

$$c_v \frac{t}{H^2} = \frac{k(1 + e)t}{\gamma_w a_v H^2} = \frac{(\text{cm sec}^{-1})(\text{sec})}{(\text{g cm}^{-3})(\text{cm}^2 \text{ g}^{-1})(\text{cm}^2)}.$$

When dimensions are inserted in this relationship, it is found to be a dimensionless ratio which may be designated as a time factor T.

Using this time factor in place of $c_v \dfrac{t}{H^2}$ in the consolidation equation, Equation 9.06d becomes

$$U\% = 100 \left\{ 1 - \sum_{N=0}^{N=\infty} \frac{8}{(2N+1)^2\pi^2} \xi - \left[\frac{(2N+1)^2\pi^2}{4} T \right] \right\}. \quad \text{(Eq. 9.07}a\text{)}$$

Because both U and T are dimensionless, a relationship exists between U and T which is independent of properties or dimensions.

9.08   RELATION BETWEEN DEGREE OF CONSOLIDATION U AND TIME FACTOR T

The value of U for any value of T can be determined from the above equation by summing all the values determined by the insertion of integral values of N from 0 to the number required to provide the degree of accuracy desired.

Values of T for each 5 per cent interval of U are given in the following table.

Table 9.08a

U%	0	10	15	20	25	30	35	40
T	0.000	0.008	0.018	0.031	0.049	0.071	0.096	0.126

U%	45	50	55	60	65	70	75	80
T	0.159	0.197	0.238	0.287	0.342	0.405	0.477	0.565

U%	85	90	95	100
T	0.684	0.848	1.127	∞

This relationship is plotted to arithmetic scales in Figure 9.08a and with U per cent to an arithmetic scale and T to a logarithmic scale in Figure 9.08b.

The semilogarithmic plot has advantages which will become apparent in the practical application of consolidation-time curves.

These curves can easily be converted to theoretical degree of consolidation-time curves for a layer of soil of given properties and dimensions.

The theoretical relationship between U and T is approximately parabolic up to a degree of consolidation of about 60 per cent. A comparison between actual theoretical values as given in Table 9.08a

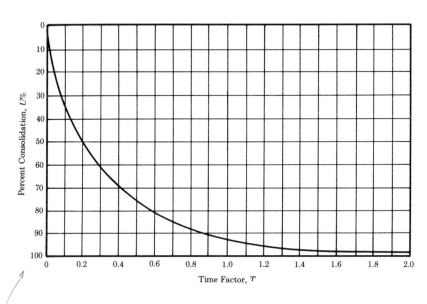

Figure 9.08a
Relationship Between Per Cent Consolidation and Time Factor
Arithmetic Scale

Same Data

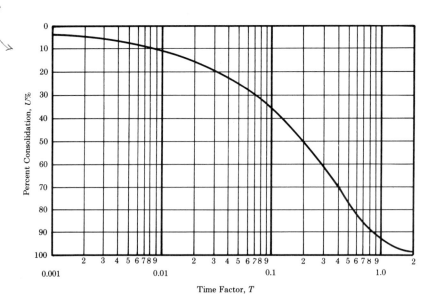

Figure 9.08b
Relationship Between Per Cent Consolidation and Time Factor
Semilogarithmic Scales

and a parabolic relationship is given in Table 9.08b.   In the parabolic relationship

$$U^2 = CT.$$ (Eq. 9.08a)

If the theoretical and parabolic relationships are assumed to be the same at $U\% = 30$, $C = \dfrac{30^2}{0.071} = 12{,}700$ from which other values in the parabolic relationship can be computed.

Table 9.08b

$U\%$	0	10	15	20	25	30	35	40
T_T	0.000	0.008	0.018	0.031	0.049	0.071	0.096	0.126
T_p	0.000	0.008	0.018	0.031	0.049	0.071	0.096	0.126

$U\%$	45	50	55	60	65	70	75	80
T_T	0.159	0.197	0.238	0.287	0.342	0.405	0.477	0.565
T_p	0.159	0.197	0.238	0.284	0.332	0.386	0.443	0.504

$U\%$	85	90	95	100
T_T	0.684	0.848	1.127	∞
T_p	0.569	0.638	0.710	0.787

Furthermore, it will be found that for values of U per cent greater than about 55 per cent, only the first term of the Fourier series of Equation 9.07a is needed to yield a sufficiently accurate result.   By substituting $N = 0$ into Equation 9.07a and rearranging the results an explicit expression for T is obtained.

$$T = 1.781 - 0.933 \log (100 - U\%)$$ (9.08b)

Within their applicable ranges, the simpler relationships represented by Equations 9.08a and 9.08b may be used in place of the more complicated Equation 9.07a.   If Equation 9.08a is used for values of U per cent lower than about 55 per cent and Equation 9.08b for values greater than 55 per cent, the maximum deviation from the theoretical relationship will not exceed about 0.5 per cent.

9.09   COMPARISON THEORETICAL CONSOLIDATION-TIME CURVE WITH CONSOLIDATION-TIME CURVE FOR IDEALIZED SATURATED CLAY

A mass of saturated clay consists of clay particles with their attached water layers and void spaces filled with free water.   At points of contact the particles are separated by attached water for which the thickness and viscosity depend upon the pressure between the grains.   The application of external load produces pressure in the free pore water, forcing it out and allowing a volume decrease and transferral of pressure to the soil grains.

This consolidation produced by squeezing out of the free pore water is referred to as primary consolidation and is the only consideration in

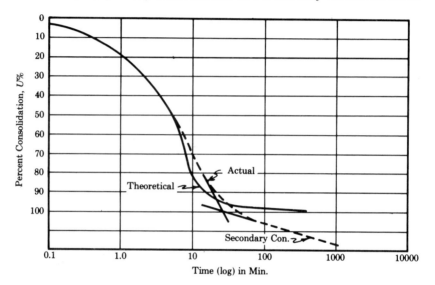

Figure 9.09a
Theoretical and Actual Consolidation-Time Curves for a Layer of Saturated Chicago Clay

the development of the theoretical consolidation-time relationship just discussed.

As pressure is transferred from the pore water to the soil grains during consolidation, the highly viscous water separating the particles at points of contact is subjected to heavy pressure.   Under this heavy pressure the highly viscous water begins to flow like seemingly brittle asphalt.   The rate at which this highly viscous water is forced from between the grains is extremely slow, continuing for years after application of load.   This very slow adjustment of grains to allow additional volume change is referred to as secondary consolidation or secondary compression.   It does not begin until primary consolidation has progressed sufficiently to produce additional pressure between the grains.   As the effect of primary consolidation decreases, the effect of secondary compression increases.

The theoretical consolidation-time curve, which does not include secondary consolidation, and the curve for an idealized saturated clay layer obtained by measurement of deformation with time almost coincide in the early portion of the consolidation cycle and separate as consolidation progresses.

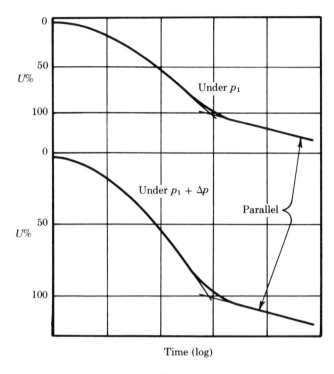

Figure 9.09b
Consolidation-Time Curves

The theoretical curve is asymptotic to 100 per cent consolidation. The actual curve continues beyond 100 per cent. When time is plotted to a logarithmic scale, the portion of the curve representing only secondary consolidation generally becomes a straight line on laboratory determined curves. This straight $U - \log t$ curve indicates a constant deceleration of the rate at which the viscous water is forced out from between the grains. Actually the secondary portion of the consolidation-time curve cannot continue indefinitely. The secondary compression must stop as the void ratio approaches zero. However, such an extremely long time is required for completion of the secondary compression that for the short period of the consolidation test the laboratory determined secondary portion of the curve may be considered as straight and continuous.

When an additional load is applied to the layer of clay after complete consolidation under an applied load, the rate at which the viscous water is forced from between contact points is the same as the rate after consolidation under the lesser load; i.e., the straight secondary portions of $U-\log t$ curves for different load increments are parallel. Such a distinction between primary and secondary consolidation cannot be made on an arithmetic plot of a consolidation-time curve. This parallel relationship contributes to the ease and accuracy of determining the change in void ratio due to complete primary consolidation caused by an added pressure.

It should be observed, however, that the slope of the secondary branch of the curve for a given load is dependent upon the history of the soil. The slope is much flatter for pressures smaller than the maximum pressure which has previously acted on the soil than it is for pressures to which the soil is being subjected for the first time. This fact is sometimes useful in helping to determine the maximum pre-consolidation pressure which is discussed in Article 9.11.

9.10 LABORATORY DETERMINATION OF CONSOLIDATION-TIME CURVE

A consolidation-time curve for a sample of clay can be constructed from data obtained by measuring the volume change of the sample at time intervals after application of a load. The volume change can be determined by measuring the amount of water which is squeezed out of the saturated sample as is commonly done in a triaxial test or by measuring with a dial indicator the deformation of a sample confined so that only vertical deformation can occur.

In the consolidation test a specimen of the soil sample is cut to fit snugly in a ring which confines the specimen laterally and loads applied in increments to the specimen through porous stones. A consolidation-time curve is obtained for each increment of applied load.

The specimen in the ring with porous stones which fit loosely inside the ring on each side of the specimen is placed in a vessel that will hold water to a depth above the top porous stone and set in a loading device with a dial indicator arranged for measuring vertical deformation as shown in Figure 9.10a. A light load is applied and the specimen flooded by filling the vessel to above the bottom of the top porous

stone.   After equilibrium has been established under this load and flooded condition, an increment of load is applied.   The dial should be read as soon as possible after application of the increment of load and the elapsed time and dial reading recorded.   After a longer period of time has elapsed, the dial should be read and the reading and the elapsed time recorded.   This pro-
cedure should be continued until after complete consolidation of the specimen under the applied incre-
ment of load.   Because the rate of volume change decreases rapidly with time, the readings should be taken at short intervals early in the consolidation period and increased in length as consolidation progresses.   If the time interval be-

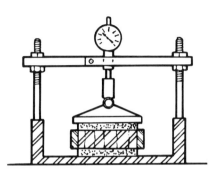

Figure 9.10a
Consolidation Test

tween readings is doubled each time, points will be spaced about an equal distance apart on the log time scale.

Dial readings are then plotted on an arithmetic scale and elapsed time on a logarithmic scale as shown in Figure 9.10b.

This dial reading-time curve can be converted to a consolidation-time curve by locating the position of zero consolidation and 100 per cent consolidation.   The scale between 0 per cent and 100 per cent is divided arithmetically to form a consolidation-time curve.   Because

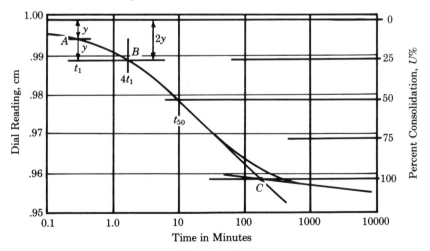

Figure 9.10b
Dial Reading-Time-Consolidation Curve

the logarithmic scale does not start at 0 and because the curve extends beyond 100 per cent consolidation, it is necessary to establish the 0 per cent and 100 per cent lines.

The theoretical consolidation-time curve is an approximate parabola up to about 60 per cent consolidation and the actual curve follows the theoretical curve approximately during the early portion of the consolidation period. Therefore, the early portion of the actual curve should be very nearly parabolic. On a semilog plot the curve is not

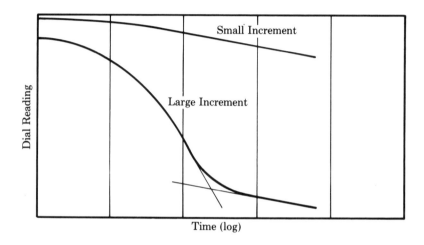

Figure 9.10c
Dial Reading-Time Curves for Large and Small Increments

parabolic but it is a parabola plotted semilogarithmically and has all the properties of a parabola so that $U^2 = Ct$. Therefore, if a point A on the curve at time t_1 has a U ordinate of y from the origin, a point B on the curve at time $4t_1$ has a U ordinate $2y$; i.e., the ordinate distance y between points whose absissas are t_1 and $4t_1$ is the same as the ordinate distance from the first point (t_1, y) to the origin. Therefore, to locate the origin of the consolidation-time curve, lay off the vertical distance between any two points on the curve that have absissas of t_1 and $4t_1$ vertically from the point at t_1 and draw a horizontal line through this point.

A. Casagrande discovered that 100 per cent consolidation can be located approximately by extending the straight portion of the curve representing secondary consolidation back to its intersection with a tangent drawn to the curve at its point of inflection. This construction is shown in Figure 9.10b.

It is obvious that the total change in height of the specimen will depend upon the intensity of the added pressure Δp which produced the deformation.   If the load increment is very small, the dial changes during consolidation will be small.   Because the time for consolidation is the same for a small increment as for a large one, the slope of the curve for a small increment may be so slight that it is impossible to distinguish between the primary and the secondary portions of the curve.   A very small increment may produce a slope of the primary portion of the curve that is less than that of the secondary portion. For a large increment the slope of the primary portion of the curve is much steeper than the slope of the secondary portion, which contributes to accuracy in determining the consolidation-time relationship. A comparison of the two curves may be seen in Figure 9.10c.   In either case, the slope of the secondary portions of the curves is the same.

9.11   VOID RATIO-PRESSURE CURVE

So far, this discussion has dealt with degree of consolidation without regard to the amount of deformation or settlement of a layer of soil subjected to additional load.   From the relationship developed, the time required for a certain percentage of the total settlement to occur can be determined, but the amount of settlement can be determined only if the total settlement is known.

Because the volume change of a soil mass occurs in the voids only, the void ratio may be used as a measure of deformation or settlement.   As shown in Figure 9.11a, if the void ratio of a mass of soil $1 + e_1$ thick is changed by an amount Δe, the unit deformation $\epsilon = \dfrac{\Delta e}{1 + e_1}$. The total settlement of a layer of clay after complete consolidation, without regard to time, caused by the addition of an increment of pressure Δp to the already applied pressure p_1 can be determined if the relationship between void ratio and pressure is known.

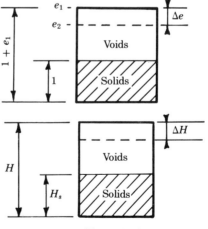

Figure 9.11a
Relationship Between Change in Void Ratio and Change in Height

A curve showing the relationship between void ratio after complete consolidation and pressure can be drawn from data obtained from the dial reading-time curves determined from a consolidation test. One point on the *e-p* curve can be obtained from each dial reading-time curve.

It is not necessary to know the exact value of *e* corresponding to a given value of *p* because deformation or settlement is measured by the change in void ratio from one value of *p* to another. The slope of the *e-p* curve between p_1 and p_2 is important. Because the time for consolidation of the same soil specimen is about the same for all load increments and because the slope of the secondary portion of the dial reading-time curves is the same, values of *e* computed from dial readings taken off the secondary portion of the dial reading-time curves at the same time when plotted against pressure will produce an *e-p* curve having the correct shape and slope. Using a dif-

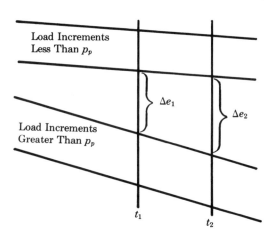

Figure 9.11b
Comparison Secondary Consolidation Curves for Pressure Less Than and Greater Than Preconsolidation Pressure

ferent time for computation of values of *e* will produce an *e-p* curve shifted slightly in position but having the same shape and slope.

Because the slope of the secondary portion of the dial reading-time curve is slightly different for pressures less than the maximum pressure under which the soil has been consolidated (preconsolidation pressure) than the slope of the secondary portion for pressures greater than the preconsolidation pressure, the shape of the junction between the recompression and virgin portions will likely be slightly different for *e-p* curves prepared by computing the values of *e* from points on the curves taken at different times. Portions of *e-p* curves prepared from points on parallel dial reading-time curves will be parallel no matter what time is used for computing Δ*e* between load increments. As shown in Figure 9.11b, the Δ*e* values between load increments below and above the preconsolidation pressure are different when computed for different times, thus causing a slightly different shape of the junction of the recompression and the virgin e-p curves prepared for different times after the beginning of consolidation. Curves prepared for different times will yield slightly different preconsolidation pressures when determined as described later. Such curve determined preconsolidation pressures are only approximate, but are useful for determining whether the soil is normally consolidated or overconsolidated. If found to be normally consolidated, the preconsolidation pressure should be taken as the known overburden pressure.

Data for plotting the e-p curve for a soil sample can be obtained by following the procedure given below.

a. Determine the initial void ratio of the specimen from the volume of the specimen at the beginning of the test (volume of confining ring), the dry weight of the specimen (obtained at end of test after specimen is oven dried), and the specific gravity of the solids.

b. After the test is set up with porous stones and loading head in contact with the soil specimen, set and record the dial reading before any load is applied so that the initial dial reading corresponds to the initial void ratio.

c. Flood the specimen and allow volume to reach equilibrium, unless data is desired for the natural water content.

d. Apply loads in increments allowing complete consolidation between increments and prepare a dial reading-time curve for each increment.

e. From the dial reading-time curves determine the dial reading corresponding to some elapsed time beyond 100 per cent consolidation; i.e., on the straight secondary portion of the $e - \log t$ curves.

f. Determine the void ratio after complete consolidation produced by the applied load by subtracting the change in void ratio from the initial void ratio. The change in void ratio is equal to the change in dial reading divided by the equivalent height of solids in the specimen.

g. After the void ratio for each of the applied loads has been determined, plot these values to form the void ratio-pressure curve.

When the e-p curve is plotted to arithmetic scales, the slope of the curve decreases with increasing pressure and decreasing void ratio as shown in Figure 9.11c. If pressure is applied up to some value p_2 and released, a certain amount of rebound or swelling will occur following a relationship somewhat as shown in Figure 9.11c. The amount of rebound depends upon the type of soil. If the specimen is again loaded in increments, the relationship between e and p for this recompression does not follow quite the same pattern as for the first compression.   Loading the soil beyond the preconsolidation pressure p_2 produces a curve which is parallel to the extension of the first compression curve but slightly below the extension.   This shift in position of the curve indicates that the recompression has produced an additional adjustment of the grains to produce a small amount of additional inelastic deformation.

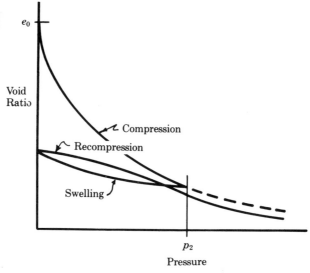

Figure 9.11c
Typical Void Ratio-Pressure Curve

Terzaghi discovered that there are certain advantages in plotting the pressure to a logarithmic scale in the e-p relationship.   When so plotted, that portion of the $e-\log p$ curve which represents recompression is curved and that portion which represents consolidation for the first time under the indicated pressures is a straight line.   This straight portion of the $e-\log p$ curve is called the virgin curve.   For soils that have been consolidated previously under a maximum pressure p_p (preconsolidation pressure), the curved recompression branch of the curve blends smoothly into the straight virgin branch.   Although the intersection of these two branches cannot be clearly distinguished, Terzaghi reasoned that the two branches must intersect at an abscissa corresponding to the preconsolidation pressure p_p.   By studying the $e-\log p$ curves of clays for which the stress history was known with reasonable assurance from geological and other evidence, Terzaghi found that the preconsolidation pressure can be determined

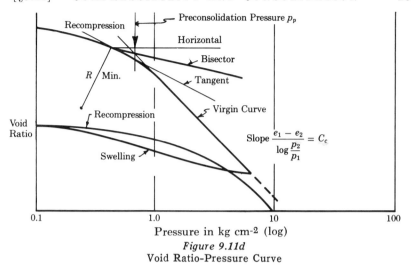

Figure 9.11d
Void Ratio-Pressure Curve

approximately from the $e - \log p$ curve as shown in Figure 9.11d by means of the following procedure:

a. Locate by eye the point on the recompression curve at which the curvature is greatest (radius of curvature least).
b. From point of greatest curvature draw a horizontal line.
c. Draw a tangent to the curve at the point of greatest curvature.
d. Bisect the angle between the horizontal and the tangent.
e. Extend the straight virgin curve back to an intersection with the bisector of the angle between the horizontal and the tangent.
f. The intersection of the virgin curve and the bisector locates the approximate preconsolidation pressure p_p.

The slope of the virgin portion of the $e - \log p$ curve is important in settlement analysis. This slope is designated C_c, the compression index, and for convenience is used as a positive number. By definition of slope

$$C_c = \frac{e_1 - e_2}{\log p_2 - \log p_1}$$
$$= \frac{e_1 - e_2}{\log \frac{p_2}{p_1}}. \qquad \text{(Eq. 9.11a)}$$

C_c is equal to the change in void ratio subtended by the curve over one cycle of the logarithmic scale.

The equation of the virgin $e - \log p$ curve is

$$e_1 - e_2 = C_c \log \frac{p_2}{p_1} \qquad \text{or} \qquad e = e_1 - C_c \log \frac{p}{p_1}. \qquad \text{(Eq. 9.11b)}$$

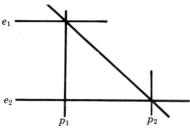

Figure 9.11e
Slope of *e*-Log *p* Curve

If p is smaller than p_1, the fraction can be reversed and the sign changed to plus. Moreover, if e_1 is taken as the ordinate of that point on the virgin curve (or its extension) corresponding to a pressure of unity, the equation becomes

$$e = e_1 - C_c \log p. \qquad \text{(Eq. 9.11c)}$$

Although the swelling curve on the $e-\log p$ plot is usually not a straight line, it is sometimes desirable to know its general slope, which is designated C_s and called the swelling index.

The discussion of the determination of volume change from $e-\log p$ curves which follows in Section 9.12 assumes that the virgin portion of the curve as shown in Figure 9.11d after recompression is parallel to an extension of the virgin curve before release of load and recompression. Theoretically, a heavy enough load could be applied to the soil to force all the water and gas from the voids and produce a void ratio approaching zero. Practically, the heaviest load that could be applied without producing disintegration of the soil particles would produce a void ratio somewhat above zero. No matter how many increments were applied, released and additional increments applied in approaching this maximum load, the void ratio produced by this maximum load would be approximately the same minimum value as that produced without release of incremental loads and recompression. Therefore, the virgin curve representing loading without release of increments and reloading and the virgin curve after release of an increment and reloading cannot be quite parallel but must intersect at the point of minimum void ratio. The point of intersection of the virgin curves is not necessarily exactly the same for any number of unloadings and recompressions.

For ordinary loads encountered in practice the error in assuming that the slope of the virgin curve before and after recompression is the same is probably insignificant in comparison with the error in assuming that the slope determined by a consolidation test on a small sample represents the slope for the aggregate soil deposit. Consolidation tests can be made on only a small amount of the soil under a loaded area. The compressibility of most of the soil must be inferred from index properties and a knowledge of the geologic history.

In the foregoing discussion, the virgin portion of the $e-\log p$ curve has been represented as a straight line. When recently deposited normally consolidated clays and clays compacted on the dry side of optimum possessing a flocculent structure and saturated are compressed, the volume change is caused by the deformation of particles, reduction of thickness of double layer water, and rearrangement of grains, all taking place simultaneously as the pressure is transferred to the soil grains.   If the contributions to volume change of the three factors occur in the proper proportion, the ratio of the log of the pressure to the void ratio is a constant and the virgin $e-\log p$ curve is a

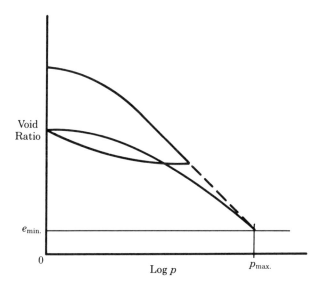

Figure 9.11f
e-Log p Curve Extended to e_{min}

straight line.   For most normally consolidated naturally deposited clays and clays compacted on the dry side of optimum and saturated the virgin $e-\log p$ curve is a straight line within the limitations of the accuracy of measurements.   If conditions exist which disturb this proportionate simultaneous contributory relationship, the virgin $e-\log p$ curve is not straight.

a. Honeycomb Structure

For some soils having a honeycomb structure, such as loess and some silts, the $e-\log p$ virgin curve is concave upward and right as shown for the loess from Grant County, Kansas.   The arched structure of these loose, honeycomb soils resists small loads with relatively little deformation.   Under a certain critical load the arches are broken down to form a more or less single grained structure with an accompanying large decrease in volume.   As more and more arches collapse and a single grained structure is produced, the resistance to deformation increases.   While the arches are being broken down the $e - \log p$

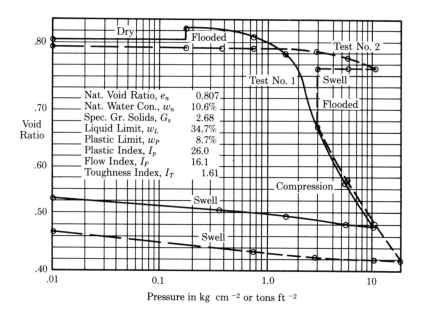

Figure 9.11g
Void Ratio-Pressure Curves for Grant Co., Kansas Loess

curve is very steep.   Under heavier loads on the single grained structure, the slope becomes less steep.

The loess for which the $e - \log p$ curves are shown in Figure 9.11g is a loose wind blown deposit of silt with a small amount of montmorillonite and calcium carbonate.   Some loess deposits contain little or no calcium carbonate.   In the dry state, the silt grains are cemented by the calcium carbonate and montmorillonite and are able to resist heavy loads without collapse of the soil structure and with little accompanying deformation.   When wet, the calcium carbonate and montmorillonite lose their cementing properties and the montmorillonite acts as a lubricant to allow collapse of the soil structure under light loads. Because the deformation is predominantly due to rearrangement of grains, there is little rebound after release of load.

The sample of this loess was taken from a test pit in a relatively dry condition. Consolidation tests were performed on two as nearly as possible identical specimens of the sample. The first test was loaded to about the overburden pressure and flooded after which increments of load were added at 48 hour intervals. In the second test, the dry specimen was loaded to 11.9 tons ft^{-2}, unloaded to 2.98 tons ft^{-2} and flooded. After flooding at 2.98 tons ft^{-2}, loads were added in increments up to 23.85 tons ft^{-2}, after which the load was released in increments. Complete swelling was allowed after each flooding and removal of load increment and complete consolidation was allowed between load increments.

These tests indicate that the dry loess will support fairly heavy loads with only small settlements and that saturation may produce sudden large settlements. This loess apparently begins to break down in the saturated state at about 1 to 1.5 tons ft^{-2}.

b. Sensitive Clays

For sensitive clays the virgin $e-\log p$ curves are also concave upward and right. When clays are subjected to pressure for a very long time after primary consolidation, the clay particles are joined at points of contact by highly viscous water. This bonding by the highly viscous water produces a high resistance to relative movement of the grains (high shear strength). When this strong bond is broken by remolding, less viscous water is drawn into the spaces between the grains and the shear strength is reduced. The ratio of the shear strengths in the natural undisturbed state and in the remolded state at the same water content is known as the sensitivity of the clay S_t.

As in the case of honeycomb structure the resistance to deformation of sensitive clays is quite high up to a certain critical shear stress which causes remolding of the soil along failure planes. Once the original strong bond is broken, the clay grains are easily displaced to form a denser structure. For pressures a little greater than the preconsolidation pressure, the $e-\log p$ curve is steep. But as the density of the soil is increased under heavier pressure, the resistance to deformation increases and the $e-\log p$ curve becomes less steep.

An example of the characteristic shape of the $e-\log p$ curve for a highly sensitive clay is shown in Figure 9.11i. The clay used for this illustration is from the St. Lawrence River Valley in the Province of Quebec taken from 30 ft below the surface.   It has a high void ratio and a water content about equal to its liquid limit.   In its natural undisturbed state the clay is quite stiff as shown in Figure 9.11h by the stress strain curve determined from an unconfined compression test.

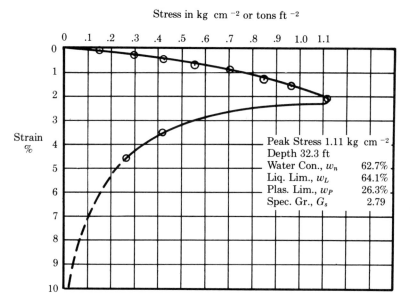

Stress Strain Curve for Unconfined Compression Test

Figure 9.11h
Stress Strain Curve for Unconfined Compression Test
St. Lawrence River Clay

The structure breaks down rapidly at a shear stress of $\dfrac{1.1}{2} = 0.55$ tons ft^{-2}.   In the remolded state, the clay has the same consistency as the liquid limit.   All clays at the liquid limit possess a shear strength of approximately 27 g cm^{-2} or 0.027 tons ft^{-2}.   These values indicate for this clay a sensitivity of about $\dfrac{0.55}{.027} = 20$.

Pertinent data for this clay are given on the diagrams.

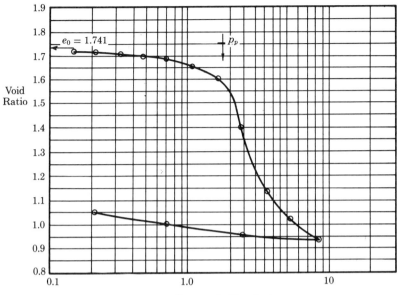

Figure 9.11i
Void Ratio-Pressure Curve for Highly Sensitive Clay
St. Lawrence River Clay

c. Overconsolidated Clay

An overconsolidated clay is one that has been consolidated under pressures greater than its present overburden. The heavy previous pressure has produced deformation by displacing the soil grains, forcing out double layer water from between the grains, and bending of elastic soil particles, probably in such proportion as to produce a linear $e - \log p$ relationship. When the earlier preconsolidation pressure was released, the elastic rebound was caused primarily by thickening of the double layer water and elastic rebound of soil particles. If the preconsolidation pressure was very great, the rearrangement of grains will have produced a more nearly parallel arrangement of grains (dispersed structure), in which case rebound after release of pressure will be accompanied by very little elastic rebound of soil particles and little rearrangement of grains toward regaining a flocculent structure.

Light loads applied to these highly overconsolidated clays force out double layer water with little movement of grains. Heavier loads produce some additional rearrangement of grains, thus causing the unit deformation produced by the heavier pressures to be greater than that produced by light pressures. This change in the factors contributing to deformation under light and heavy pressures causes the $e-\log p$ curve for heavily overconsolidated clays to be curved slightly convex upward and right.

The Pauls Valley, Oklahoma clay illustrated in the example of Sec. 9.20L was deposited by the Washita River and has been subjected to many cycles of drying and wetting. During dry periods, desiccation has consolidated the clay under capillary pressures of about 4.5 tons ft^{-2}. Probably this clay has not been subjected to greater pressure than that caused by desiccation. This conclusion is based upon knowledge of the geologic history of the clay and the straight $e-\log p$ curve beyond the preconsolidation pressure. Almost all the original undisturbed deposits of these desiccated Permian clays extending from southern Kansas to northern Texas indicate a preconsolidation pressure of 3.5 to 4.5 tons ft^{-2}, and an $e-\log p$ curve slightly convex to pressures of over 30 tons ft^{-2}. The geologic history of the clays indicates previous overburden pressures of 20 to 30 tons ft^{-2}.

The clays mentioned above were deposited during the Permian Period in an inland sea. The climate was semiarid. Torrential rains occurring between droughts produced torrents flowing down arroyos and gulleys carrying sand, silt, and clay with considerable amounts of limonite and hematite into the sea and producing heterogeneous deposits of clays, silty clays, sandy clays, and sands, all stained yellow and red by the iron oxides. The sands were cemented into yellow and red sandstones of various hardnesses by the iron oxides. The sea dried up three times toward the end of the Permian Period leaving thick deposits of gypsum and salt above the clays and sandstones. Deposits made during the Mesozoic Era and later ages covered the Permian deposits to a depth of 400 to 600 ft. An uplift to the east (probably the Ozark uplift) tilted the earlier deposits downward to the west. The younger Rocky Mountains cut through the earlier deposits without affecting the slope of the deposits under consideration. Erosion has since removed the 400 to 600 ft of overburden so that the Permian deposits are exposed in central Oklahoma. In addition to preconsolidation under the 20 to 30 tons ft^{-2}, the surface clays have been subjected to thousands of cycles of saturation and drying. This desiccation has apparently produced preconsolidation pressures of 3.5 to 4.5 tons ft^{-2}. A typical $e-\log p$ curve for these desiccated overconsolidated clays is shown in Figure 9.11j.

Clays that have been consolidated under pressures considerably below the maximum pressure applied in a consolidation test but greater than their overburden pressures exhibit $e - \log p$ curves similar in shape to those for normally consolidated clays. These clays, however, have been through an extra cycle of loading and unloading which

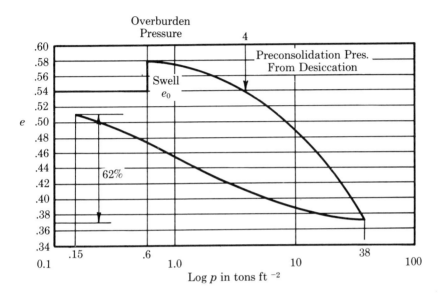

Figure 9.11j
Void Ratio-Pressure Curve
Permian Red Clay

causes the virgin portion to slope slightly less steeply than the curve for the clay if it had been normally consolidated.

d. Compacted Clays

The $e - \log p$ curves for compacted clays usually show a curved portion up to a hypothetical preconsolidation pressure produced by the compactive effort. The shape of the curve beyond this hypothetical pressure depends to some extent upon the manner in which the compacting was done and the water content of the soil as compacted. Compaction produced by dynamic action of the falling hammer tends to produce a dispersed (parallel) structure in clays when compacted wet of optimum and a flocculent (random) structure when compacted dry of optimum. Compaction by static load tends to allow a flocculent structure to be maintained to a much greater extent than dynamic compaction.

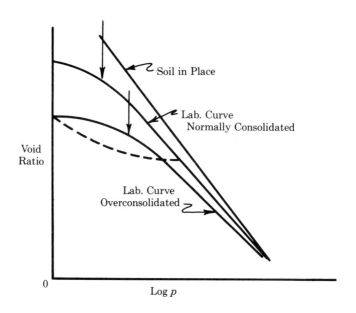

Figure 9.11k
Comparison e-Log p Curves for Soil in Place and Laboratory Curves
for Normally Consolidated and Overconsolidated Clay

The shape of the $e-\log p$ curves for those clays which retain a flocculent structure generally show a curved portion for light pressures and a straight portion for heavier pressures, similar to the curves for normally consolidated clays. For clays compacted wet of optimum the $e-\log p$ curves tend to be shaped somewhat like those for overconsolidated clays.

The compressibility of clays compacted wet of optimum is greater under light pressures than that of clays compacted to the same void ratio on the dry side of optimum. Under heavy pressures, clays compacted dry of optimum are more compressible than those compacted to the same void ratio wet of optimum.

e. Effect of Disturbance

Disturbances of a soil sample amount in effect to a partial remolding of the soil. If the soil in its undisturbed state has a low void ratio

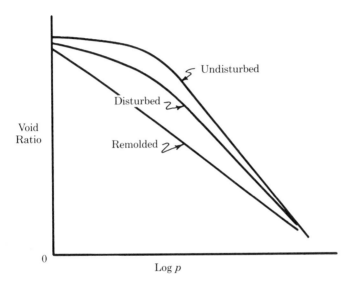

Figure 9.11l
Comparison e-Log p Curves for Saturated Clay in Undisturbed, Disturbed, and Remolded States

(dense) and is not completely saturated, disturbance probably increases the void ratio of that portion of the soil that is disturbed, making it more compressible than the original undisturbed soil. The result of this disturbance on the $e - \log p$ curve is to cause it to be shifted to the left and become steeper under light pressures than the curve for the completely undisturbed soil. As the disturbed soil is compacted by the applied pressure, the $e - \log p$ curves for the disturbed and the undisturbed soils approach a common void ratio and pressure.

If the clay is completely saturated when disturbed, the void ratio probably will be little affected by the disturbance. The void ratio of a completely saturated clay is little changed by remolding at the same water content. There exists the same amount of soil solids and water in either state. Disturbance of the soil breaks down the structure making the disturbed soil more compressible under light loads. Although the void ratios of the clay in the undisturbed and the disturbed states are approximately the same under zero pressure, the void ratio at the lowest pressure indicated on the log scale will probably be less than the void ratio of the undisturbed soil at that pressure because of the greater compressibility of the disturbed soil.

Figure 9.11l shows a comparison of the $e - \log p$ curves for a saturated clay in the undisturbed, disturbed, and completely remolded states.

9.12   COMPARISON OF LABORATORY e-p CURVE WITH e-p RELATIONSHIP FOR SOIL IN PLACE

a. Normally Consolidated Clay

A normally consolidated clay is a clay that has been consolidated under its present overburden but has never been subjected to a greater pressure.

An undisturbed soil sample no matter how carefully taken is never completely undisturbed. A sample of soil that has been consolidated under several feet of overburden when removed from the ground is relieved of its overburden pressure, which allows the soil to swell. A specimen of this soil in a consolidation test is subjected to recompression up to the overburden pressure. When a load is added to the soil in place without removing the overburden, the soil is subjected to the greater pressure for the first time without recompression.

The laboratory curve has the characteristic shape consisting of a curved recompression portion and a straight virgin portion. On the other hand, the relationship followed by the soil in place is that indicated by the virgin portion only, starting at the preconsolidation pressure. Because the laboratory specimen has been recompressed and the soil in place has not, the laboratory curve is slightly lower than the curve for the soil in place as shown in Figure 9.12a.

When the pressure on the soil in place is increased from p_0 to p_2, the void ratio is reduced from e_n to e_{2n} and the unit deformation of the soil is $\dfrac{e_n - e_{2n}}{1 + e_n}$ or $\dfrac{C_c}{1 + e_n} \log \dfrac{p_2}{p_0}$. The void change $\Delta e = e_n - e_{2n}$, which is a measure of the deformation of the soil in place, is the same as the void change $\Delta e = e_{1L} - e_{2L}$ indicated by the virgin portion of the curve extended over the range of pressure under consideration. (See Figure 9.12a). The deformation of the soil in place cannot be predicted from the change in void ratio from p_0 to p_2 taken directly from

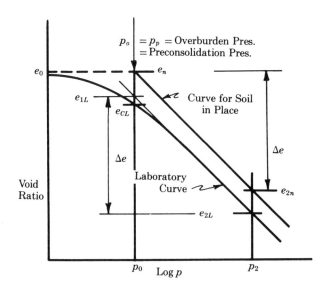

Figure 9.12a
Laboratory and Soil in Place *e-p* Curves for Normally Consolidated Clay

the laboratory curve. The error indicated by the difference between e_{1L} and e_{CL} in some cases might be appreciable.

Therefore, in estimating the settlements of a layer of normally consolidated clay, the void change should be computed from the equation of the virgin curve and not from void ratios taken directly from the laboratory curve; i.e., $\Delta H = H \dfrac{C_c}{1 + e_n} \log \dfrac{p_2}{p_0}$ and not $\Delta H = H \dfrac{e_{CL} - e_{2L}}{1 + e_{CL}}$. Since the two curves are nearly parallel, the slopes C_c are almost the same. For practical purposes because of the resistance to swelling provided by capillary forces when the soil sample is removed from the ground, the initial void ratio of the soil sample e_n and the void ratio of the soil in place consolidated under its present overburden e_0 may be assumed as equal. The error introduced by using $1 + e_0$ instead of $1 + e_{1L}$ in determining deformation is insignificant. Therefore, the laboratory determined e-p relationship may be used for making reliable estimates of settlements due to consolidation of normally consolidated clays.

b. Overconsolidated Clay

An overconsolidated clay is a clay that has at some past time been consolidated under a pressure greater than the present overburden pressure.

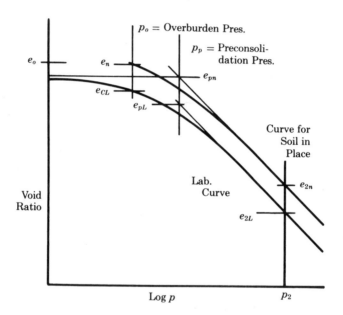

Figure 9.12b
Laboratory and Soil in Place e-p Curves for Overconsolidated Clay

In the laboratory, the specimen of overconsolidated clay is recompressed from zero pressure to the preconsolidation pressure. When a load is added to the load of the overburden on a soil in place, the soil is recompressed from the overburden pressure to the preconsolidation pressure. As shown in Figure 9.12b, the void change of the soil in place for overconsolidated clay is $e_n - e_{2n}$. This indicated void ratio change is nearly the same as read on the laboratory curve as $e_{CL} - e_{2L}$. Probably the displacement of the laboratory curve below the curve representing the e-p relationship for the soil in place is greater above the preconsolidation pressure p_p than below, but any error introduced by using the laboratory curve is negligible.

Therefore, computations for estimating the settlement of a layer of overconsolidated clay should be made from values of the void ratio taken from the curve rather than by computation from the equation of the virgin curve. Because the two curves are for practical purposes parallel, values of e taken from the laboratory curve can be used without appreciable error. Frequently, the settlement produced by recompression is so small compared to the total settlement that it can be neglected and settlement computed from the virgin curve from p_p to p_2. The unit deformation in this case is $\epsilon = \dfrac{e_{pL} - e_{2L}}{1 + e_{pL}} = \dfrac{C_c}{1 + e_{pL}} \log \dfrac{p_2}{p_p}$. The error in this procedure is illustrated by the distance $e_{CL} - e_{pL}$ or $e_n - e_{pn}$. For highly overconsolidated clays loaded to less or slightly greater than the preconsolidation pressure, most of the settlement will be caused by recompression, in which case the recompression portion of the settlement should not be neglected. In this case, however, the total settlement will be small.

9.13   TOTAL SETTLEMENT COMPUTATION

In this article settlement is assumed to be due to consolidation of the soil under static load. It does not include settlements caused by vibration of cohesionless soils.

Before a settlement analysis can be made the thickness of the soil layers must be determined by exploration. For accurate analysis, undisturbed samples of the different soil strata must be taken and a consolidation test made on a sample of each of the different soils. Usually, the soils underneath an area vary so much that it is impractical to sample and test all the variations. Practically, only a few consolidation tests are made. Estimates are made of the properties of the soils not tested directly for compressibility. Aids to the judgment in making these estimates of properties are natural void ratio, natural water content, liquid limit and plastic limit used in comparison with these properties of the tested samples. A fairly accurate estimate of C_c can be made from the natural void ratio from a rational development made by Nishida, $C_c = 0.54(e_n - 0.3)$. The natural water

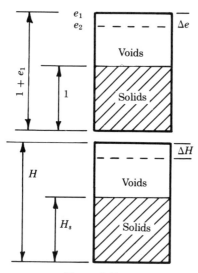

Figure 9.13a
Relation Between Change in Void Ratio and Settlement

content w_n for clays taken from below the water table (saturated clays) can also be used for estimating the slope of the virgin $e-\log p$ curve. Using 2.70 for G_s, the Nishida relationship becomes $C_c = 0.54(2.7w_n - 0.3)$. For normally consolidated clays, C_c can be estimated from the empirical relationship based upon the liquid limit given by Terzaghi and Peck as $C_c = 0.009(w_L - 10)$, in which w_L is per cent water content at the liquid limit.

For greater accuracy in making a settlement analysis, thick strata of soils are divided into thin layers. Present overburden pressures p_1 and the corresponding e_1 must be determined or estimated for each layer. The average pressure on each layer produced by the added load is determined and added to p_1 to find the value of p_2. With the information described, the total settlement of a layer of clay under an added load can be computed.

The unit deformation of the clay layer is

$$\epsilon = \frac{e_1 - e_2}{1 + e_1} = \frac{C_c}{1 + e_1} \log \frac{p_2}{p_1}.$$

The total settlement of a layer of clay H thick is

$$\Delta H = H \frac{C_c}{1 + e_1} \log \frac{p_2}{p_1}.$$

Example of Settlement Analysis

Assume a soil profile consisting of 2 strata of homogeneous clays above an impervious layer and with a drainage layer of incompressible sand separating the strata as shown in the tabulation of Figure 9.13b. Undisturbed samples were taken from the positions shown and consolidation tests made from which values of C_c were determined. The water table is 4 ft below the surface.

Assume that a very large area is to be loaded on the surface with a uniform load of 1500 lb per sq ft.

Physical Properties of Clay Layers
Normally Consolidated

<table>
<tr><td>Upper Layer</td><td>Lower Layer</td></tr>
<tr><td>One point on e-p curve</td><td>One point on e-p curve</td></tr>
<tr><td>$e_c = 1.02$ $p_c = 1.0$ kg cm^{-2}</td><td>$e_c = 0.73$ $p_c = 1.6$ kg cm^{-2}</td></tr>
<tr><td>$G_s = 2.74$ $C_c = 0.36$</td><td>$G_s = 2.70$ $C_c = 0.24$</td></tr>
</table>

Solution

The calculations are carried out on the tabulated form of Figure 9.13*b* in accordance with the following procedure:

a. Divide the upper and lower clay strata into convenient layers for the desired degree of accuracy. In the upper layer the water table is a convenient plane for division. The lower 13 ft stratum is arbitrarily divided into 6 ft and 7 ft layers.

b. Determine the average unit weight of the soil in each stratum. The unit weights are computed in Col. 5 of the tabulation.

c. Determine the overburden pressure p_1 at middepth of each layer. See Col. 6.

d. Compute the value of e_1 corresponding to the overburden pressure p_1, and record in Col. 8. These values of e_1 are computed from the relationship $e_1 = e_c + C_c \log \dfrac{p_1}{p_c}$. The value of e_1 for $p_1 = 232$ lb ft^{-2} at middepth of the top layer is $e_1 = 1.02 + 0.36 \log \frac{2000}{232} = 1.02 + 0.33 = 1.356$. For practical purposes, 1 kg cm^{-2} = 1 ton ft^{-2} = 2000 lb ft^{-2}. The void ratio at middepth of the bottom layer where $p_1 = 1853$ lb ft^{-2} is $e_1 = 0.73 + 0.24 \log \frac{3200}{1853} = 0.73 + 0.057 = 0.787$.

e. Divide C_c by $1 + e_1$ and record in Col. 10. For the first layer

$$\frac{C_c}{1 + e_1} = \frac{0.36}{2.356} = 0.153.$$

f. Multiply the value of $\dfrac{C_c}{1 + e_1}$ in Col. 10 by the thickness of the layer H and record in Col. 11. Because $\dfrac{C_c}{1 + e_1}$ is dimensionless, H can be in any units. ΔH will be in the same units.

g. Determine and record in Col. 12 the pressure produced at middepth of each layer by the added load. In this case, because the loaded area is large compared to the depth of the compressible layers, the increase in pressure Δp may be considered as constant for the full depth. Under a small loaded area the total added load is distributed to a larger area at depth which makes the unit pressure Δp decrease with depth. Computation for determining pressure at depth under different loading conditions is discussed in *Theoretical Soil Mechanics*, Terzaghi and elsewhere in the literature.

h. Add Δp to initial overburden pressure p_1 and record as p_2 in Col. 13.

i. Determine log p_2/p_1 and record in Col. 14. p_2 and p_1 can be in any convenient units provided they are both in the same units.

j. Multiply the values in Col. 14 by $H \dfrac{C_c}{1 + e_1}$ of Col. 11 to determine the contribution to settlement of the different layers. Record in Col. 15.

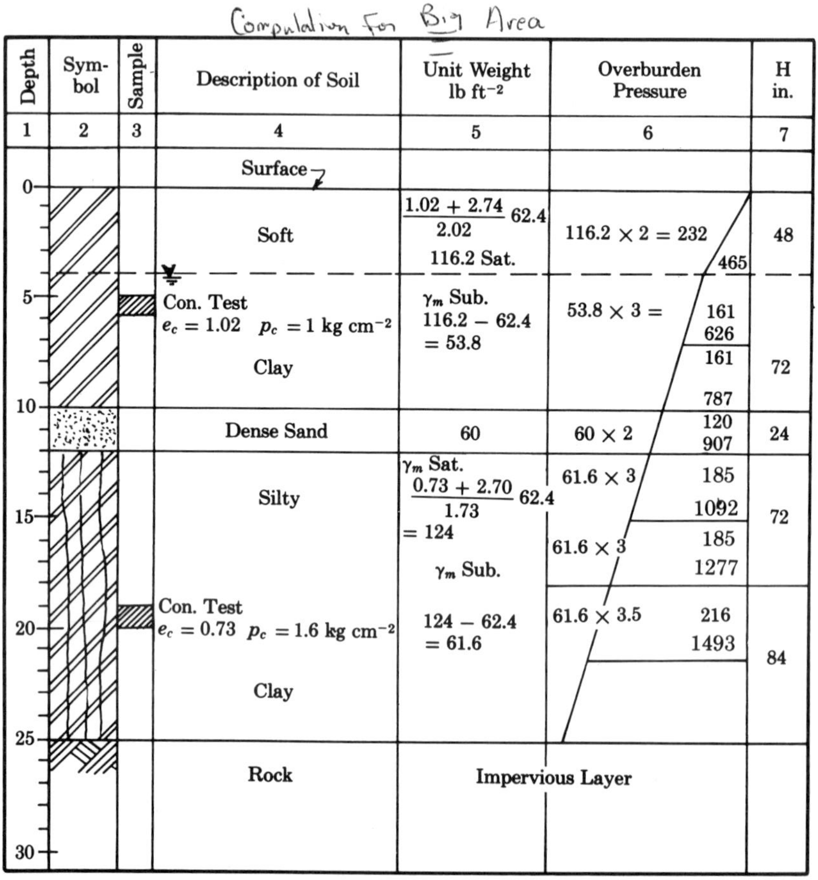

Computation For Big Area

Depth	Sym-bol	Sample	Description of Soil	Unit Weight lb ft^{-2}	Overburden Pressure	H in.
1	2	3	4	5	6	7
0—			Surface			
			Soft	$\dfrac{1.02 + 2.74}{2.02}$ 62.4 116.2 Sat.	$116.2 \times 2 = 232$ 465	48
5—		Con. Test $e_c = 1.02$ $p_c = 1$ kg cm^{-2} Clay	γ_m Sub. $116.2 - 62.4$ $= 53.8$	$53.8 \times 3 =$ 161 626 161	72
10—			Dense Sand	60	60×2 120 907	24
15—			Silty	γ_m Sat. $\dfrac{0.73 + 2.70}{1.73}$ 62.4 $= 124$ γ_m Sub.	61.6×3 185 1092 61.6×3 185 1277	72
20—		Con. Test $e_c = 0.73$ $p_c = 1.6$ kg cm^{-2} Clay	$124 - 62.4$ $= 61.6$	61.6×3.5 216 1493	84
25—			Rock	Impervious Layer		
30—						

Figure 9.13b
Sheet No. 1
Settlement Analysis

k. Add the contributions of the individual layers to obtain the total settlement produced by the added load.

This summation is the total estimated settlement after complete consolidation of all the layers. No consideration in this analysis has been given to the time required for the settlement to occur.

The preceding example of settlement analysis is for the simplest condition that could exist. It is presented here in order to illustrate the use of information obtained from a consolidation test and to illustrate how the computations for a settlement analysis can be organized. Often deposits of soft clays which will allow large settlements are fairly uniform over a large area. Usually, however, natural soil deposits are not as uniform as assumed in the example. A fill of uniform

e_1	C_c	$\dfrac{C_c}{1 + e_1}$	$H\dfrac{C_c}{1 + e_1}$	Δp	p_2	$\log \dfrac{p_2}{p_1}$	$H\dfrac{C_c}{1 + e_1}\log\dfrac{p_2}{p_1}$ $\dfrac{}{\Delta H}$ in.
8	9	10	11	12	13	14	15
Surface							
1.356		0.153	7.35	1500	1732	0.874	6.43
	0.36						
1.201		0.163	11.75	1500	2126	0.532	6.25
							(12.68)
0.826		0.131	9.43	1500	2594	0.374	3.52
	0.24						
0.787		0.134	11.3	1500	2993	0.302	3.40
							6.92
						Total Settlement	19.60

Figure 9.13b
Sheet No. 2
Settlement Analysis

thickness applied over a very large area could provide the approximate loading conditions assumed in the example.   The usual loading conditions consist of small areas of different sizes, shapes, and distribution supporting different total loads and applied at different depths beneath the surface.   The determination of unit pressures at different points in the soil caused by loads on these small areas is a highly complicated procedure and beyond the scope of this text.

Estimates of unit pressures caused by loads applied to the soil are based upon a solution developed by Boussinesq for a point load applied to the surface of an isotropic, elastic body semi-infinite in extent. Westergaard also developed the relationship between applied load and unit pressure at any point for a point load applied at the surface of a semi-infinite layer of soil which is restrained against lateral deformation.   The Westergaard solution was developed for application to varved clays.

Influence tables for the Boussinesq relationship were computed by Glennon Gilboy and published in a paper in *Proceedings of Am. Soc. C. E.*   Application of the Boussinesq and Westergaard solutions were developed for loaded areas and influence values computed by N. M. Newmark.   R. E. Fadum computed influence tables and curves for the application of Boussinesq and Westergaard solutions to loads applied at a point, uniformly along a line of finite length, and uniformly distributed over rectangular and circular areas.   Influence tables for the Boussinesq solution for point load, uniformly distributed load on a rectangular area, and uniformly distributed load on a circular area are included in the appendix of *Theoretical Soil Mechanics,* Terzaghi.

9.14   TIME-SETTLEMENT RELATIONSHIP

The relationship between per cent consolidation U per cent and time factor T is given in Table 9.08a.   If the time factor for a given layer of soil at a given time is known, the value of U per cent is known and the settlement at time t is equal to U times the total settlement ΔH. The problem in time settlement analysis is to determine the value of T for a given time t or to determine the time for a given value of T.

Since $T = c_v \dfrac{t}{H^2}$, if c_v for the soil is known, then for a layer of clay H thick the value of c_v/H^2 is approximately constant and the relationship between T and t is known.   In this case, H is the drainage thickness, full thickness of a layer drained one side only, and one-half the thickness of a layer drained both sides.

The coefficient of consolidation c_v can be determined from one of the time-dial reading curves obtained from a consolidation test. The time-dial reading curve should be converted to a time-consolidation curve as shown in Figure 9.10b. A curve for one of the larger load increments and a fairly large deformation should be chosen for this purpose. From this consolidation-time curve, the time for some per cent consolidation should be determined. The time for 50 per cent consolidation is commonly used because it occurs in a steeper part of the curve and can be more accurately determined than can much smaller or much larger values of U. 50 per cent consolidation occurs early enough that the secondary consolidation has not appreciably distorted the curve and late enough that the percentage of error in reading the time and dial readings after application of the increment of load is much smaller than for the earlier readings.

The value of H in the consolidation test is one-half the thickness of the specimen.

$$c_v = \frac{H^2 T}{t}. \quad \text{For } U\% = 50, \ T = 0.2 \text{ (approx.)}$$

Then

$$c_v = \frac{0.2 H^2}{t_{50}}.$$

For example, if the thickness of the specimen is 1 in. or 2.54 cm and if t_{50} is 1.2 min or 72 sec,

$$c_v = \frac{(0.2)(0.5^2)}{1.2} = 0.0416 \text{ in.}^2 \text{ min}^{-1} \text{ or } \frac{(0.2)(0.5^2)}{72} = 0.000695 \text{ in.}^2 \text{ sec}^{-1}.$$

$$\text{or} = \frac{(0.2)(1.27^2)}{1.2} = 0.269 \text{ cm}^2 \text{ min}^{-1} \text{ or } \frac{(0.2)(1.27^2)}{72} = 0.0448 \text{ cm}^2 \text{ sec}^{-1}.$$

Thus, the value of c_v can be obtained in any convenient units, but in its use these units must be preserved consistently.

Once c_v is known for the soil, a time-settlement curve can be prepared for a single layer of that soil during the consolidation period from the relationship $t = \frac{H^2}{c_v} T$. When two or more strata of the same or different soils contribute to the settlement, time-settlement curves must be prepared separately for each of the strata and the total settlement at a given time found by adding the settlements of the contributing layers at that time. In this sense, a stratum or layer of soil is a layer bounded by drainage layers on both sides or by a drainage layer on one side and an impervious layer on the other.

Example of Time-Settlement Analysis

Assume the same soil and loading conditions as used for the example of total settlement in Article 9.13 and the following additional data.

Upper Layer	Lower Layer
Hgt. of Spec. = 2.54 cm.	Hgt. of Spec. = 1.62 cm.
t_{50} = 4.6 min.	t_{50} = 3.1 min.
$c_v = \dfrac{(0.2)(1.27^2)}{4.6} = 0.07$ cm^2 min^{-1}.	$c_v = \dfrac{(0.2)(0.81^2)}{3.1} = 0.0423$ cm^2 min^{-1}.
$c_v = \dfrac{0.07}{2.54^2} = 0.0108$ in.^{2} min^{-1}.	$c_v = \dfrac{0.0423}{2.54^2} = 0.00655$ in.^{2} min^{-1}.

Computation for a solution can be tabulated as follows:

Table 9.14a

		Upper Layer $H = 60''$ drained 2 sides $\Delta H = 12.68''$, $\dfrac{H^2}{c_v} = 333,000$			Lower Layer $H = 156''$ drained 1 side $\Delta H = 6.92''$, $\dfrac{H^2}{c_v} = 3,720,000$		
$U\%$	T	Settlement $U \times 12.68$	t min.	t months	Settlement $U \times 6.92$	t min.	t months
10	0.008	1.27	2,670	0.062	0.69	29,750	0.69
20	0.031	2.54	10,300	0.238	1.38	115,200	2.66
30	0.071	3.80	23,700	0.549	2.08	264,000	6.10
40	0.126	5.07	42,000	0.972	2.77	468,000	10.85
50	0.197	6.34	65,600	1.520	3.46	734,000	16.98
60	0.287	7.60	95,600	2.22	4.15	1,068,000	24.65
70	0.405	8.86	135,200	3.13	4.84	1,508,000	34.82
80	0.565	10.15	188,200	4.36	5.54	2,100,000	48.60
85	0.684	10.80	228,000	5.28	5.86	2,540,000	58.70
90	0.848	11.40	282,100	6.54	6.23	3,150,000	72.80
95	1.127	12.05	375,000	8.69	6.57	4,190,000	96.70

The settlement of a single stratum of clay at a given time or the time required for a certain amount of settlement to occur can be estimated from the parabolic relationship between U and T, provided U is less than 60 per cent.

$$\frac{U_1^2}{U_2^2} = \frac{T_1}{T_2} = \frac{c_v \dfrac{t_1}{H_1^2}}{c_v \dfrac{t_2}{H_2^2}}$$

For the same soil c_v is constant, so the relationship may be written $\dfrac{U_1^2}{U_2^2} = \dfrac{t_1 H_2^2}{t_2 H_1^2}$, or stated in words, the square of the degree of consolidation is directly proportional to the time and inversely proportional to the square of the thickness if drained 1 side or the square of one-half the thickness if drained 2 sides.

For example, the time required for 40 per cent consolidation or 5.07 in. settlement of the upper stratum of clay of the preceding

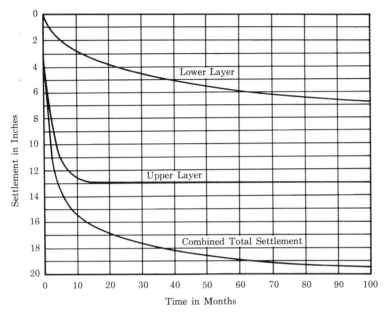

Figure 9.14a
Time-Settlement Curve

example can be estimated by comparing the consolidation speci-men with the stratum of clay from the parabolic relationship as,

$\dfrac{40^2}{50^2} = \dfrac{(t)(0.5)^2}{(4.6)(60)^2}$ from which $t = \dfrac{1600}{2500}\dfrac{(4.6)(60)^2}{(0.5)^2} = 42{,}400$ min. If

the time required for 4 in. settlement is desired, $U = \dfrac{4}{12.68} = 31.6\%$

and $t = \dfrac{(31.6)^2}{2500}\dfrac{(4.6)(60)^2}{(0.5)^2} = 26{,}500$ min. If the settlement at the end

of 6 months is desired, $t = 259{,}200$ min. and

$$U = \sqrt{\dfrac{(2500)(0.5)^2(259{,}200)}{(4.6)(60)^2}} = 99\%$$

or 12.5 in. settlement.

By comparison with values in Table 9.14a, it can be seen that this parabolic relationship gives quite accurate results for low degrees of consolidation, but for the 6 months calculation of settlement there is considerable error because at 6 months the degree of consolidation is considerably greater than 60 per cent. For degrees of consolidation greater than about 60 per cent, the relationship between U and T as given in Table 9.08a must be used or, as a substitute, the approximate relationship of equation 9.08b, $T = 1.781 - 0.933 \log (100 - U)$. For the 6 months settlement of the above illustration, the approximate relationship for T and degree of consolidation above 60 per cent may be used as follows:

$$T = T_2 \frac{tH_2{}^2}{t_2H^2} = \frac{(0.2)(259,200)(0.5^2)}{(4.6)(60^2)} = 0.784$$

and $\quad \log (100 - U) = \dfrac{(1.781 - 0.784)}{0.933} = 1.069$

$$100 - U = 11.7 \text{ and } U = 88.3\%$$
$$\Delta H = 0.883 \times 12.68 = 11.2 \text{ in. settlement.}$$

This value of the settlement of the upper layer at 6 months agrees fairly well with the settlement as computed by using the relationship between U and T of Table 9.08a listed in Table 9.14a or with the settlement curve of Figure 9.14a.

9.15 LABORATORY TESTS FOR COMPRESSIBILITY

Three types of tests may be made in the laboratory for determining the compressibility of soils.   These tests are classified according to the amount of lateral restraint applied to the test specimen as vertical load is applied.

The unconfined compression test is made by loading a small column cut from the soil with no lateral restraint applied during testing. Vertical deformation is measured with a dial indicator.

The triaxial compression test is made on a column of the soil subjected during test to an all around confining pressure.   In this test, any confining pressure desired can be applied to the specimen.   The specimen is free, except for some lateral restraint at the top and bottom surfaces of contact with the loading device, to be deformed both laterally and vertically by the applied stresses.

In the consolidation test, as already described, lateral deformation is prevented by confining the test specimen in a relatively unyielding ring.

9.16   UNCONFINED COMPRESSION TEST

The unconfined compression test is primarily a strength test and is described in detail in the chapter on shear strength.   It is the simplest of the tests for compressibility.   Considerable information concerning the deformation characteristics of cohesive soils can be obtained from an unconfined compression test.   The shape of the stress-strain curve for an unconfined compression test tells a great deal about the history of clays and how they may be expected to deform under load.

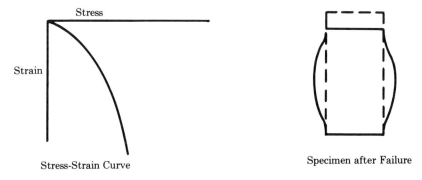

Stress-Strain Curve   Specimen after Failure

Figure 9.16a
Recently Worked Clay in Unconfined Compression

The following discussion applies to wet clays taken from below the water table or in a capillary zone.

A freshly deposited clay that has been consolidated under its present overburden or a remolded clay has a stress-strain relationship which builds up gradually to a maximum stress as shown in Figure 9.16a. Deformation continues at this maximum stress.   Creep occurs in such a clay at smaller stresses than the maximum.   The shape of the stress-strain curve and the maximum stress are dependent to some extent upon the rate of load application.   At failure the specimen bulges without a distinct failure surface.

A clay that has been subjected to pressure for a very long time (hundreds of years) is bonded at points of contact of the clay particles with highly viscous asphalt-like water.   Such a clay has built into it a high initial strength which is rapidly reduced at deformations beyond a critical value to approximately the maximum strength of the remolded specimen at the same water content.   Up to this initial strength the ratio of stress to strain is much higher than for the freshly deposited or reworked clay, and the creep is relatively low. This stress-strain characteristic is illustrated in Figure 9.16b. Failure of such a clay usually occurs along a distinct surface.

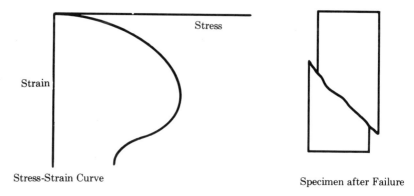

Stress-Strain Curve

Specimen after Failure

Figure 9.16b
Sensitive Clay in Unconfined Compression

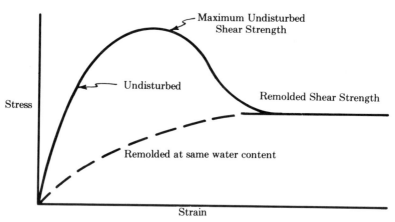

Figure 9.16c
Stress-Strain Relationship for Sensitive Clay

A comparison of the stress-strain relationship for an old normally consolidated clay in the undisturbed state and in the remolded state at the same water content is shown in Figure 9.16c. The ratio of the high initial shear strength to the shear strength of the remolded clay at the same water content is referred to as the sensitivity of the clay. Unconfined compression tests made on undisturbed and remolded specimens of a sample of clay give information concerning the sensitivity of the clay and relationship between shearing stress and deformation. The shape of the stress-strain curve from an unconfined compression test can give information concerning the amount of disturbance the clay has been subjected to during sampling or preparing the test specimen.

Skempton and Northey classify the sensitivity of clays as follows:

Insensitive clays...............	1
Low-sensitive clays..............	1–2
Medium-sensitive clays..........	2–4
Sensitive clays..................	4–8
Extra-sensitive clays............	8
Quickclay.....................	16

9.17   TRIAXIAL COMPRESSION TEST

The triaxial test is almost exclusively a shear strength test. It is sometimes called a triaxial shear test. The triaxial compression test is discussed in detail in the chapter on shear strength.

9.18   CONSOLIDATION TEST

The consolidation test is devised to simulate as nearly as possible the soil in place when subjected to an increased pressure from added load. Each piece of soil in its natural state in the ground is prevented from deforming laterally, when subjected to vertical stress and deformation, by the adjacent surrounding soil which is subjected to the same stress conditions. In the consolidation test, this assumed idealized condition is simulated by confining the test specimen in a ring which prevents lateral deformation. Porous stones placed above and below the specimen allow drainage. In most cases for loading conditions less than those that cause failure, the assumption that lateral deformation is prevented in the soil in place appears to be fairly accurate. There may be conditions of unsymmetrical lateral dimensions, loading conditions, or involving changes in type of soil that might cause this assumption to be in error to some extent.

Detailed procedure for making the consolidation test follows. Tests should be made by or under the direct supervision of an engineer who knows the conditions under which he wishes the soil to be tested. He should not feel obligated to follow A.S.T.M. or any other standard testing procedure if some other procedure will give him better information for his particular purpose.

a. Soil Samples

If the object of the test is to determine the properties of the soil in place, soil samples should be as little disturbed as possible. Any sample removed from its natural position is disturbed to some extent, at least to the extent of a change in the state of stress. The greatest disturbance in the sample occurs, generally, at and near the surface where the soil structure has been broken down by the cutting implement and shear distortion caused by friction along the sides of the sampler. Samples for testing should be large enough that the test specimen can be taken from the interior of the sample where the disturbance is least. All samples should be protected to preserve their natural water content.

Under suitable conditions samples taken carefully by hand from a test pit suffer least disturbance. This method of taking samples is expensive and feasible only for soils near the surface above the water table. One of the simplest and best methods of taking a sample from a test pit is to expose a block of soil of the desired size and shape, or one that will stand without damage, leaving the block to stand on its bottom as shown in Figure 9.18a. The block should be trimmed to

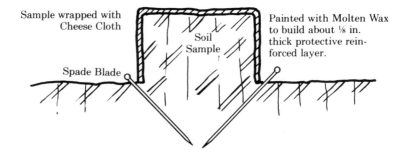

Sample wrapped with Cheese Cloth

Soil Sample

Painted with Molten Wax to build about ⅛ in. thick protective reinforced layer.

Spade Blade

Figure 9.18a
Hand Sample in Test Pit

the desired size and shape and wrapped with one or two layers of cheese cloth. Molten wax should be applied to the cheese cloth covered surface with a paint brush to a thickness of about $\frac{1}{8}$ in. If greater reinforcement is needed, the waxed covering can be wrapped with another layer of cheese cloth and wax. The sample can be loosened from its bed by inserting the blade of a spade below the block and prying it out. The sample can then be turned over, the excess soil cut off, and the bottom reinforced and coated with molten wax to form a completely protected sample.

Relatively undisturbed samples of soils that will not stand alone can often be taken from a test pit in a sharpened cylinder. A column is cut as high as will stand without damage and slightly larger than the tube. The sharpened tube is then pushed over the trimmed portion of the sample cutting off the excess soil as the cylinder is pushed down. Then more of the sample is exposed and trimmed below the cylinder and the cylinder pushed down again. This procedure is repeated until a sample of the desired length is obtained. Great

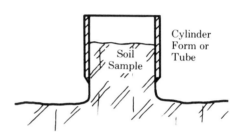

Figure 9.18b
Soil Sample in Cylinder

care must be exercised to see that the sample is not damaged as the cylinder is pushed over it. The ends of the sample should be recessed about $\frac{1}{8}$ in. and covered with a layer of molten wax.

When a sample is to be taken of fissured clay from below the water table in a test pit in which the water has been lowered by pumping, it may be necessary to drive or push a short 6 in. diameter sharpened cylinder the full depth of the sample into the soil. The water in the fissures of the clay coming to the surface at the sides of an excavated block causes the clay to crumble. This may occur even in hard clay having a high shear strength in its natural position.

The recovery of undisturbed soil samples from bore holes is made in some sort of sampling spoon which is forced into the soil at the bottom of the hole. The sample is usually preserved in the tube which receives and retains it.

The simplest of sampling spoons consists of a seamless thin walled steel tube secured to a head as shown in Figure 9.18c. These tubes vary in size from 2 in. to 4 in. O.D. and are often called Shelby tubes. The tube is fastened to the head with set screws arranged so that the tube and sample can be removed from the head. Holes through the head allow water to drain from the drive pipe to prevent a column of water from forcing the sample from the tube as it is withdrawn from the hole. In some cases a ball check in the head helps to create a partial vacuum at the top of the sample to compensate for the vacuum formed at the bottom as the sample is withdrawn from the soil. Samples are preserved in the tubes by covering the ends with molten wax. In order to keep disturbance to a minimum, the sharpened

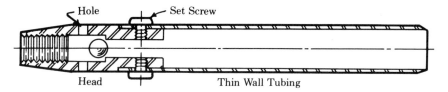

Figure 9.18c
Shelby Tube Sampler

edges of the tubes should be rolled inward slightly to reduce friction along the sides of the tubes. Friction along the sides of the tubes having the same diameter as the cutting edges often causes extreme distortion and makes pushing the sample out of the tube difficult. Sometimes, side friction causes so much distortion that the length of the sample entering the tube is less than the penetration of the sampler into the soil. Remolding of the soil next to the tube walls may cause the soil to stick to the tubes so tightly that a snugly fitting piston used to push the sample from the tube penetrates through the sample leaving a thin layer of clay adhering to the walls of the tube.

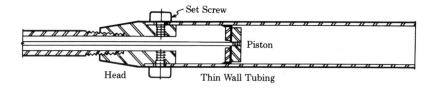

Figure 9.18d
Piston Sampler

A number of variations of the thin wall sampler have been devised for special purposes. One called a piston sampler is provided with a piston which fits snugly inside the thin wall tube and is operated by a rod extending up through the drive pipe. The piston is positioned at the bottom of the tube initially. It is then held stationary by clamping the piston rod in a fixed position while the tube is pushed into the soil at the bottom of the hole. During the taking of the sample the piston is pulled up with the sampler and carries the weight of the column of water above the sample and creates a vacuum between the piston and the sample which helps to hold the sample in the tube. A disadvantage of the piston sampler is the difficulty in holding the piston stationary as the sampler is pushed into the soil. Another disadvantage of the piston sampler is that the samples either must be removed from the tube in the field where the possibility of disturbance is great or a new tube provided for each sample. Sampling tubes should be sharpened, the cutting edge rolled inward slightly and

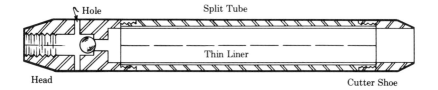

Figure 9.18e
Split Spoon Sampler

reamed to size, the upper end drilled for attachment to the sampler head, and the inner surface coated to prevent corrosion. Obviously such tubes are expensive. Dr. Jorj Osterberg has improved the piston sampler to simplify the holding of the piston stationary during the taking of the sample and has provided thin liners so that samples may be removed from the sampler inside the liner and protected and stored in the liner. This improved piston sampler is known as the Osterberg sampler. The piston sampler is particularly useful in soft clays.

Another sampling device is known as a split spoon sampler. It consists of a permanent tube split lengthwise and screwed into a driving head and a cutting shoe as shown in Figure 9.18e. A thin tube liner is often provided into which the sample enters and in which the sample is preserved after being removed from the sampler.

Other more complicated samplers have been devised. Some have a loop of fine wire in a groove in the cutting shoe for cutting off the sample after penetration and before extraction of the sampler. Some are provided with a compressed air outlet in the cutting shoe for applying air pressure to the bottom of the sample to help hold it in the sampler. These complicated samplers are made in the larger sizes for obtaining undisturbed samples of soft clays for consolidation tests.

One of the most successful samplers, especially for hard clays, is the Denison barrel sampler. It was originally devised and used for exploration of the site of the Denison Dam on Red River between Durant, Oklahoma and Denison, Texas. It is now used quite generally by the Corps of Engineers, U.S. Army; the U.S. Bureau of Reclamation; and the U.S. Soil Conservation Service. This sampler consists of an inner split spoon type sampler with a thin liner and core catcher and an outer rotating barrel with teeth on the cutting edge. As the outer barrel is rotated, water is forced through the drill stem down between the outer barrel and the sampler, out under the cutting edge and up between the outer barrel and the sides of the hole bringing the soil cuttings to the top of the hole. As the outer barrel advances, the sampler is pushed into the soil. The liner is 24 in. long and 6 in. in diameter. With this sampler, almost continuous samples can be taken and preserved in the liners in 24 in. sections. Such sampling is very expensive and is commonly used only on large projects for which the expense is justified. This sampler is described by H. L. Johnson in *Civil Engineering*, Volume 10, 1940.

The amount of disturbance of samples taken with samplers from bore holes is dependent upon the ratio of the area of the sampler walls to the total cross sectional area of the sampler and the method of forcing the sampler into the soil. Driving the sampler with successive blows of a hammer produces the most disturbance. Pushing the sampler into the soil continuously and rapidly produces the least disturbance. Samples for accurate laboratory testing of compressibility and strength should be at least 4 in. in diameter. This large sampler requires a large bore hole which adds materially to the cost of exploration. Two inch Shelby tube samples cannot be considered as undisturbed, but if nothing better is available, they can be used for consolidation and shear tests. In cased holes in soft clay, the sample should be taken just below the bottom of the casing. If the hole is drilled ahead of the casing before the sample is taken, the soft clay may partially fill the drilled hole below the casing and result in a highly disturbed sample.

Samples kept for extensive periods should be stored in a humid room to prevent loss of natural water.

An exhaustive study of soil sampling has been made by Dr. M. Juul Hvorslev. The Swedish Geotechnical Institute has for some years rather continuously engaged in the design and testing of improved sampling devices.

b. Fitting Specimen in Ring

The first step in making a consolidation test is to fit a specimen of the soil in the confining ring with as little disturbance of the soil structure as possible. Different methods have to be used for different soils.

Soft clays without gravel intrusions can be cut with a fine high strength wire pulled tight in a coping or hack saw frame. A portion of the sample 1 or 2 inches larger than the diameter of the ring and somewhat longer than the height of the ring should be cut from the sample.

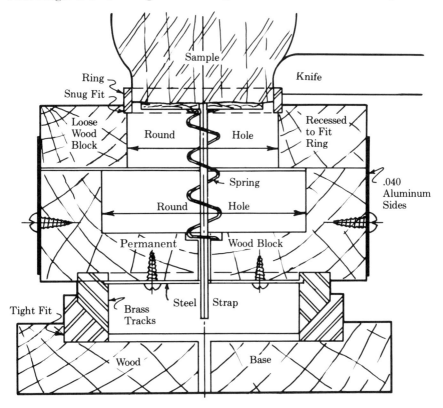

Figure 9.18f
Section Through Lathe for Fitting Consolidation Specimen to Confining Ring
Half Size

The ring can be held in a hand operated lathe while cutting the specimen to fit the ring. A section through such a lathe is shown in Figure 9.18f.

The lathe consists of a wood base about 8 in. square, a brass ring track in two parts, a rotating frame, a loose wood block recessed to hold the confining ring in position, and a spring lifted sample support. The sample support is a disc of hardwood or plywood about $\frac{3}{8}$ in. thick secured to a $\frac{3}{8}$ in. diameter dowel rod. The spring should be just strong enough to provide some support for the specimen but not enough to lift the specimen and ring. A loose block is needed for each size of ring. The ring should fit into the recess tight enough that it will not slip as the ring and sample are turned against the cutting knife. An excellent knife can be made from an old power operated hack saw blade.

In fitting the specimen in the ring, the portion of the sample, as described above, is placed on the ring and shaved off immediately above the ring until it will just start into the ring. The specimen is then cut to fit snugly and accurately in the ring by cutting off the excess sample with a knife provided with a shoulder that allows the sharpened edge of the knife to extend exactly to the inside edge of the ring. As the lathe is slowly turned by hand, the knife is held on the ring against the sample shaving off the excess soil until the shoulder comes against the ring all around. The knife is then removed and the sample pushed gently into the ring a short distance, being careful not to disturb the specimen. Alternate trimming and pushing of the sample is continued until the specimen extends a distance below the bottom of the ring to eliminate the bottom portion of the sample that does not quite fit the ring.

That portion of the sample that projects above the ring is then cut off, being careful not to remove any of the specimen below the top of the ring. With a straight edge, the excess soil is shaved off to form a smooth plane surface exactly in the plane of the top surface of the ring. This smooth surface should be covered with a clean glass plate, and the ring with the specimen removed from the lathe and turned upside down on the glass plate. After the excess soil is removed from this side and the face planed to a smooth surface, it should be covered with another glass plate to prevent drying out.

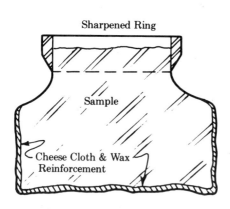

Figure 9.18g
Fitting Fissured Clay into Sharpened Ring

Sometimes, small pebbles are encountered as the soil is being fitted into the ring. They should be carefully picked out and the void filled with soil cuttings on a small spatula. Pebbles found in the upper and lower faces of the specimen can be treated in the same manner.

Hard clays cannot be cut with a wire saw but can usually be trimmed down to a proper size with a sharp butcher knife and fitted into the ring by trimming to exact size with the special knife. Fissured clays are sometimes difficult to fit into a consolidation ring because they are pulled apart by the cutting knife. Sometimes, a specimen of such a clay can be fitted into a ring with one end sharpened to a cutting edge. The sharpened edge is pushed a short distance into the sample after which the soil below is trimmed to a diameter slightly larger than the ring, so alternating until the ring is filled.

Occasionally it may be desirable to test a specimen of a thin wall tube sample without removing it from the tube. In this case, a short section of the tube and sample is cut to a length equal to the desired height of the specimen, being careful to maintain parallel ends and to disturb the soil as little as possible.

Confining rings are often made of brass. High strength machinable plastics are sometimes used. Steel and aluminum rings may corrode during a test, causing the specimen to adhere to the sides of the ring, and are not suitable materials for this purpose.

c. Setting Up the Test

Either before the specimen is placed in the ring or at the end of the test, the inside diameter and the height of the ring are measured with a vernier caliper to the nearest 0.01 cm and the ring weighed to the nearest 0.01 g. These measurements and the weight are recorded on the Water Content and Computation Sheet.

After the specimen has been fitted to the ring, the ring with the specimen is weighed to the nearest 0.01 g. The glass plates may or may not be included in this weighing. If they are included in this weighing, their weight must be included in the tare, along with the weight of the ring.

A porous stone is then placed on each side of the specimen, and the stones and specimen are placed in a vessel which is provided with a support for holding an indicator dial and which allows flooding of the specimen. The porous stones should be medium or fine grade about $\frac{1}{16}$ in. smaller in diameter than the inside of the ring and about $\frac{3}{8}$ in. thick. Before placing the stones in contact with the specimen, they should be moistened to prevent absorption of pore water and resulting shrinkage of the specimen by capillary forces.

The test may be run in either a floating ring or a fixed ring. In the floating ring test both top and bottom stones are smaller than the ring so that the ring is supported only by friction along the sides of the specimen as shown in Figure 9.18h. In the fixed ring test the bottom stone is larger than the ring and supports the ring and specimen. In the consolidation test deformation of the specimen is resisted by the vertical stress in the soil and by friction along the contact surface of the ring and specimen.   Frictional resistance is less in the floating ring test than in the fixed ring test because both stones can enter the ring as deformation occurs.   With some soils, the specimen will not support the ring during the test which makes necessary the use of a fixed ring. Also, if the specimen is dry or if it is allowed to dry out during the test and is subsequently flooded, capillary forces pull some of the soil from the specimen between the stones and the ring and deposit it on the bottom of the flooding vessel.   This loss of soil can be reduced by the use of a fixed ring.

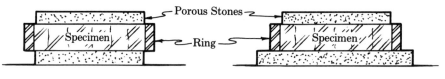

| Figure 9.18h | Figure 9.18i |
| Floating Ring | Fixed Ring |

After carefully centering the stones so they will enter the ring, the specimen and stones are carefully placed in the center of the vessel. Brass is a satisfactory material for the vessel.   Aluminum has been found to corrode badly.   The bottom of some aluminum vessels have been pitted to such an extent that the bottom stones are broken during testing.   The vessel should be large enough to allow the fingers between the bottom stone and the sides of the vessel for the largest specimen or bottom stone used in testing.   Usually, a 4 in. diameter specimen is as large as ever desired or practical.   A vessel 9 in. inside diameter should be sufficient for all tests.   Provision must be made for the reception of standards to support the dial indicator.   A suggestion for vessel, standards, and dial holder is given in Figure 9.18j.   These details are believed to be self explanatory.   If the temperature of the laboratory is likely to vary over a wide range, the variation in length of the standards may cause erratic results.   Invar steel standards have been found to be satisfactory under all conditions.   Temperature variation is not desirable but sometimes cannot be avoided.

After placing the specimen and stones in the center of the vessel, the setup is ready to be placed in the loading frame for testing.

d. Loading Frames

Any system that applies a known load to the specimen and follows the deformation without excessive friction and which provides for measurement of deformation can be used for consolidation testing.

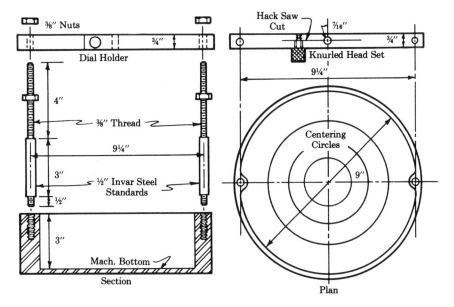

Figure 9.18j
Holding Vessel for Consolidation Test

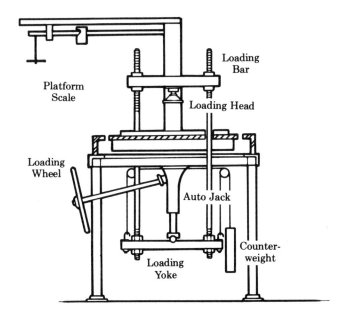

Figure 9.18k
Platform Scale Loading Frame

One of the simplest loading frames is made with an ordinary platform scale of about 3000 lb capacity supported on a substantial frame and provided with a 2 ton screw jack which operates a counterweighted loading yoke extending through small holes in the platform. A steel reinforcing plate is usually necessary to prevent excessive deflection of the platform.

A number of other arrangements for loading the specimen have been devised using mechanical advantages of from 8 to 1 to 100 to 1. In

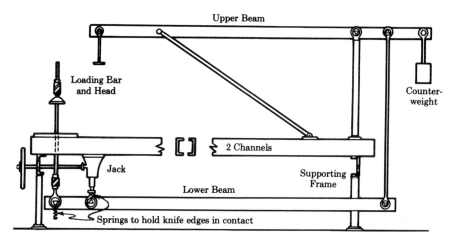

Figure 9.18l
Two Beam Loading Frame

any case, the loading yoke must be counterweighted so that the load on the specimen can be accurately controlled from zero to the maximum capacity of the machine. In some frames using a single beam, the loading weights are applied on the opposite side of the machine from where the readings are taken. Loading of those frames using low load to weight ratios is inconvenient because of the difficulty in handling the heavy weights, which for the heavier loads must be placed in two lifts. The use of two beams with greater load to weight ratios enable the use of light easily handled weights and are easily arranged for loading from the same position as the readings are taken. The disadvantage of the two lever system is greater friction and difficulty in adjusting the height of the loading head and leveling the top beam as deformation occurs.

A sketch of a two beam loading frame is shown in Figure 9.18l. These frames can be mounted alternately facing in opposite directions on a supporting frame so that a considerable number of frames can be mounted in a relatively small space.

e. Arrangement for Loading and Measuring Deformation

The loading frame is provided with a loading head supported on the loading bar with provision for self adjustment of the head to the surface of the specimen. Allowance for adjustment of the head is often provided by supporting the head on the bar in a ball joint with 4 tension springs.

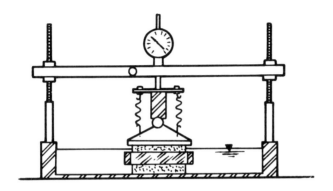

Figure 9.18m
Floating Ring Consolidation Test

The setup of specimen, ring, and stones arranged for a floating ring or a fixed ring test in the vessel is carefully centered under the loading head below the loading bar and the beam adjusted with the jack until the head is in contact with the specimen.

The standards are then screwed snugly into position and the dial holder adjusted to the proper height, leveled, and secured firmly in place with nuts above and below on the standards. A dial indicator reading directly to thousandths cm, or an equivalent in inches, is placed in the holder bar, adjusted to the approximate dial reading desired and fastened in place with the knurled head set screw. The dial should be fastened tight enough that it will not slip but not tight enough to cause the movable stem to bind. The exact initial reading is set by turning the dial face to the hand. This initial reading is recorded on the Water Content and Computation Sheet. (See section IX-L, Chapter XII.)

f. Testing the Specimen

Means must be provided for preventing loss of moisture from the specimen by evaporation.

Soft saturated clays that do not swell when submerged can be flooded before application of a load increment by pouring water into the vessel to above the bottom of the top porous stone. In general, this procedure should be used only when the soil exhibits little or no tendency to swell under zero load. If it is likely that the soil will swell under these conditions, evaporation can be largely prevented by surrounding the ring and stones with moist paper towels and arranging a sheet of rubber or plastic over the vessel. The towels are removed and the specimen flooded after the loading has progressed far enough to prevent swelling of the soil.

Load increments should be applied gently and as nearly as possible when the second hand of a clock arranged for easy viewing is at zero. The loading time should be decided upon before application of the load and the load and initial time recorded on the Data Sheet.

Because the rate of volume change decreases rapidly with time, the first dial reading should be made as soon as possible after application of the load. Just prior to reading, the face of the dial should be tapped lightly with the eraser end of a pencil to make sure that the stem is in contact with the loading bar. A convenient elapsed time for the first reading is 6 sec or 0.1 min. Because the time is to be plotted to a logarithmic scale, points on the log scale will be approximately equally spaced if the elapsed time is doubled for subsequent readings; i.e., readings at 6 sec, 15 sec, 30 sec, 1 min, 2 min, 4 min, 8 min, 15 min, 30 min, 1 hr, 2 hr, 4 hr, 8 hr, and each 8 hours thereafter. It is not necessary that the readings be made at these times, but it is necessary that the exact time of reading be recorded. Care should be exercised to see that dial readings and elapsed times are correctly and accurately recorded on the Data Sheet.

As consolidation progresses after application of the first load increment, the dial reading-time curve should be plotted to ascertain when sufficient time has elapsed to extend beyond 100 per cent consolidation. It is desirable that all load increments remain on the specimen for the same length of time so that the effect of secondary compression will be essentially the same for each consolidation period.

Because the pressure is plotted to a logarithmic scale, points on the log scale will be approximately equally spaced if each successive increment doubles the load. Good dial reading-time curves are produced by large load increments. Good void ratio-pressure curves are produced by a large number of points, i.e., a large number of increments. Doubling the load incrementally is a satisfactory compromise between these two conflicting requirements for good curves.

The size of the first load depends upon the soil and what the investigator is seeking. For soft clays investigated from zero load, a first load that produces a unit pressure of 0.05 kg cm^{-2} or 100 lb ft^{-2} is often desirable. A total load of 3 or 6 kg is often used as the first increment. Weights are commonly provided for loads of 3, 6, 12, 24, 48, 96, 192, 384, 768, and 1536 kilograms.

After consolidation under the heaviest load to be applied during the test, the load is removed in increments allowing the same time for swelling as was allowed for consolidation. Readings should be taken in the same manner as during consolidation. The load may be removed at a faster rate than it was applied, usually in only 3 or 4 steps.

Sometimes in the study of desiccated clays, it may be desirable to load the specimen to some pressure without flooding after which the specimen is flooded and the swelling under this pressure determined. The specimen can then be subjected to a consolidation test as described above and unloaded to determine the amount of elastic recovery.

g. Dial Reading-Time Curves

A dial reading-time curve is plotted on 5 cycle semilog paper for each increment of loading and unloading. The dial reading is usually plotted in cm to some convenient scale and the time in min on the log scale. One of these dial reading-time curves representing a fairly large deformation should be converted to a consolidation-time curve as described in Article 9.10 and the time for 50 per cent consolidation determined.

h. Removing the Specimen

At the end of the test after swelling is completed under zero load and the final reading taken and recorded, the dial, holder and standards are removed, the loading bar and head is raised and the vessel with the specimen removed from the loading frame. The porous stones and ring with the specimen are carefully lifted out of the vessel and the stones removed from the specimen. Any soil clinging to the stones should be scraped off with a small spatula and placed on the specimen. Excess water should be carefully removed from the ring with a paper towel or other absorbent material. Immediately, the ring and specimen are weighed to the nearest 0.01 g and this weight recorded on the Water Content and Computation Sheet.

The ring and specimen are then placed in an evaporating dish and set in a drying oven at 105° C. After the specimen is thoroughly dry, the dried specimen and ring are removed from the oven, cooled in a desiccator and weighed to the nearest 0.01 g. This dry weight should be recorded in the proper place.

i. Calibrating Machine and Porous Stones

In the consolidation test as described, load is applied to the specimen through the loading bar and head and the porous stones. The deformation read on the dial includes the deformation of these parts as well as the deformation of the specimen. In order to determine the deformation of the specimen, it is necessary to determine for each load the deformation of these parts without the specimen and subtract the deformation of the parts from the total deformation.

To determine the deformation of the machine parts and the stones, the porous stones are placed in the vessel without the ring and specimen, placed in the loading frame with the loading head in contact with the stones and the dial set to an initial reading under zero load. Load increments are then applied and removed in exactly the same order as they were applied to the specimen and corresponding dial readings taken and recorded. These loads can be applied and removed as rapidly as readings can be taken.

j. Computations

The computations indicated on the Water Content and Computation Sheet are self-explanatory. The water content at the start and at the end of the test is used for computing the degree of saturation at the beginning and end of the test. The initial void ratio is determined as the ratio of the equivalent height of the voids to the equivalent height of the solids in the specimen.

During a consolidation test lasting from a week to a month and the application of several load increments, errors may be caused by slipping of the dial in its holder, by jostling of the machine, or by other accident. During the several consolidation periods practically all the entrapped air should be driven from the specimen and the degree of saturation should be 100 per cent. A degree of consolidation at the end of the test close to 100 per cent is evidence that no such accident has occurred during the test and that the dial changes from start to finish are an accurate measure of the deformation. The degree of saturation is computed as the ratio of the equivalent height of water in the specimen to the equivalent height of the voids.

Values used for preparing data for the e-p curve are equivalent height of solids in the specimen H_s and initial void ratio e_0.

$$H_s = \frac{W_s}{AG_s} \text{ and } e_0 = \frac{H_{v1}}{H_s},$$ in which W_s is weight of the dry specimen, A is area of the ring and specimen, G_s is specific gravity of solids, H_{v1} is equivalent height of voids at start of test, and H_s is equivalent height of solids in specimen.

k. Preparing Void Ratio-Pressure Curve

Unit pressure is determined by dividing the total applied load by the area of the specimen.

For determining void ratios e corresponding to unit pressures p the change in void ratio for each load is added to or subtracted from the initial void ratio e_0. The change in void ratio is equal to the ratio of the change in height of the specimen ΔH after complete consolidation under the given pressure to the equivalent height of solids H_s. The change in height of the specimen is the change in dial reading from the initial reading less the deformation of the machine parts and porous stones as determined in the machine calibration. Dial readings for computing these changes in void ratio are taken from the dial reading-time curves at the same time on the secondary compression portion of the curves. These computations are made on the Summary and Computation Sheet.

The e-p curve is then plotted on 3 cycle semilog paper with void ratio to some convenient scale on the arithmetic ordinate and the pressure on the logarithmic scale. Pressures are usually in kg cm^{-2} or tons ft^{-2}. For practical purposes, it may be assumed that 1 kg cm^{-2} is equal to 1 ton ft^{-2}.

l. Computing Soil Properties

Information is now available for computing the coefficient of compressibility a_v, the compression index C_c, the coefficient of consolidation c_v, and the coefficient of permeability k.

$$a_v = \frac{\Delta e}{\Delta p},$$

and can be determined from values on the Summary and Computation Sheet,

where

Δe is the difference in void ratio from one value of p to the next,

and

Δp is the difference in pressure over the same interval.

The value of a_v is not a constant but becomes smaller with increase in pressure and decrease in void ratio. It is used for computing the coefficient of permeability and should be chosen for a void ratio at which the coefficient of permeability is desired.

C_c is the slope of the virgin $e - \log p$ curve and is equal to the difference in void ratio subtended by the straight virgin curve or its extension over one cycle of the log scale.

$$c_v = \frac{TH^2}{t} = \frac{0.2H^2}{t_{50}},$$

where, in this case, H is one-half the original height of the specimen,

and

t_{50} is the time for 50 per cent consolidation as determined on a consolidation-time curve.

$$k = \frac{c_v \gamma_w a_v}{1 + e_m}.$$

The coefficient of permeability thus determined is the range for which a_v is known. The void ratio e_m is the average over the range for which a_v is determined.

m. Effect of Size and Thickness of Specimen

Because the time for the same degree of consolidation is approximately proportional to the square of the thickness of the specimen, thick specimens require much longer time for consolidation and are, therefore, subject to greater influence of secondary compression than thin ones. The thickness of the specimen is determined largely by the type of soil. Specimens of soft smooth clays can be only $\frac{1}{2}$ in. thick while those of clays containing small pebbles must be thicker. Seldom is a specimen thicker than 1 in. necessary.

The percentage of error caused by friction between the specimen and the ring is dependent upon the ratio of the contact area between specimen and ring to the cross sectional area of the specimen. Therefore, the thinner the specimen compared to its area, the less is the error because of friction.

9.19 BRIEF INSTRUCTIONS FOR CONSOLIDATION TEST PROCEDURE

A brief step by step procedure for carrying out a consolidation test is given in the corresponding section in the chapter on Laboratory Testing.

9.20   TEST DATA

Test data, curves, and computation of properties are given in the chapter on Laboratory Testing.

REFERENCES

Hough, B. K., *Basic Soils Engineering*, New York: The Ronald Press Company, 1957, Chapter 5.

Leonards, G. A., "Engineering Properties of Soils," *Foundation Engineering*, New York: McGraw-Hill Book Company, Inc., 1962, Chapter 2.

Peck, Ralph B., Walter E. Hanson, and Thomas H. Thornburn, *Foundation Engineering*, New York: John Wiley and Sons, Inc., 1953, Art. 2.8, Chapter 2.

Sowers, G. F., "Shallow Foundations," *Foundation Engineering*, Edited by G. A. Leonards, New York: McGraw-Hill Book Company, Inc., 1962, Chapter 6.

Lo, K. Y., "Secondary Compression of Clays," *Journal Soil Mechanics and Foundations Division*, Proc. Am. Soc. C. E., Vol. 87, No. SM4, (August, 1961).

Scott, Ronald F., "New Method of Consolidation—Coefficient Evaluation," *Journal Soil Mechanics and Foundations Division*, Vol. 87, No. SM1, (February, 1961).

PROBLEMS

9.1 The terms "Liquidity Index" and "Sensitivity" are frequently used in soil mechanics.

(*a*) Define the two terms.

(*b*) State what practical significance may be attached to their numerical values.

9.2 A certain saturated clay deposit has a water content of 400 per cent and a specific gravity of solids of 2.42. Assuming the water table to be initially located at the ground surface, determine the unit pressure acting on a horizontal plane at a depth of 30 feet. What would the pressure be on this plane after the water table is lowered 25 feet by pumping from deep wells?

9.3 (*a*) The water table is at a depth of 6 feet in a thick deposit of clean, coarse sand having an average void ratio of 0.65. Determine the effective pressure at a depth of 15 ft beneath the ground surface.

(*b*) Determine the effective pressure at a depth of 15 ft in the soil
described above if conditions are altered as follows:
(1) the ground water table coincides with the surface, and
(2) water rises to an elevation of 10 ft above the surface in a
piezometer installed with its porous tip 15 ft below the surface.

9.4 A deposit of soft, organic clay ($G_s = 2.3$) in a swampy area overlies a
thick stratum of coarse sand. The clay deposit is 25 ft thick and has
an average natural water content of 160 per cent. A piezometer
inserted into the coarse sand shows that an excess pressure exists which
causes the water surface in the piezometer tube to rise to 5 ft. above the
ground surface. Determine the *effective* stress (in units of lb/sq ft) at
the base of the clay stratum.

9.5 (*a*) For the soil profile shown in the figure, estimate appropriate values
for G_s for the various strata and draw a diagram which shows the
variation, with depth, of the effective pressure on the horizontal
plane. Note on the diagram the effective pressures at depths of
4, 9, 18, and 30 feet beneath the ground surface.

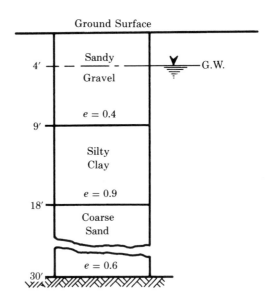

(*b*) Suppose that the conditions were as shown, except that the water
rises to a height of 5 feet above the ground surface in the tube of a
piezometer which has been installed in the coarse sand stratum.
Plot the effective pressures for this condition (in red) on the same
diagram plotted for part (*a*).

9.6 Derive an expression for the total decrease in thickness of a normally consolidated clay stratum in terms of the thickness of the layer, the average initial void ratio, the slope of the virgin compression curve, and the initial and final pressures which act on the soil.

9.7 Laboratory temperatures are usually somewhat warmer than the temperatures of soils in the ground. *Explain* what effect, if any, this will have on the rate of consolidation of clay as compared to the rate predicted from the results of laboratory tests.

9.8 The following data were obtained from a consolidation test on an undisturbed clay sample, with a test temperature of 26° C:

$$p_1 = 1.5 \text{ kg cm}^{-2} \qquad e_1 = 1.30$$
$$p_2 = 3.0 \text{ kg cm}^{-2} \qquad e_2 = 1.18$$

The sample was 1 inch thick, was drained on both sides, and 50 per cent consolidation was reached in 20 minutes. Compute for this range of load increment, the average coefficient of permeability reduced to 20° C, and expressed in units of 10^{-9} cm sec^{-1}.

9.9 List 4 factors which may influence the results of laboratory consolidation tests in such a way as to affect our interpretation of the results.

9.10 In a consolidation test on an undisturbed clay specimen, the following results were obtained:

Pressure kg/cm^2	Void Ratio
0.2	0.953
0.4	0.948
0.8	0.938
1.6	0.920
3.2	0.878
6.4	0.789
12.8	0.691
3.2	0.719
0.8	0.754
0.2	0.791
0.0	0.890

Plot these results on both arithmetic and semilogarithmic graphs, and determine the equations for the virgin compression curve and the swelling curve on the semilogarithmic plot.

Estimate the stress to which this clay had been preconsolidated.

9.11 Consolidation tests on undisturbed samples from a 20 ft thick stratum of clay gave the pressure-void ratio relationship shown in Problem No. 9.10. The average existing overburden pressure on this clay stratum is 1.2 kg/cm². It is proposed to construct a building which will increase the average pressure on the clay to 2.7 kg/cm².

Estimate the decrease in thickness of the clay stratum caused by full consolidation under the building load, assuming:

(a) that the preconsolidation pressure as determined in problem 9.10 has existed prior to the present overburden pressure, and

(b) that the clay has been preconsolidated only under its existing overburden, i.e., that the entire deformation is governed by the compression index C_c of the virgin curve.

Show these estimates of Δe on an $e - \log p$ plot.

9.12 A clay layer 40 feet thick, having an initial void ratio of 1.00, will be reduced to a void ratio of 0.94 under the weight of an extensive gravel fill which increases the load by about 0.5 tons/sq ft. How much will the surface of the fill settle?

9.13 The compression index C_c for a certain normally consolidated clay soil is found to be 0.25. If the average natural void ratio of the soil in a layer 20 feet thick is 1.15, determine the settlement of the surface when the average load on the soil is increased from 0.5 to 2.5 tons/sq ft.

9.14 Two points on the virgin compression curve for a normally consolidated clay are:

$$e_1 = 1.00, \qquad p_1 = 0.5 \text{ kg/sq cm}$$
$$e_2 = 0.90, \qquad p_2 = 2.5 \text{ kg/sq cm}$$

If the average overburden pressure on a 30 foot layer of the soil is 0.75 kg/sq cm, compute the decrease in thickness of the layer under an average *pressure increase* of 2.5 kg/sq cm.

9.15 A surface stratum of clay 20 feet thick is underlain by a stratum of coarse sand which outcrops on a nearby stream bank. The clay is to be loaded by placing an extensive gravel fill over the area. The weight (1 ton/sq ft) of the fill will result in a reduction of the void ratio of the clay from an average initial value of 1.06 to a final average value of 0.84. Previous experience with this soil has shown it to have a coefficient of permeability of 54×10^{-9} cm/sec at a void ratio of 1.14. Compute

(a) the total settlement of the surface, and

(b) the coefficient of permeability corresponding to the average void ratio during the period of consolidation.

(c) Assuming the answer to part (b) to be $k = 50 \times 10^{-9}$, compute the time required for 50 per cent consolidation of the clay stratum.

9.16 A sample of clay 3 cm thick is tested in a consolidation apparatus (drained both sides), and 30 per cent consolidation was reached in 10 minutes. How long will it take a layer 30 meters thick to reach the same degree of consolidation if drained on the upper surface only? How long will it take to reach 50 per cent consolidation?

9.17 The following data were obtained from a consolidation test on an undisturbed sample of organic silt:

$$p_1 = 1.4 \text{ kg/sq cm} \qquad e_1 = 1.37$$
$$p_2 = 2.0 \text{ kg/sq cm} \qquad e_2 = 1.25$$

The coefficient of permeability in this range of loading was determined by a direct permeability test to be 215×10^{-9} cm/sec at a temperature of 23° C. Assuming that the pressures shown above correspond, also, to the average initial and final pressures imposed on a 20 ft stratum of this soil in the field, compute

(a) the time required for the stratum to reach 50 per cent consolidation, and

(b) the settlement at this time (i.e. $U\% = 50$).

NOTE: The stratum is drained only at its upper surface, and the average ground temperature is 8° C.

9.18 The coordinates of two points on the virgin compression curve of a certain silty clay are

$$p_1 = 2 \text{ kg/sq cm} \qquad e_1 = 0.77$$
$$p_2 = 10 \text{ kg/sq cm} \qquad e_2 = 0.61$$

The time required for 50 per cent consolidation of a laboratory specimen 2 cm thick was 7.5 minutes. In the field, a 10 foot stratum of this clay, having an average void ratio of 0.75, lies between two strata of coarse sand. The average pressure on the clay is increased from 2.4 to 4 kg sq cm. Determine

(a) the total decrease in thickness of the layer (inches),

(b) the time in days for 50 per cent consolidation in the field,

(c) the time in days for 1 inch of settlement to occur, and

(d) the settlement at the end of 3 months.

9.19 The following data were obtained from a consolidation test on an undisturbed clay sample:

$$p_1 = 1.65 \text{ kg cm}^{-2} \qquad e_1 = 0.895$$
$$p_2 = 3.10 \text{ kg cm}^{-2} \qquad e_2 = 0.782$$

The average value of the coefficient of permeability of the clay in this range is 3.5×10^{-9} cm sec^{-1}. The clay layer undergoes a change in pressure from 1.65 kg cm^{-2} to 3.10 kg cm^{-2}. By utilizing the known theoretical consolidation curve (numerical or graphical relation between per cent consolidation and time factor), compute and plot (with the settlement in inches as ordinates, and the time in years as abscissas) the decrease in thickness with time for a 30 ft layer of this clay which is drained:

(a) on the upper surface only,

(b) on the upper surface and at a depth of 10 ft by a thin horizontal sand layer, providing free drainage.

9.20 The time for 50 per cent consolidation of a clay sample 2.54 cm in height was found to be 10.3 min.   The total settlement due to added load on a 20 ft layer of this clay was estimated as 6.42 inches.   If the layer is drained on one side only, estimate the settlement at the end of one year.

9.21 Laboratory tests of a normally consolidated clay specimen 2 cm thick indicate a compression index of 0.15 and a time of 7.5 min for 50 per cent consolidation.   Over a large area of the ground surface a 25 ft stratum of this same clay is to be subjected to loading during which the stress increases from an average of 0.5 tons/sq ft to an average of 1.5 tons/sq ft.   If the average initial void ratio of the clay is 0.7, estimate the settlement which will take place during the first three years after loading.   The clay stratum is underlain by impervious rock.

9.22 In a laboratory consolidation test, a soil sample 1 inch thick reached 50 per cent consolidation in 8 minutes under a given load.   A layer of this soil 40 feet thick (drained top and bottom) is to be subjected to the same loading.   The *total* decrease in thickness of this layer will be about 20 inches.   How long will it take to settle 6 inches?

9.23 For the clay described in Problem No. 9.14, the time required for 50 per cent consolidation of a laboratory specimen 2 cm thick was found to be about 4.2 minutes for the range of loading in which we are interested.   Determine the time required to produce a 6 inch settlement of the surface of the 30 foot layer under the *pressure increase* of 2.5 kg/sq cm.

CHAPTER X

Shrinkage

10.01 GENERAL

One of the important physical properties of soil in which the engineer is interested is volume change caused by change in water content. When water is evaporated from a saturated compressible soil, a volume decrease or shrinkage occurs. This phenomenon is observable on the surface of dried up ponds.

Shrinkage of clays caused by loss of pore water and subsequent swelling caused by absorption of pore water has produced enormous damage to structures built on or of clays subjected to variable moisture conditions. The evaluation of shrinking and swelling amounts and pressure intensities under existing conditions is of primary concern to the engineer in the design of structures.

281

Shrinkage can occur only in compressible soils. Sands and gravels have the same volume when dry as when inundated. When moistened, these cohesionless relatively incompressible soils occupy a volume greater than that of either the dry or inundated state. Shrinkage of clays and bulking of sands is caused by tension in the pore water reacting against the soil grains.

10.02   STRESS IN PORE WATER AND SHRINKAGE

Imagine a compressible soil consisting of tiny grains with capillary pore spaces between the grains. These capillary tubes are interconnected and variable in size somewhat as shown in Figure 10.02a.

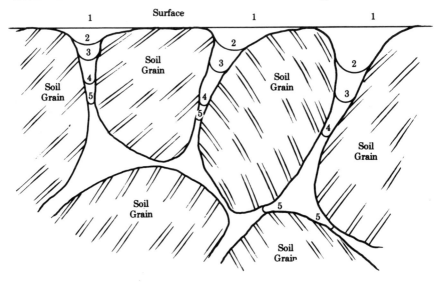

Figure 10.02a
Idealized Section Through Soil

When the pore spaces are completely filled with water and there is free water on the surface of the soil, the meniscus is a plane surface as shown by position 1 and the tension in the water is zero. As evaporation removes water from the surface, a meniscus begins to form in each of the pores at the surface with a resulting tension in the water. Because the stress in the water is distributed equally throughout the water in the communicating pore spaces and tubes, the radii of all the menisci are equal. The menisci react against the soil grains producing a compression between the grains somewhat as though the soil mass were loaded with elastic bands stressed in tension. If the soil is compressible, this compression between the grains causes a deformation or shrinkage of the volume.

At some time after evaporation has started the menisci will have receded to some position as shown at 2 in Figure 10.02a, with a fully developed meniscus in the largest pore. At this stage the stress in the pore water is $u_2 = 2T_s/R_2$ and the soil is compressed by this stress by an amount dependent upon the compressibility of the soil. As further evaporation occurs, the fully developed meniscus in the largest pore recedes from position at 2 to a smaller diameter at 3 and develops a meniscus with a smaller radius equal to R_3. All other menisci will have radii equal to R_3 and the stress in the water will be increased accordingly. This increased stress compresses the soil still further.

As evaporation continues the menisci continue to recede to smaller portions of the tubes, the radii continue to decrease, the tension in the water continues to increase, and the compression between the soil grains and the resultant shrinkage continues to increase.

Eventually, the menisci will have reached the smallest diameter in the largest of the communicating pores as shown by position 5 in Figure 10.02a. At this point the radii will have become the smallest possible when the menisci are fully developed and the stress in the water will have become a maximum. Since there are no pores of smaller diameter in all passages, further recession of the menisci produces no further increase in the stress and therefore no further shrinkage.

This lower limit of volume change as evaporation takes place is called the shrinkage limit and is expressed as water content. It is usually designated w_{SL} or S_w. It should be observed that at the shrinkage limit the soil is saturated, but there can be no free water on the surface. The void ratio has been reduced by the amount of water evaporated during the shrinkage process. Since the smallest pore diameters lie immediately beneath the surface, the limit of shrinkage is reached just as the surface becomes dry and may be observed by a change in color of the surface to a lighter shade.

There may be a slight additional shrinkage as compressible soil is completely dried out from the shrinkage limit. This additional shrinkage was determined by Terzaghi to be less than 5 per cent of the total shrinkage. For practical purposes this may be ignored.

When free water is again applied to compressible soil which has been shrunk by evaporation of the pore water, the menisci are destroyed, the tension in the water becomes zero, and the pressure between the

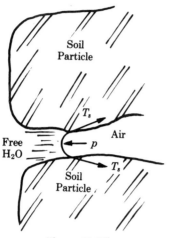

Figure 10.02b
Meniscus on Surface of Dry Clay

grains is relieved. If part of the deformation was elastic, there is an elastic expansion or swelling of the soil. Although swelling of the soil cannot take place until water has been pulled into the pores, it is the attraction for the dipolar water molecules by the negative charges on the surface of the soil grains which pulls the water into the pores and allows the expansion to occur.

When a dried out lump of clay is dropped into a vessel of water, the lump quickly slakes into a pile of individual particles. In this case the water pulling itself along the surfaces of the soil particles into the pores of the soil forms menisci which react against the air in the pores of the dry soil. At the surface the water is not stressed and the surface tension can be equated to the pressure in the air.

$$\frac{p\pi d^2}{4} = T_s \pi d \text{ or } p = \frac{4T_s}{d} = \frac{0.3}{d} \text{ g cm}^{-2}$$

when d is in cm. The entrapped air in the extremely small pores of dry clay can be subjected to enormous pressure. This pressure in the entrapped air breaks loose small bits of soil on the surface. As the bit of soil is broken loose the pressure in the air is suddenly released and the expanding air throws the small piece of soil from the surface with explosive force. A dried out lump of clay in its undisturbed state having a honeycomb or flocculent structure with a large volume of entrapped air slakes much more rapidly when immersed than the same clay after having been remolded and dried.

10.03 SHRINKAGE LIMIT

As has already been stated, the shrinkage limit is expressed as the water content at which there is no further volume decrease with further evaporation of pore water.

At the beginning of the shrinkage process the expanded soil may have a void ratio equal to some value e_0 and an effective pore diameter

d_0 capable of producing a pressure between the soil grains p_0. As evaporation occurs and this pressure is applied to the soil and the volume is reduced, the effective pore diameter is reduced. This reduction in pore diameter increases the potential stress. The increased stress is capable of still further reducing the volume, which still further increases the capacity for reducing the volume. But this capacity for increasing the shrinkage does not continue indefinitely.

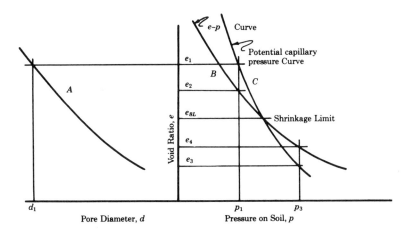

Figure 10.03a
Relationship of Pore Diameter, Void Ratio, and Pressure as Shrinkage Progresses

The relationship between pore diameter and void ratio can be shown by a curve such as the one marked A on the left side of Figure 10.03a. The relationship between void ratio and pressure capable of being produced by the capillary pore water may be as shown by curve C in Figure 10.03a. If the saturated compressible soil with a void ratio e_0 is loaded with a unit pressure p_1, the void ratio will be reduced to some value e_1. Further increase in pressure produces further decrease in void ratio. But, the coefficient of compressibility, slope of e-p curve a_v, decreases as p becomes larger. This relationship between e and p is shown by curve B in the figure.

Refer to Figure 10.03a and consider a soil in which at some time the pore diameter is d_1. The corresponding void ratio is e_1 and the pressure between the grains produced by the stress in the pore water as shown by curve C is p_1. As shown by curve B the pressure p_1 is capable of compressing the soil to a void ratio e_2. Therefore, evaporation from the soil at e_1 produces further shrinkage. Now consider the soil with a void ratio e_3. The pore water at this void ratio is capable of exerting a pressure of p_3 between the grains as shown by curve C. But as shown by the e-p curve, this pressure p_3 is not capable of compressing the soil to void ratio e_3. Therefore, there can be no further shrinkage with further drying out. It can be readily seen that the intersection of the void ratio-pressure curve B with the void ratio-capillary pressure curve C marks the shrinkage limit.

10.04 DETERMINATION OF SHRINKAGE LIMIT

Because there is little change in volume from the shrinkage limit after complete drying out, the shrinkage limit can be determined from the volume of an oven dried specimen. If the total volume, the specific gravity of the solids, and the weight of the dry specimen are known, the shrinkage limit can be obtained by assuming the voids of the dry soil to be filled with water and determining the ratio of the weight of this water to the weight of the dry solid.

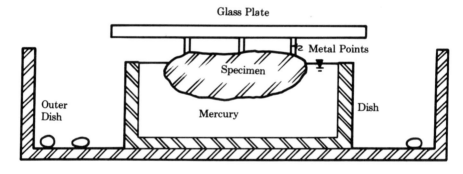

Figure 10.04a
Apparatus for Determining Volume of Soil Specimen

The total volume of the dry specimen can be determined by displacement in mercury. The displacement can be made in a dish or vessel with a smooth top such as shown in Figure 10.04a. A glass plate larger than the dish is provided with 3 metal points for forcing the specimen into the mercury. In making the determination of the total volume of the dried specimen, the dish is first set in a larger dish and filled with mercury. The cover plate with the metal points is placed over the dish and pushed down against the top of the dish, forcing the mercury displaced by the points to overflow into the larger dish.　The cover is then lifted off and the mercury emptied out of the outer dish.　A dry specimen of soil of known weight considerably smaller than the inside of the dish is placed on the surface of the mercury and pushed down by the points on the glass cover until the cover is in contact with the top of the dish forcing a volume of mercury equal to the volume of the specimen to overflow into the outer dish. The mercury which is caught in the outer dish is weighed and the volume of the specimen computed as follows:

$$V_{SL} = \frac{W_q}{G_q}$$ in which V_{SL} = volume at shrinkage limit, W_q = weight of displaced mercury, and G_q = specific gravity of mercury (about 13.6).

Having the total volume of the dry soil specimen V_{SL}, the weight of the dry soil W_s, and the specific gravity of the soil solids G_s, the shrinkage limit can be determined as follows:

$$V_v = V_{SL} - V_s = V_{SL} - \frac{W_s}{G_s \gamma_0}.$$

$$W_w = V_v \gamma_w.$$

$$w_{SL} = \frac{W_w}{W_s} = \frac{\left[V_{SL} - \frac{W_s}{G_s \gamma_0} \right] \gamma_w}{W_s}.$$

In the metric system γ_0 and γ_w may be assumed to be equal to unity for practical purposes.

The foregoing determination can be made for either a remolded or an undisturbed soil.

Another method of determining the shrinkage limit for remolded soils in which the specific gravity of the solids is not required was devised by the Bureau of Public Roads (then Public Roads Administration).　In this method a soil sample is mixed with water to approximately the liquid limit and placed in a vessel of known volume V_1 and its weight W_1 determined.　The sample is oven dried and the dry weight W_s determined.　The volume of the dry sample, which is the same as the volume at the shrinkage limit V_{SL}, is determined by displacement in mercury.　The relationship which exists for these weights and volumes and the shrinkage limit may be seen on the diagram shown in Figure 10.04b.

If volume be plotted against weight as water is evaporated from a sample of saturated compressible soil, the line showing the relationship is a straight line as shown on Figure 10.04b.　When the metric system is used and the horizontal and vertical scales are the same, the line makes 45° with the axes.　At the shrinkage limit the volume ceases to change as water is evaporated, so from that point the volume remains constant while the weight continues to decrease, making the line show-ing the relationship between volume and weight horizontal below the shrinkage limit.

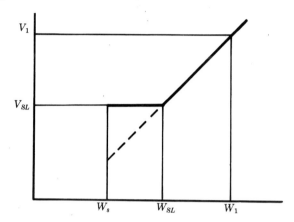

Figure 10.04b
Relationship Between Volume and Weight

By definition

$$w_{SL} = \frac{W_w \text{ at } S.L.}{W_s}$$

$$= \frac{W_{SL} - W_s}{W_s}.$$

From Figure 10.04b

$$W_{SL} = W_1 - (W_1 - W_{SL})$$
$$= W_1 - (V_1 - V_{SL})\gamma_w.$$

The water content at the shrinkage limit is

$$w_{SL} = \frac{W_1 - (V_1 - V_{SL})\gamma_w - W_s}{W_s}.$$

As a by-product of this BPR method of determining the shrinkage limit for remolded soils, the specific gravity of the soil solids can be evaluated.   By definition

$$G_s = \frac{W_s}{V_s \gamma_0}.$$

From Figure 10.04b it can be seen that

$$V_s = V_1 - \frac{W_1 - W_s}{\gamma_w}.$$

Then

$$G_s = \cfrac{W_s}{V_1\gamma_0 - \cfrac{W_1\gamma_0}{\gamma_w} + \cfrac{W_s\gamma_0}{\gamma_w}} \text{ or } \frac{W_sG_w}{V_1\gamma_w - W_1 + W_s}.$$

In determining the volume and weight of a dry specimen by displacement in mercury, the dry specimen should be weighed before immersion in the mercury. During immersion tiny particles of mercury enter the surface pores increasing the weight of the specimen.

Another method of determining the volume of a fairly large specimen of cohesive soil is to dip the specimen of known weight into molten wax being sure that a continuous waterproof coating of wax covers the specimen. The volume of the wax covered specimen is then found by immersing it in water in a graduated cylinder, noting the displacement. By weighing the wax covered specimen, the weight and volume of the wax can be determined and the volume of the specimen obtained by subtracting the volume of the wax from the displaced volume.

10.05 SIGNIFICANCE OF SHRINKAGE

The shrinkage limit alone is of comparatively little significance. Its relationship to other plasticity limits is not always the same. Reliable information concerning shrinkage can best be obtained by direct measurement.

Volume change of soils with change in water content is of importance because of its effect upon the stability of structures of the soil and upon structures resting on the soil. In those localities and conditions where the soil is continuously below the water table or the evaporation is so slow that the capillary fringe remains near the surface, shrinking and swelling are of relatively little importance. In semi-arid regions where clay exists near the surface, the clay shrinks during dry periods forming cracks that extend some distance below the surface; and during wet periods the clay swells closing the shrinkage cracks. This opening and closing of shrinkage cracks causes the clay to become fissured. Sometimes dust falls into the shrinkage cracks producing a soil consisting of blocks of over-consolidated (by desiccation) clay separated by thin layers of silt.

This filling of shrinkage cracks with dust in clay backfills against retaining walls is responsible for the gradual forcing of the walls out of plumb. In a period of years, repeated cycles of shrinking, filling the cracks with dust, and swelling may cause complete failure of the retaining wall unless the wall has been designed to resist the passive earth pressure of the backfill.

Swelling of desiccated clays has caused enormous damage to light buildings and other structures on these clays. For a discussion of the type of damage and suggestions for reducing the damage from this cause, the reader is referred to the papers presented at the Soil Mechanics Conference on Expansive Clays held at Golden, Colorado in 1959.

10.06 MEASUREMENT OF SWELLING PRESSURES

The engineer is more interested in the relationship between amount of swelling and pressures against which the swelling occurs, than in the limited significance of the water content at the shrinkage limit. Some idea of this relationship can be obtained from the swelling curves obtained from consolidation tests. From the $e-\log p$ curve for the Pauls Valley clay in Section 9.20L in the chapter on Laboratory Testing, the swelling after release of pressure from consolidation under 23.63 tons ft^{-2} is found to be about $\dfrac{0.555 - 0.521}{1.521} = 2.2\%$ under 5 tons ft^{-2}, and $\dfrac{0.608 - 0.521}{1.521} = 5.7\%$ under 1 ton ft^{-2}, and $\dfrac{0.687 - 0.521}{1.521} = 10.8\%$ under zero pressure.

Another method used to study the relationship between swelling and pressure consists in fitting a specimen of the clay with a water content well below the shrinkage limit in a consolidation test arrangement. A predetermined pressure p_1 is applied to the dry specimen, after which the specimen is flooded and allowed to swell against the applied pressure. After swelling is complete, the specimen is subjected to the standard consolidation test. An $e-\log p$ curve is obtained as shown in Figure 10.05a.

Theoretically, the pressure which reduces the void ratio of the saturated (swelled) clay to that of the dry soil is the pressure that will prevent swelling and is equal to the pressure created by the capillary water. The relationship between swelling and pressure can be obtained from a series of these tests on different specimens of a dry sample of clay swelled under different pressures.

The greatest difficulty in making these tests is fitting the dried out clay into the consolidation ring. Sometimes, it is necessary to fit the clay into the ring while moist and let it dry out in the ring. Shrinkage reduces the size of the specimen until it no longer fits the ring. Filling this space so that the specimen is confined during swelling under the applied load has been done fairly successfully by wrapping the specimen with metal foil and filling the space between the foil and the ring

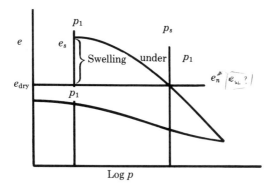

Figure 10.05a
e-Log p Curve

with leadite. Also adjustable rings have been used. In order to prevent excessive loss of soil by capillary action during flooding, the test must be set up for the fixed ring arrangement.

REFERENCES

Conference, *Theoretical and Practical Treatment of Expansive Soils*, Golden, Colorado: Quarterly of the Colorado School of Mines, October, 1959, Vol. 54, No. 4. Papers by R. E. Means, T. William Lambe and Robert V. Whitman, Raymond Dawson, W. G. Holtz, Chester McDowell.

Seed, H. B., J. K. Mitchell, and C. K. Chan, "Studies of Swell and Swell Pressure Characteristics," *Highway Research Board Bulletin No.* 313, 1962.

Seed, H. B., Richard J. Woodward, Jr., and Raymond Lundgren, "Prediction of Swelling Potential for Compacted Clays," *Journal Soil Mechanics and Foundations Division*, Proc. Am. Soc. C. E., Vol. 88, No. SM3, (June, 1962).

Lambe, T. William, *Soil PVC Meter—The Character and Identification of Expansive Soils*, Washington, D.C.: Technical Studies Report FHA-701, Federal Housing Administration, December, 1960.

PROBLEMS

10.1 Using an appropriate sketch, give a detailed description of the process of shrinkage of clay.

10.2 Why does a clay shrink when exposed to drying? Why does shrinkage stop when the color of the clay changes from dark to light?

10.3 Assume that the water in the voids of a soil sample has been replaced by another liquid having the same specific gravity as water but having a greater surface tension. *Explain* the effect (if any) which this substitution would have on the shrinkage limit of the soil.

10.4 The weight and volume of a clay specimen after oven drying were measured, and the specific gravity of the grains was determined. Explain clearly, with the aid of a sketch, how the shrinkage limit of the clay can be determined from this information and derive the formula for the shrinkage limit in terms of known quantities.

10.5 The volume and weight of a saturated clay specimen are measured once when its water content is about equal to the liquid limit and again after the specimen has dried out. With the aid of a suitable sketch, derive expressions for the shrinkage limit of the clay and the specific gravity of the solid matter in terms of the known quantities.

10.6 When remolded at a consistency roughly corresponding to that at the liquid limit, 40.2 g of soil were required to fill a 25 cc mold. After thorough drying, the soil pat weighed 24.0 g and had a volume of 12.8 cc. Determine the shrinkage limit of this soil.

10.7 An undisturbed block of saturated clay has an initial volume of 25 cc and weighs 45.5 g. After drying, the volume is 15 cc and the weight 32 g. Determine the *shrinkage limit* of the clay and the *specific gravity* of solids.

10.8 A saturated sample of undisturbed inorganic clay occupies a volume of 17.2 cc and weighs 29.5 g. Oven drying at about 105° C reduces the volume of the sample to 10.45 cc and the weight to 19.5 g. For this soil compute
(*a*) the initial void ratio,
(*b*) the average specific gravity of the mineral particles, and
(*c*) the shrinkage limit.

10.9 An oven-dried clay specimen weighs 103.6 g and has a volume of 52.9 cc. The specific gravity of the soil particles is 2.74. Determine the shrinkage limit, the unit weight at the shrinkage limit, and the dry unit weight of the clay.

10.10 A certain highly plastic clay has a dry unit weight (accurately determined from an air-dried specimen of the soil) of 135 lb/cu ft. Determine the shrinkage limit of this clay. (Estimate $G_s = 2.75$).

10.11 An oven-dried specimen of inorganic clay weighs 19.5 g and has a volume of 10.45 cc. Based on reasonable assumptions, compute the shrinkage limit of this soil.

10.12 An oven-dried specimen of clay, weighing 273.6 g, was found to have a volume of 116.3 cc. The soil possessed a very high dry strength. Make an approximate determination of the shrinkage limit of this soil.

10.13 A specimen of dried clay weighs 136.8 g and has a volume of 65.2 cc. The dried soil has only a moderately high strength. Determine an approximate value for the shrinkage limit.

10.14 A pat of very fat clay, when dried, has a volume of 25 cc and weighs 56 g. Estimate the shrinkage limit of this clay.

CHAPTER XI

Stress–Deformation and Strength Characteristics

11.01 GENERAL

The shear strength is the most difficult to comprehend and one of the most important of the soil characteristics. Shear strength *per se* is not difficult to understand; however, for soils the relationships become vastly complicated by many factors extraneous to the soil itself and by unavoidable differences which exist between the laboratory specimen and the soil in the ground. These remarks are not made with the intention of creating apprehension and misgivings in the student just beginning a study of soil mechanics. However, in the interest of clarity, an understanding of the shear strength of soils must be developed gradually, beginning in the simplest possible manner. The student, therefore, should not be surprised to learn, as the chapter unfolds, that some of the concepts first presented are of limited practical value. Yet, they are fundamental to a basic understanding of soil behavior.

The stress-deformation and strength characteristics of soils, along with the compressibility discussed in Chapter IX, provide the information most frequently needed in practical problems of soil mechanics. The methods used to determine these properties in the laboratory must be understood in detail in order to permit the intelligent application of laboratory results to field conditions. The soils engineer who attempts to get along without this understanding is like unto the blind man whose seeing eye dog has departed taking with him the man's cane.

11.02   STATES OF STRESS

A brief review of the relationships existing among the stresses acting on various planes which pass through a point in a stressed body will serve as an introduction to the basic mechanics to be used in this chapter. The relationships given apply equally well to steel, concrete, wood, soil, or any other substance because they are independent of the physical properties of the material.

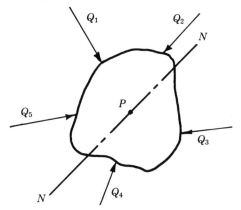

Figure 11.02a
Stressed Body

The state of stress at point P in a body acted on by a system of forces Q may be described by reference to the stresses on one or more planes N, which pass through the point P (see Figure 11.02a). In general, it will be found that the resultant stress on any plane has an obliquity θ. The resultant R may be resolved into components which act along the plane and perpendicular to the plane as shown in Figure 11.02b. The component σ perpendicular to the plane is called the normal stress, and the component τ or s acting tangentially to the plane is called the shearing stress.

If plane N, passing through the stressed body, is rotated in space, it will be discovered that there are three mutually perpendicular planes on which the shearing stress τ is zero. The normal stress acting on one of these planes is algebraically the largest normal stress acting on any plane. On another of the planes the normal stress is algebraically the smallest, while on the third plane the normal stress has a value intermediate between the stresses acting on the other two planes. The three planes on which τ is zero are called principal planes and the normal stresses acting on these planes are referred to as principal stresses.

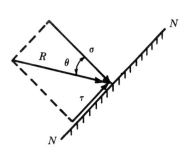

Figure 11.02b
Stress on a Plane

The principal stresses are denoted σ_1, σ_2, and σ_3, and are called, respectively: major, intermediate, and minor principal stresses.

In soil mechanics it is generally sufficient to consider only a plane state of stress because of a primary concern with the maximum shearing stress which acts on any plane. This maximum shearing stress occurs on a plane that is perpendicular to the plane of the intermediate stress σ_2. It was stated above that the major and minor principal stresses also act on planes perpendicular to the plane of the intermediate principal stress. In Figure 11.02c, if the plane of this sheet represents the plane on which σ_2 acts, lines drawn on the sheet may be used to represent planes perpendicular to the sheet and to the plane of σ_2.

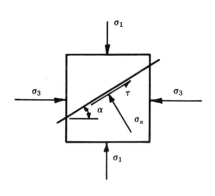

Figure 11.02c
Planar Stress

In the figure, which represents a very small prism of the material, horizontal lines represent planes on which the major principal stress acts, and vertical lines represent planes on which the minor principal stress acts. Although the orientation chosen is arbitrary, the principal planes must be mutually perpendicular. If σ_1 and σ_3 are known, the stresses acting on any other plane (such as the plane shown in the sketch which makes an angle α with the plane of the major principal stress) can be computed.   Conversely, if the stresses acting on any two planes are known, the principal stresses may be computed and the angle between these planes and the planes of the principal stresses σ_1 and σ_3 may be determined.

In ordinary mechanics it is customary to designate σ_1 as the largest tensile stress and σ_3 as the smallest tensile (or largest compressive) stress.   However, in soil mechanics the opposite convention is customarily used because of the nature of the stresses commonly imposed on soils.   Thus, the greatest compressive stress is σ_1 and the smallest compressive stress is σ_3.

Now, consider the equilibrium of the infinitesimal triangular prism constituting the upper left hand portion of the element in Figure 11.02c as shown in Figure 11.02d.   Solution of the two equations of equilibrium, $\Sigma F_V = 0$ and $\Sigma F_H = 0$, taking into consideration the relative areas of the surfaces in the x, y, and z planes and assuming a thickness perpendicular to the paper as unity, yields the equations

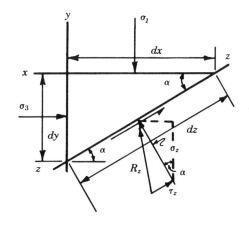

Figure 11.02d
Equilibrium Prism

$$\sigma_3 \, dy = \sigma_z \, dz \sin \alpha - \tau_z \, dz \cos \alpha$$

and $\qquad \sigma_1 \, dx = \sigma_z \, dz \cos \alpha + \tau_z \, dz \sin \alpha.$

Inserting values of dx and dy in terms of dz and solving these equations simultaneously yields the relationships

$$\sigma_z = \frac{\sigma_1 + \sigma_3}{2} + \frac{\sigma_1 - \sigma_3}{2} \cos 2\alpha. \qquad \text{(Eq. 11.02a)}$$

$$\tau_z = \frac{\sigma_1 - \sigma_3}{2} \sin 2\alpha. \qquad \text{(Eq. 11.02b)}$$

It is important to remember that these equations are the result of the application of mechanics, having nothing to do with the properties of the material. The theory of elasticity deals with materials having particular properties.

11.03 MOHR'S CIRCLE OF STRESS

Equations 11.02a and 11.02b are the parametric equations of a circle. In 1895 Otto Mohr of Germany suggested the use of this circle for the graphic representation of the state of stress. The use of this simple but powerful tool has proved to be a major convenience in stress

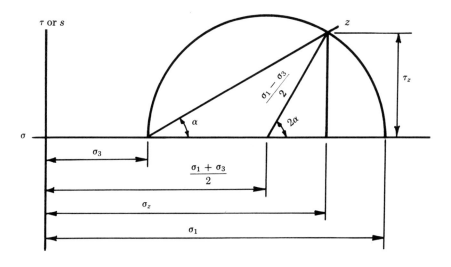

Figure 11.03a
Mohr's Circle of Stress

analysis. A complete understanding of the properties of Mohr's circle of stress is a necessity for the soils engineer. Exactly the same relationship exists for moments and products of inertia as for normal and shearing stresses.

As shown in Figure 11.03a, if a vertical axis along which shearing stress τ or s is measured and a horizontal axis along which normal stress σ is measured are drawn, and σ_3 and σ_1 are laid off from the origin on the σ axis, and a circle drawn with its center midway between σ_1 and σ_3 on the σ axis and a radius of $\dfrac{\sigma_1 - \sigma_3}{2}$; Equations 11.02a and 11.02b are satisfied for every point on the circle thus drawn. In

Figure 11.03a, only one-half the circle is drawn.   If a line representing plane z making an angle α with the σ axis is drawn from the σ_3 intersection of the circle with the σ axis, a line drawn parallel to the τ axis from the intersection of the z plane and the circle to the σ axis represents the value of the shear stress τ_z on the z plane.   From the sketch

$$\tau_z = \frac{\sigma_1 - \sigma_3}{2} \sin 2\alpha. \qquad (11.02b)$$

The distance from the origin along the σ axis to the intersection of the

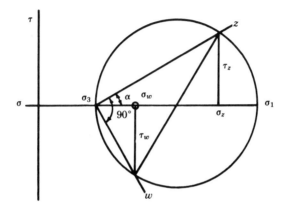

Figure 11.03b
Stresses on Perpendicular Planes

τ line with the σ axis represents the value of the normal stress σ_z on plane z.   From the sketch

$$\sigma_z = \frac{\sigma_1 + \sigma_3}{2} + \frac{\sigma_1 - \sigma_3}{2} \cos 2\alpha. \qquad (\text{Eq. } 11.02a)$$

Note that when 2α exceeds $90°$, $\cos 2\alpha$ is negative and the second term of Equation 11.02a becomes negative.   If the angle α is constructed at the point σ_3, 0 on the circle, as shown, Mohr's circle of stress is a graphical solution of the equations of equilibrium.   The coordinates of any point on the circle represent the stresses on some plane inclined at an angle α to the plane of the major principal stress.

Figure 11.03b shows that certain requirements concerning the shearing stresses are fulfilled.   For example, the shearing stresses τ are seen to be zero on the principal planes.   Moreover, the shearing stresses on any two perpendicular planes at α and $\alpha - 90°$ are equal and of opposite algebraic sign as required by consideration of equilibrium.   It is also apparent that the maximum shearing stress, which is equal to the radius of the circle, must occur on two planes which are inclined at 45° to the principal planes.

A state of pure shear is defined as the state of stress occurring on any plane which is subjected to shearing stresses in the absence of normal stress. In soil testing a state of pure shear is difficult to achieve because of the relatively low tensile strength of soils. The only practical way of creating pure shear in the laboratory is illustrated in

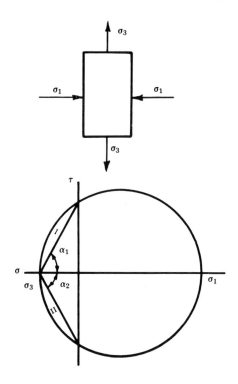

Figure 11.03c
State of Pure Shear

Figure 11.03c. An element of material subjected to tensile stress in one direction and to compressive stress in the other is shown, together with the corresponding Mohr's circle of stress. For this state of stress there are two planes on which a state of pure shear exists. These are the planes I and II corresponding to the two points on the circle for which the abscissa σ is zero. If the major and minor principal stresses are equal in magnitude but of opposite sign, a state of pure shear exists on the planes of maximum shearing stress; i.e., α_1 and α_2 are 45°.

It will be shown later that the plane of failure does not necessarily coincide with the plane of maximum shearing stress. The location of the critical (failure) plane depends upon the properties of the material. In general, failure occurs as a result of the combination of normal and shearing stresses in a specific manner.

In the case of substances which are incapable of resisting any shearing stress, the principal stresses must always be equal. For example, the state of stress existing at depth H in a body of water is represented by $\sigma_1 = \sigma_2 = \sigma_3 = \gamma_w H$. The Mohr circle is therefore a point on the σ axis, situated at a distance $H\gamma_w$ to the right of the origin. This condition is known as the hydrostatic state of stress.

11.04 SIGN CONVENTION

In books on mechanics, various rules are given for designating the nature or the direction of stresses and for the orientation of planes. A particular sign convention for shearing stresses must be adhered to if such rules are to be applied successfully. Some persons consider tension as a positive stress while others use compression as positive. In soil mechanics compression is commonly considered as a positive stress and is plotted on the positive side of the τ axis in the Mohr's circle analysis. Shear is commonly considered as positive when the external shearing force is downward on the right of a prism and upward on the left. A disadvantage of this rule is that the sign changes when the prism is viewed from the opposite direction.

Rules for determining the orientation of planes and the direction of stresses in the Mohr's circle analysis are complicated and hard to remember. Unless one has a faultless memory, arbitrary rules for this purpose are practically useless. The general manner of orientation of a specific plane of stress may be determined by application of common sense. The specific angle of orientation may be found by using trigonometry in conjunction with a crude sketch of the Mohr circle. The algebraic sign of the shearing stress is, in this case, of little consequence.

Consider, for example, the state of stress illustrated in Figure 11.04a(1). In this example σ_y is assumed to be greater than σ_x. τ_{xy} and τ_{yx} are equal and of opposite sense of rotation. Suppose that it is desired to determine the orientation of the plane of the major principal stress σ_1. In this case, the shearing stresses tend to deform the element in such a way that the diagonal AC shortens while the diagonal BD lengthens. AC is, therefore, referred to as the compression diagonal

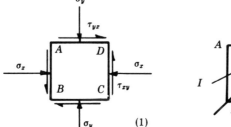

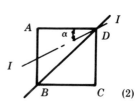

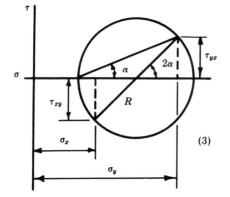

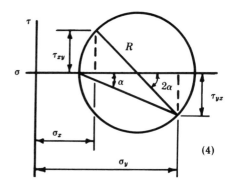

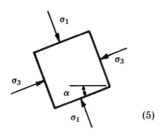

Figure 11.04a
Orientation of Planes

and BD as the tension diagonal. The plane of the major principal stress must be so oriented that it lies somewhere between a plane perpendicular to the compression diagonal and one perpendicular to the direction of the larger normal stress σ_y. Plane I-I fulfills this condition. Any plane lying outside the angle ADB is influenced more by the smaller stress σ_x than by σ_y and could not possibly be the plane of the major principal stress. A similar plane lying inside the angle DAC is relatively influenced more by the tension diagonal and, again, could not be the plane of the major principal stress.

Thus, the angle has been restricted to some angle smaller than 45°. Figures 11.04a(3) and (4) illustrate that there is only one absolute value possible for this angle and that it does not matter what sign convention is used for the shear stress.   The magnitude of α may now be readily determined as

$$\alpha = \frac{1}{2}\left[\tan^{-1}\frac{\tau}{\dfrac{\sigma_y - \sigma_x}{2}}\right]$$
(Eq. 11.04a)

which is always less than 45°. The magnitude of the principal stresses may also be easily found because the radius R of the circle is given by the expression

$$R = \frac{\sigma_y - \sigma_x}{2\cos 2\alpha} = \frac{\sigma_y - \sigma_x}{2} \leftrightarrow \tau$$
(Eq. 11.04b)

and
$$\sigma_1 = \frac{\sigma_y + \sigma_x}{2} + R$$
(Eq. 11.04c)

and
$$\sigma_3 = \frac{\sigma_y + \sigma_x}{2} - R.$$
(Eq. 11.04d)

The state of stress may now be fully represented by the reoriented element of Figure 11.04a(5).   Similar processes may be used for different initial conditions, such as, when the principal stresses are known and it is desired to find the normal and shearing stresses acting on some other plane.

Another condition for equilibrium that may be of assistance in the orientation of planes and stresses is that the sum of the normal stresses on any set of perpendicular planes through a point in a stressed body is a constant for any given state of stress; i.e., $\sigma_y + \sigma_x = \sigma_1 + \sigma_3$. This fact is the basis of a circle of stress and circle of inertia method of analysis.   In this method the diameter of the circle of stress is equal to the sum of the normal stresses on perpendicular planes and the origin of the coordinate system lies on the circle.   This method using the sum of the normal stresses has certain advantages over the Mohr's circle method, except that it does not combine the state of stress with properties of materials as does the Mohr's circle method.

The procedure indicated in the preceding discussion is recommended as fostering an insight and feeling for stress relationships.   A clear understanding and picture of the orientation of planes and stresses and of the nature of the stresses is as important as a knowledge of the magnitude of the stresses.   Often it is more important to know how than how much.

Strictly graphical solutions using Mohr's circle are sometimes of considerable convenience in the solution of soil mechanics problems. Neither the graphical nor the algebraic method of analysis possesses an advantage over the other with regard to accuracy.   The accuracy of algebraic methods depends upon the number of significant figures used in the computation and that of graphical methods depends upon the scale and care in making measurements.   Graphical methods possess the advantage of presenting a visual picture of existing conditions.

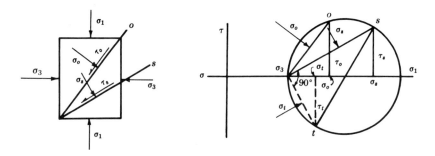

Figure 11.04b
Orientation of Planes and Stresses
Principal Stresses Known

When they are employed for the solution of problems involving the state of stress, it is necessary to adopt a rule of signs for the shearing stress as well as for the normal stresses.   In these cases it is sometimes convenient to employ an arbitrary rule for orientation of planes and for determining the stresses acting on the planes.   Such a rule for the application of Mohr's circle of stress follows.

When principal stresses and their planes are known and the stresses on other planes are desired, the orientation is inherent in the proof that the circle of stress is a graphical statement of the state of stress.   In this case, if the σ and τ axes are drawn parallel to the planes of the principal stresses and the circle of stress drawn, lines drawn through the circle from the σ_3 intersection with the σ axis parallel to the plane on which the unknown stresses are desired, determine the stresses acting on this plane as shown in Figure 11.04b.   The stresses acting on the perpendicular plane are also determined as shown by plane t which is perpendicular to plane s.   Compressive stresses are shown to the right of the origin and tensile stresses to the left.   The rule applies when the major and minor principal stresses are reversed or if either or both are tension. The direction of shearing stresses can be seen by inspection.

If the normal and shearing stresses on any set of perpendicular planes through a point in a stressed body are known, the principal stresses and the planes on which they act can be determined as follows:

Draw the σ and τ axes parallel to the planes x and y of the known stresses. Lay off the normal stresses on the σ axis. Lay off from σ_y

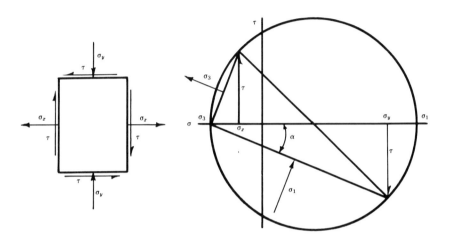

Figure 11.04c
Orientation of Planes and Stresses
Principal Stresses Not Known

the shear stress in the direction of the external shear on the right and from σ_x the shear stress in the direction of the shear on the left of the prism. From the extremities of these two shear lines draw a diameter of the circle and construct a circle through these extremities. The magnitude and nature of the principal stresses are now known. A line drawn from σ_3, 0, to the circle at the extremity of the shear line from σ_y is parallel to the plane on which the major principal stress acts and a 90° line to the circle and the shear line from σ_x is parallel to the plane of the minor principal stress. If the shear acts in the opposite direction, the value of α is the same but the plane of σ_1 will lie on the opposite side of the σ axis. The rule applies whether σ_y is larger or smaller than σ_x.

11.05   COULOMB'S LAW

In the second half of the 18th century C. A. Coulomb, a French military engineer and physicist, carried out tests on friction and in 1776 published an important paper on the subject.   Coulomb concluded that shearing resistance (i.e., that particular shear stress referred to as shear strength) follows an empirical law.   His hypothesis was primarily based on tests made on sand because he was interested in determining the pressures acting against retaining walls built in connection with the construction of military fortifications.   Although Coulomb recognized that there is a fundamental difference in the behavior of cohesionless and cohesive soils, he considered that the relationship between shear strength and normal stress could, in either case, be represented as a straight line.   The general relationship postulated by Coulomb is

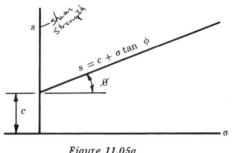

Figure 11.05a
Coulomb's Law

$$s = c + \sigma \tan \phi \qquad \text{(Eq. 11.05a)}$$

in which s is the shear stress on the plane of failure at failure.   The empirical constants, c and ϕ, represent, respectively, the s axis intercept and the slope of the straight line of Equation 11.05a, as shown in Figure 11.05a.

Through the years Coulomb's equation has formed the basis for the solution of various problems in soil mechanics, and it remains one of the most important tools.   It is, perhaps, unfortunate that the empirical constants, c and ϕ, have come to be called *the* cohesion and *the* angle of internal friction.   These appellations have undoubtedly caused many engineers to suppose that the constants are used to express the invarient properties of a particular soil.   In actuality, they are mathematical constants obtained empirically under a given set of conditions and may be wholly unrelated to the physical properties of the soil.   Failure to recognize this fact may lead to serious blunders in the use of Equation 11.05a for the solution of practical problems.   This is not the fault of the equation.   The terms cohesion and angle of internal friction have been established by long continued usage, and no attempt will be made to avoid their use in subsequent paragraphs. Their real significance, however, must be kept in mind.

The general validity of Equation 11.05a may be demonstrated by a simple experiment with clean sand. Because clean sand has no cohesion ($c = 0$), its entire resistance to shear depends upon the pressure acting between the grains. A long thin rubber tube filled with sand has practically no resistance to deformation. If the air is evacuated from the inside of the tube so that the sand grains are forced together by the pressure of the atmosphere on the outside of the tube, the sand offers a substantial resistance to deformation. The same resistance to deformation may be mobilized by means of fluid pressure applied externally.

11.06 THEORIES OF FAILURE

The failure of materials is by no means a simple phenomenon. The basic causes of failure are still imperfectly understood in spite of the many theories that have been advanced to explain it from a fundamental standpoint. Some of the theories seem to explain quite well the failure of certain materials but seem to be inapplicable to others. Fortunately, in the field of practical soil mechanics a simple theory of failure has, so far, satisfied needs. This theory was first expressed by Coulomb and later generalized by Mohr. It is concerned with failure by shear.

In soils, shear failures are produced by compression, and not by tension as is the case with some other materials. For example, a tensile test of mild steel produces a localized necking down which is accompanied by the formation of Lüder lines indicating shear displacement in the region of failure. The failure surface takes the form of a truncated cone whose surface slopes at about 45° with the plane of the applied tension, and on which the shear is maximum. On the other hand, a tension test of a soil specimen never produces a shear failure, but only a failure in tension. Even an unconfined compression test is likely to result in bulging of the specimen accompanied by the opening of tension cracks parallel to its longitudinal axis. This tension failure produced by such a test cannot be considered representative, because the soil in the ground is laterally confined and unable to fail in this manner. So, although tension failures sometimes occur in the laboratory, the reliable strength of a soil is its shear strength.

The difference between Coulomb's and Mohr's concepts of the failure condition is simply represented by the sketches in Figure 11.06a. Coulomb considered that the relationship between shear strength and normal stress could be adequately represented by a straight line.   The generalized Mohr theory also recognizes that the shear strength depends on the normal stress, but indicates that the relation is not linear.   The Mohr strength line, also called rupture line, or envelope, also is valid in a region of tension and, as shown by the dashed circle, still represents a shear failure if σ_3 is a relatively low tensile stress.

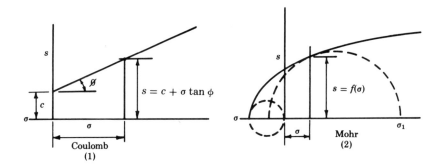

Figure 11.06a
Failure Hypotheses

However, at the extreme left the strength line is tangent to the vertical, representing a tensile failure on a plane of principal stress. The full circle corresponds to failure during a uniaxial tensile test.

The strength line in Figure 11.06a(2) roughly corresponds to that obtained from tests of plain concrete.   The tensile strength is only a small fraction of the compressive strength, while the compressive strength is noticeably dependent upon the confining pressure σ. Where the strength line crosses the s axis, it has, for concrete, a slope of about 45°.   Sometimes, because of improper capping of test cylinders or because of the wedging action of large stones near the ends, a compression test of concrete results in splitting of the specimen.   Because this splitting corresponds to a failure in tension, the strengths in these cases are lower than those found for specimens which show a definite shear failure.

In the usual triaxial test or unconfined compression test, σ_1 acts in a vertical direction while σ_2 and σ_3 act horizontally.   In both tests $\sigma_2 = \sigma_3$; in the unconfined compression test they are zero.   These failure hypotheses indicate that the compressive strength as controlled by shearing resistance depends directly upon the normal stress.

Therefore, it is reasonable to expect to find differences in the compressive strengths of specimens tested under varying values of the intermediate principal stress σ_2 if the minor principal stress σ_3 is held constant.

It is difficult to provide a truly intermediate value of σ_2 in the soils laboratory. However, it is possible to perform the triaxial test in such a way that $\sigma_2 = \sigma_1$. A test of this type is called an axial extension test and is described below. The results of an axial extension test, when combined with those of the usual triaxial compression test, provide information about the soil behavior under the two extreme values possible for σ_2.

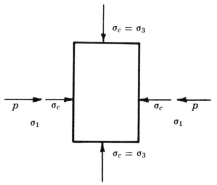

Figure 11.06b
Axial Extension Test

In an axial extension test the specimen is subjected to an all around confining pressure σ_c, as indicated in Figure 11.06b. If the confining pressure (usually supplied by liquid) is increased by an increment p while holding the vertical stress constant, both the major principal stress and the intermediate principal stress equal $\sigma_c + p$, while the minor principal stress remains constant at σ_c. It is apparent that the average pressures between the soil grains are likely to be somewhat higher in this test than those that are produced in a triaxial test of the usual type. It is conceivable that the compressive strength would be higher when $\sigma_2 = \sigma_1$ than it would be for $\sigma_2 = \sigma_3$, and has been shown to be

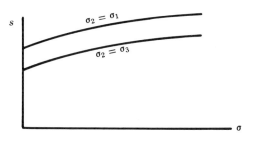

Figure 11.06c
Effect of Intermediate Principal Stress

true for some materials. The resulting strength envelopes for these two conditions of loading have somewhat the relationship shown in Figure 11.06c.

Except for certain types of tests on fine grained soils, the variation in strength due to changes in the intermediate principal stress is, for soils, small. Since the ordinary triaxial tests ($\sigma_2 = \sigma_3$) produce results which are on the side of safety, the effect of the intermediate principal stress may, for practical purposes, be ignored when dealing with coarse grained soils.

There is no doubt that for precise investigations of materials having closely controlled properties a theory of failure of the general form $s = f(\sigma_1,\sigma_2,\sigma_3)$ is necessary. There are a number of such theories, one of the best known being that of R. von Mises. These have no proved importance in connection with soils. The simple relationship between s and σ as given by the Mohr envelope seems to be satisfactory for soils.

11.07   THE MOHR ENVELOPE

It was indicated in the preceding article that the envelope of a series of Mohr's circles for states of stress at failure in compression tests made under varying conditions of confinement is, in general, a curved line. Such a line is often referred to as the strength line. It is customary in soil mechanics to use the terms strength line, rupture line, and strength envelope interchangeably. For some materials, and frequently for small ranges of confining pressures, the envelope may for practical purposes be approximated as a straight line.

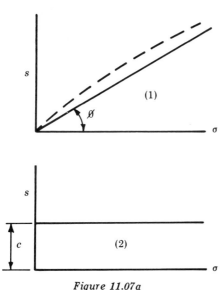

Figure 11.07a
Strength Lines

The strength line for an ideal pure friction material is illustrated by the full line in Figure 11.07a(1). This line passes through the origin and has the constant slope ϕ. It has been repeatedly demonstrated that, for practical purposes, clean sand behaves in this way. The relationship is closely approximated by loose sands tested under normal ranges of pressure. Under high pressures the relationship is affected by crushing of the grains. Dense sands exhibit a slightly curved strength line, indicated by the dashed line in Figure

11.07a(1). For working purposes it is generally taken as a straight line having a slope which best corresponds to that of the actual envelope over the particular range of pressure involved.

Figure 11.07a(2) represents purely cohesive (plastic) material, for which the strength line is parallel to the σ axis. The strength of such a material is independent of the normal stress acting on the plane of failure. Steel is a material which approximates this relationship. Even for steel, however, the strength line has a slope of about 1.5° in the compression range and curves downward sharply in the tension range. The angle which the plane of failure makes with the plane of the major principal stress is for steel only slightly larger than 45°. For clay soils, the strength line may be almost anything between the two extremes indicated in Figure 11.07a. For some test conditions ϕ may be zero and it appears that the strength is independent of the normal stress. When tested in a different manner, the same clay may have a strength line identical with that of the ideal frictional material. Suitable variations of the test procedure produce other strength lines, indicating the existence of both cohesion and internal friction. All this does not necessarily imply that clays do not possess definite values for what might be termed the true cohesion and the true angle of internal friction. Cohesion and angle of internal friction are discussed at some length later. However, the difficulty of confidently determining these physical constants is at once apparent. Fortunately, the solution of practical problems in soil mechanics does not depend upon a knowledge of these constants, but rather upon a knowledge of the strength due to the combined effects of cohesion and friction.

The strength theory upon which the Coulomb and Mohr strength lines are based indicates that definite relationships exist among the principal stresses, the angle of internal friction, and the inclination of the failure plane. Figure 11.07b shows the relationship between a particular strength line and one of the infinite number of corresponding Mohr strength circles. The point of tangency P must represent a state of stress corresponding to the failure conditions. Then, α_f is the angle, in the test specimen, between the plane of the major principal stress and the plane of failure. Since the angle PCO is the supplement of the angle $2\alpha_f$, by equating the value of this angle in terms of ϕ from triangle PCO to its value in terms of α_f from triangle PCB, it is found that

$$\alpha_f = 45° + \frac{\phi}{2} \qquad \text{(Eq. 11.07a)}$$

which indicates the fixed relationship between the location of the failure

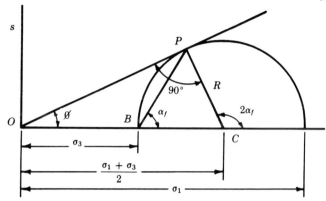

Figure 11.07b
Properties of the Strength Line

plane and the angle of internal friction of the material. If the full
Mohr circle is drawn, together with the strength line corresponding to
failure under shearing stresses of the opposite sense to those represented by the line OP, the diagram shows that two sets of failure planes
exist. The planes of
one set intersect those
of the other set at an
angle of $90° + \phi$, both
sets lying at an angle of
$45 + \dfrac{\phi}{2}$ to the plane of
the major principal
stress and at $45 - \dfrac{\phi}{2}$ to
the plane of the minor
principal stress. This
relationship exists even
if part of the resistance
is due to cohesion. For
purely cohesive mate-
rials ϕ is zero and the

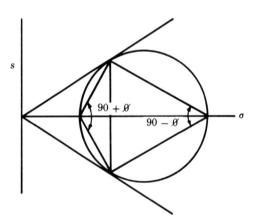

Figure 11.07c
Relationship of Planes of Failure

failure surfaces are inclined at 45°. If the strength line is a curve,
Equation 11.07a still applies provided that ϕ is taken as the slope of
the curve at its point of tangency to the strength circle.

An additional important relationship can be derived from Figure
11.07b. From the figure it can be seen that

$$R = \frac{\sigma_1 + \sigma_3}{2} \sin \phi = \frac{\sigma_1 - \sigma_3}{2}$$

from which

$$\sigma_1 = \sigma_3 \frac{1 + \sin \phi}{1 - \sin \phi}$$

and

$$\sigma_3 = \sigma_1 \frac{1 - \sin \phi}{1 + \sin \phi}. \qquad \text{(Eq. 11.07}b\text{)}$$

If use is made of trigonometric identities for the right hand terms, Equations 11.07b can be stated in the following convenient form:

$$\sigma_1 = \sigma_3 \tan^2 \left(45 + \frac{\phi}{2}\right)^\circ$$

$$\sigma_3 = \sigma_1 \tan^2 \left(45 - \frac{\phi}{2}\right)^\circ. \qquad \text{(Eq. 11.07}c\text{)}$$

The preceding equations apply only in the case of cohesionless soils.

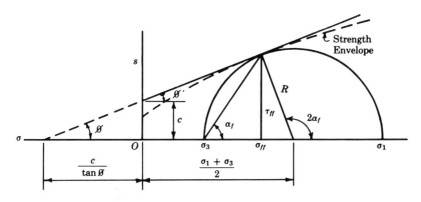

Figure 11.07d
General Strength Line Relationships

The derivation of similar equations for cohesive soils is only slightly more complicated. The geometry is shown in Figure 11.07d.

$$R = \left[\frac{c}{\tan \phi} + \frac{\sigma_1 + \sigma_3}{2}\right] \sin \phi = \frac{\sigma_1 - \sigma_3}{2}$$

from which

$$\sigma_1 = 2c \frac{\cos \phi}{1 - \sin \phi} + \sigma_3 \frac{1 + \sin \phi}{1 - \sin \phi}$$

and   (Eq. 11.07d)

$$\sigma_3 = -2c \frac{\cos \phi}{1 + \sin \phi} + \sigma_1 \frac{1 - \sin \phi}{1 + \sin \phi}.$$

Applying trigonometric identities these equations assume the following convenient form.

$$\sigma_1 = 2c \tan\left(45 + \frac{\phi}{2}\right) + \sigma_3 \tan^2\left(45 + \frac{\phi}{2}\right)$$

$$\sigma_3 = -2c \tan\left(45 - \frac{\phi}{2}\right) + \sigma_1 \tan^2\left(45 - \frac{\phi}{2}\right)$$

(Eq. 11.07e)

These equations apply to the general Coulomb Law $s = c + \sigma \tan \phi$ or to a tangent drawn from the circle of stress for failure under a particular state of stress where it meets the strength envelope of the soil. Obviously, one test cannot determine both c and ϕ. It can also be seen that the values of c and ϕ determined from two tests may depend upon the magnitudes of the confining pressures.

For cohesionless soils where $c = 0$, Equations 11.07e become identical with Equations 11.07c. It is interesting to note that for cohesionless soils having reasonably high friction angles, a small increase in σ_3 produces a relatively large increase in σ_1. This indicates that the strength of such soils can be substantially increased by only slight increase in confining pressure. It is made equally clear by the equa-

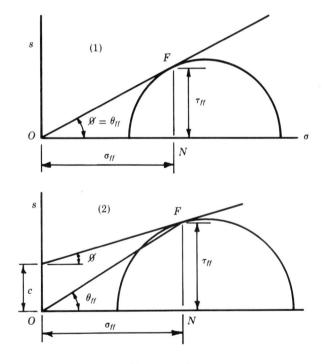

Figure 11.07e
Angle of Obliquity

tions that for purely cohesive materials σ_1 and σ_3 differ by a constant amount; i.e., the size of the strength circle is independent of the minor principal stress.   The compressive strength of a purely cohesive material is unaffected by changes of the confining pressure.   Figure 11.07a makes the preceding facts obvious.

The resultant stress R_{ff} on the plane of failure at failure will in general be inclined to a normal to the plane by an angle of obliquity θ as explained in article 11.02.   The angle of obliquity of the resultant stress on any plane is $\tan^{-1} \tau/\sigma$, when τ and σ are, respectively the shearing and normal stresses acting on the plane.   In Figure 11.07e point F on the Mohr circle represents a state of stress corresponding to the failure condition.   The coordinates of this point are σ_{ff} and τ_{ff} and the angle FON is the angle of obliquity.   It is readily seen that for cohesionless soils the maximum possible value for the angle of obliquity is the angle of internal friction ϕ.   This maximum value of the angle of obliquity for cohesionless soils occurs only on the plane of failure.

For cohesive soils as shown in Figure 11.07e(2) the angle of obliquity of the resultant stress on the failure plane is always greater than ϕ. The angle FON changes with variations of σ_{ff}.   Thus, for cohesive soils the angle of obliquity θ_{ff} may have any value between ϕ and $90°$. When $\sigma_{ff} = 0$, $\theta = 90°$ because the entire resistance to sliding is due to cohesion.   As σ_{ff} increases, the difference between θ_{ff} and ϕ approaches zero because an increasingly greater proportion of the total resistance is due to friction.

11.08   COMMON LABORATORY STRENGTH TESTS

The shear strength of soils can be determined in the laboratory in several different ways.   Of the many types of shear tests that have been devised, only a few are in general use.   Three of the most commonly used tests are described now in general terms, because a clear understanding of the principles of these tests will greatly facilitate the study of the shear strength of soils.   A detailed explanation of the laboratory equipment and procedures used for making some of these tests is given later.

a. The Direct Shear Test

One of the oldest of the shear tests, historically, is the direct shear test, probably so called because the stresses on the plane on which failure occurs are controlled directly throughout the test. This test was used extensively in the early development of modern soil mechanics. It is still used in some laboratories in routine investigations and, occasionally, even in research. Through the years the design of the direct shear apparatus has been improved in order to make the test more useful and reliable. A. Casagrande made notable contributions to the improvement of the test.

The direct shear apparatus consists essentially of a box divided horizontally into two sections. The lower section is attached to a fixed

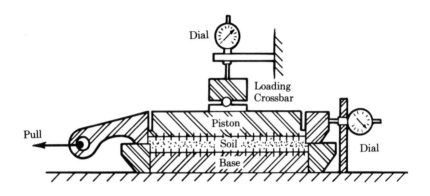

Figure 11.08a
Direct Shear Apparatus

frame, while the upper section is free to move both horizontally and vertically. The movements may be measured by means of dial gages. Figure 11.08a shows the apparatus schematically in cross section. A thin soil specimen is enclosed in the box, resting on a fixed base and covered by a piston which fits closely in the upper section of the box. Depending upon the nature of the soil being tested, either metal gratings or rough porous stones are embedded in the interior faces of the base and piston. The normal load on the failure plane is applied through the piston by means of a crossbar and loading yoke. The horizontal (shearing) force is applied to the upper section of the box, along the plane of separation between the upper and lower sections. As the upper section is displaced, the area of the shearing surface decreases slightly.

The state of stress on the failure plane is known at all times during the test. From a series of these tests, each made under a different magnitude of normal stress, σ_{ff} and the corresponding τ_{ff} are used to plot points which directly locate the strength line for the soil. The magnitude of the principal stresses and the location of the principal planes are not generally known during the test since both vary systematically as the test progresses. For the failure condition, this information can be quickly obtained by reference to the derived strength line and a corresponding strength circle.

The direct shear test is subject to some limitations which, from a practical standpoint, restrict its usefulness. As mentioned above, the prefailure stress conditions are not completely known, thus limiting the extent to which the data can be used in studying the stress-deformation characteristics of the soil. Also, flow of water to or from the soil specimen cannot easily be controlled or measured. The moisture content of the soil varies in an unknown manner during the test, necessitating the use of soil moisture data obtained before and after the test. Moreover, it is hardly practicable to measure pore pressures during the test. Finally, in spite of the gratings or rough porous stones used to transfer the shearing stress to the soil, it is practically impossible to obtain a uniform distribution of the shearing strain (and stress) along the failure plane. This latter defect may in sensitive soils result in a progressive failure which obscures the true nature of the soil.

It is probably fair to conclude that the direct shear test is really satisfactory only for investigating the strength of cohesionless soils, remolded clays, or quite insensitive clays.

b. The Triaxial Test

The strength test most commonly used in the research laboratory today is the triaxial compression test. As a result of the steady improvement of testing equipment and techniques, the test is now used in routine investigations as an aid to design. For a discussion of the equipment and procedures employed in one outstanding laboratory, the reader is referred to the excellent book on the subject by Bishop and Henkel. While the details of design and physical appearance of the equipment vary from laboratory to laboratory, the basic principles employed are shown schematically in Figure 11.08b.

The test utilizes a cylindrical soil specimen whose ratio of length to diameter is generally within the range of two to three. The test specimen is enclosed in an impermeable membrane which is an inch or so longer than the specimen. The specimen is then placed in position and the membrane securely bound to the base and cap. Next, the apparatus is assembled as shown in the figure and the pressure chamber filled with a liquid (usually water) through which hydrostatic pressure is applied to the specimen. Axial load is applied through the piston,

acted on by the loading crosshead.   Vertical movement of the crosshead is measured by means of a dial gage.   Drainage of water from
the pores of the soil may or may not be permitted during some phase of
the testing.   If drainage occurs, the quantity drained is measured in
the attached burette.   In actual practice, the equipment is frequently

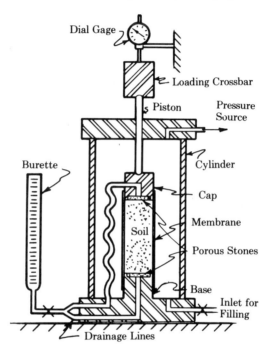

Figure 11.08b
Triaxial Test Apparatus

modified to permit separate control of the drainage from the two ends
of the specimen and to permit the measurement of pore pressure at
the ends.

The behavior of the specimen during testing is decisively influenced
by the condition of drainage permitted.   The manner of this influence
will be discussed in subsequent articles; but because of it, three basic
types of triaxial tests for soils have come to be recognized.   The adoption of these basic types was recommended by A. Casagrande, who was
the first to give a clear explanation of the factors which are responsible
for the different results obtained from the tests.   In general, the
differences are explained on the basis of the relationship between
applied pressure, effective stress, and pore water stress, as outlined in
Chapter IX.

Before defining the three basic tests, it should be pointed out that application of the all around (chamber) pressure and application of the axial load generally form two separate stages of the test. The stress caused by the axial load is commonly called the deviator stress since it is the difference between σ_1 and σ_3, or the diameter of the Mohr stress circle. The magnitude of the deviator stress at failure is referred to as the compressive strength. The tests are broadly classified according to the drainage conditions which exist during the two stages of the test. Descriptions of the three basic types of triaxial tests follow.

(1) Drained Test

This is a test in which drainage of the specimen is permitted during both stages of loading. Full consolidation first occurs under the applied chamber pressure. Following this initial consolidation, the deviator stress is applied and increased so slowly that no significant pore pressure is built up while the specimen is under test. These tests are sometimes called consolidated slow tests and are often designated S tests or Q_s tests.

(2) Consolidated-Undrained Test

In this test, complete consolidation is allowed under the applied chamber pressure before any axial load is applied through the piston. No drainage is permitted during application of the deviator stress, a circumstance generally giving rise to excess pore pressures. This test is sometimes called a consolidated quick test and has been variously referred to as a Q_c test, an R test, or a CU test.

(3) Undrained Test

This test has been called an unconsolidated quick test. In this test no drainage is permitted during either stage of the test. The specimen is brought to failure with no change in water content from its initial condition. Excess pore pressures commonly exist at all times during the test. The test is sometimes designated a Q test or a Q_q test.

Most of the shortcomings mentioned for the direct shear test are absent in the triaxial test. The state of stress is known at all times during the test; flow of water to or from the specimen can easily be measured and controlled, and pore pressures can be measured. Some non-uniformity of strain, particularly at large strains, is caused by lateral restraint at the ends of the specimen due to friction between the end faces and the cap and base. The effect of this restraint probably becomes insignificant within a rather short distance from the ends. If the test specimen is prepared so that it has an L/D ratio of two to three, the test results are not likely to be affected, because the failure plane does not generally touch the cap or base.

c. Unconfined Compression Test

The unconfined compression test, which is often designated a U/C test or sometimes a Q_u test, is one of the simplest and most widely used of the laboratory strength tests for cohesive soils. On account of the extensive correlation of the results of these tests with the actual strength of the soil in the ground, it is rather common to rely exclusively on the unconfined compression strength in the solution of practical problems involving the shear strength of the soil. This statement applies, however, only to intact clays. An unconfined compression test gives little or no reliable information concerning the strength of fissured clays in their undisturbed state in the ground.

Fundamentally, the unconfined compression test may be regarded as an undrained triaxial test in which the ambient pressure is zero. Since no pressure chamber is needed, the simplest sort of arrangement may be used for measuring the axial load and deformation. The specimen is normally brought to failure within a loading period of five to ten minutes. As in most strength tests, either the rate of loading or the rate of strain may be controlled.

The conditions of non-uniform strain discussed under the triaxial test also apply to the unconfined compression test. Change of moisture content due to evaporation or migration of pore water during the test is probably of little practical consequence if the total test time does not exceed ten minutes.

11.09   PROPERTIES OF COHESIONLESS SOILS

The simplest soils, from the standpoint of explaining and understanding their behavior under applied load, are the clean, granular materials having few particles smaller than about 0.06 mm. Clean sands are considered to be those which have no dry strength. This is likely to be the case so long as not more than about three per cent of the particles pass the No. 200 sieve. Unless otherwise stated, the soils discussed in this article are also assumed to be homogeneous, isotropic materials. A homogeneous material is one which exhibits identical properties at every point throughout its mass.   An isotropic material is one for which identical characteristics exist on every plane passing through any point in the mass.   Thus, it is easy to imagine the existence of a homogeneous, anisotropic material; but the converse is not true unless we refer to a microscopic element of a non-homogeneous mass.

a. Internal Friction

It has been previously indicated that the shear strength of cohesionless soils depends entirely upon the resistance developed by internal friction.   Therefore, it becomes necessary to inquire closely into the nature of the soil property called internal friction.

The frictional resistance to relative displacement of two plane surfaces in contact has been shown to be directly proportional to the normal pressure between the surfaces.   The maximum resistance that can be developed is given by the expression $s = f\sigma$, in which f is the coefficient of friction.   Experiments with many materials have shown that f depends upon the type of materials in contact and the surface characteristics of the contiguous planes.   In particular, it has been discovered that large differences exist between the coefficients of friction for clean and for contaminated surfaces.   In experiments using metals it has been found that truly clean surfaces are physical rarities, the most careful techniques being required to assure their existence. Surface molecules of the metal tend to snatch the moisture from the air instantly unless the tests are carried out in a vacuum.   The coefficient of friction is much higher when the moisture film is absent.   It may be concluded that the highest values of f correspond to those conditions under which mineral to mineral contact is established between the sliding surfaces.

More than thirty-five years ago K. Terzaghi investigated the frictional resistance of glass, using a loaded watchglass in contact with a glass plate.   He found that the angle of frictional resistance ϕ (note, $f = \tan \phi$) was about 12° under small loads, but that it jumped to about 45° under higher loads.   The pressure at which the transition took place was not well defined.   Subsequent microscopic examination of the glass revealed scratches and tiny globules of molten glass where sliding had occurred under heavy loads.   The surface was not marred where sliding had taken place under the light loads.   No special precautions had been taken that would have precluded the existence of thin films of surface moisture taken from the air.   It would appear that true mineral contact can be established even between contaminated surfaces if the normal stresses are sufficiently high.

It is possible that an illustrative natural example of the preceding described phenomenon is furnished by the St. Peter sandstone formation found in the central part of North America.   The St. Peter sandstone is an aggregate of rounded quartz grains of fairly uniform size.   It has only slight cohesion, as indicated by the fact that grains can be dislodged by blowing strongly on the surface.   Nevertheless, tests on this sandstone show a very steep curved strength envelope

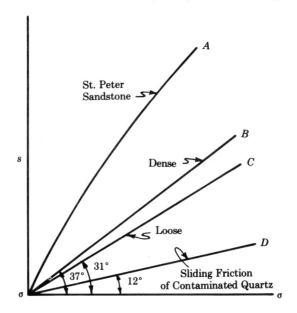

Figure 11.09a
Frictional Resistance of St. Peter Sandstone and Its Constituent Grains

represented by line A in Figure 11.09a.   Even at high normal pressures the envelope has a slope of 60°.   If the tiny cohesion of the sandstone is destroyed, and if tests are carried out on the resulting soil, the strength lines for the material in the dense and loose states are about as shown by lines B and C in the figure.   The void ratio corresponding to the dense state is about the same as that of the undisturbed sandstone. The large difference in the slopes of line A and lines B and C cannot be accounted for by a difference in cohesion, for in each case the lines begin essentially at the origin.   It seems logical to suppose that true mineral contact in the undisturbed sandstone is responsible for the higher angle of internal friction.   It has been hypothesized that this condition resulted from heavy overburden pressure which existed at some time during the geological history of the formation.

It is interesting to observe that the angle of internal friction for the dense sand is greater than that of the loose sand and that both are substantially greater than the angle of friction between contaminated quartz surfaces. The latter is represented by line D in Figure 11.09a on the basis of Terzaghi's experiments with glass. It seems that the angle of internal friction of granular materials depends upon something besides the frictional resistance to sliding on the surfaces of contact.

The resistance above that offered by friction must be the result of interlocking of the grains. The importance of the interlocking effect increases with decreasing void ratio, but it may be seen from the figure that a substantial part of the resistance of loose sand must be due to interlocking of the grains. Thus, it would seem appropriate to refer to ϕ as the angle of shearing resistance, had not the use of the term angle of internal friction become so firmly established.

It is not surprising that the effect of interlocking is large, for some grains must be lifted and rolled over others as sliding occurs along the failure planes. Since the motion of individual particles has a component normal to the plane of failure, a considerable amount of the work required to produce failure must be used in overcoming the resistance which the normal force offers to this motion. An analysis of the data from a direct shear test of sand can provide a rough indication of the quantity of energy so expended.

b. Factors Affecting the Angle of Internal Friction

The discussion in the preceding paragraphs permits at least a qualitative evaluation of the principal factors which affect the angle of internal friction. Quantitative data must usually be obtained experimentally, often indirectly.

It is probably safe to assume that under ordinary conditions the grains of a cohesionless soil are covered with at least a thin film of moisture. The mineral grains consist largely of quartz and feldspar with the former usually predominant. Therefore, it is uncommon to find that the angle of internal friction varies appreciably on account of differences in mineralogical properties.

The particles of a granular soil may vary in size and shape. The aggregate may vary in gradation and in relative density. The effect of such variations on that portion of the shearing resistance due to interlocking is apparent after a study of the experimentally determined values of ϕ for a number of different granular materials in both the loose and dense states as shown in Table 11.09a. In the loosest and densest states possible, spheres of equal size have void ratios of 0.91 and 0.35, respectively. The difference in void ratios for these two states is greater than that for any natural granular soil. A careful study of this and other similar data indicates that the grain size itself (for example, as expressed by the effective grain size D_{10}) has an unimportant influence on the angle of internal friction. On the other hand, the factors listed below appear to influence ϕ significantly.

Table 11.09a
Angle of Internal Friction of Non-Cohesive Soils
After A. Casagrande

No.	General Description	Grain Shape	D_{10} mm	C_u	Loose e	Loose ϕ deg.	Dense e	Dense ϕ deg.
1	Ottawa standard sand	Well rounded	0.61	1.2	0.70	28	0.53	35
2	Sand from St. Peter sandstone	Rounded	0.16	1.7	0.69	31	0.47	37*
3	Beach sand from Plymouth, Mass. (Fill for Boston Army Base)	Rounded	0.18	1.5	0.89	29	—	—
4	Silty sand from Franklin Falls Dam Site, N. H.	Subrounded	0.03	2.1	0.85	33	0.65	37
5	Silty sand from vicinity of John Martin Dam, Colo.	Subangular	0.04	4.1	0.65	36	0.45	40
6	Slightly silty sand from the shoulders of Ft. Peck Dam, Mont.	Subangular to subrounded	0.13	1.8	0.84	34	0.54	42
7	Screened glacial sand, Manchester, N. H.	Subangular	0.22	1.4	0.85	33	0.60	43
8	Sand from beach of hydraulic fill dam, Quabbin Project, Mass.	Subangular	0.07	2.7	0.81	35	0.54	46
9	Great Salt Lake fill-sand	Angular	0.07	4.5	0.82	38	0.53	47
10	Artificial, well graded mixture of gravel with sand No. 7 and No. 3	Subrounded	0.16	68.0	0.41	42	0.12	57
11	Well graded, compacted crushed rock	Angular	—	—	—	—	0.18	60

* The angle of internal friction of the undisturbed St. Peter sandstone is larger than 60°, and its cohesion so small that slight finger pressure or rubbing, or even stiff blowing at a specimen by mouth, will destroy it.

Friction angles for the dense specimens apply to relatively small normal pressures for which the tests were conducted. For high normal pressures these angles are considerably smaller, as indicated by the usual curvature of the strength envelopes for dense, granular materials.

(1) Degree of Density, Void Ratio, or Porosity

For a given material there is a marked increase in ϕ as the void ratio decreases. This is probably the most influential of all the factors.

(2) Grain Shape

The value of ϕ increases substantially with increasing angularity of the grains.

(3) Coefficient of Uniformity, C_u

Soils having higher coefficients of uniformity (better gradation) have higher values of ϕ. Such soils also tend to have lower void ratios (though not necessarily greater degrees of density) than do more uniform soils.

It is possible that some other factors have a minor effect upon the angle of internal friction. For example, some studies have shown that submerged soils have angles of internal friction about one degree smaller than those of the dry soils. The angle of internal friction, especially of dense sands, is dependent on the confining pressure which is reduced by submergence.

The results of laboratory tests on fine grained, saturated cohesionless soils are likely to be influenced by the development of excess pore pressures, to the end that the data will not yield a reliable value for ϕ unless interpreted in terms of effective stress (stress between the grains). An appreciation of all these causes of variation in ϕ is important to the soils engineer. For instance, the bearing capacity of soils, which depends on the shear strength, is quite sensitive to small changes in the angle of internal friction.

The angle of internal friction of granular soils is usually determined in the laboratory from a series of direct shear tests or triaxial tests. In the case of triaxial tests, the difficulties of testing increase under low confining pressures so that the envelope or strength line is often poorly defined near the origin. In carefully conducted studies at Harvard University, Chen found the envelope to be generally a straight line having a positive intercept on the s axis, falsely indicating a slight cohesion. This error decreased with increased care in performing the tests.

Because it is sometimes impracticable to conduct a careful and thorough laboratory investigation of the soils from a particular location, empirical charts similar to the one shown in Figure 11.09b may be used in the solution of practical problems. Charts of this type should be constantly studied and supplemented with additional information as more dependable data become available. Information so acquired is likely to be more reliable than that obtained from poorly conducted laboratory tests. In using the chart, it is only necessary to determine the angle of internal friction corresponding to one value of the void ratio. The angle of internal friction corresponding to other void ratios may then be estimated from the chart. A simple procedure for determining the initial pair of values consists of determining the void ratio for a very loose state obtained by gently pouring the sand

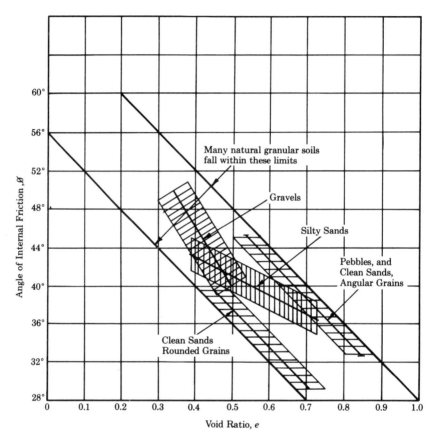

Figure 11.09b
Range of Angle of Internal Friction of Natural, Granular Soils as a Function of the Void Ratio.   After A. Casagrande

into a volumetric container. The corresponding value of ϕ may be found by measuring the angle of repose assumed by the sand when it is poured similarly onto a plane surface. The term angle of repose is sometimes incorrectly used as a synonym for angle of internal friction. The angle of internal friction for a given material varies with the void ratio and also, especially for dense sands, with the confining pressure. The angle of repose is that particular value of ϕ for the sand in a loose state under zero confining pressure. Although values for the angle of repose of clays have sometimes been published, the term has no meaning when applied to cohesive soils.

Experience has shown that the angle of internal friction for cohesionless soils is rarely less than 28° and seldom exceeds about 45°.

c. Stress-Deformation Characteristics

The behavior of a soil specimen during the entire period of loading which precedes failure constitutes an interesting and important study. It does not particularly matter whether the stress is plotted as a function of the strain, or the strain as a function of the stress.   In practice, the two procedures seem to be about equally divided.   In this discussion, the strain is generally plotted as a function of the stress.   In some tests the stress is controlled, whereas in others the strain is controlled. The stress most generally plotted is the deviator stress $(\sigma_1 - \sigma_3)$ although the effective stress ratio $(\bar{\sigma}_1/\bar{\sigma}_3)$ is sometimes used, especially for undrained shear tests.

Before continuing the study of cohesionless materials, the following brief review of the behavior of some idealized materials is interposed.

Stress-strain curves for the ideally plastic and ideally brittle materials are shown in Figure 11.09c.   In the ideally plastic material, the behavior is elastic up to a certain magnitude of stress beyond which the deformation continues under unchanged stress, which is by definition plastic behavior. The ideally brittle mate-

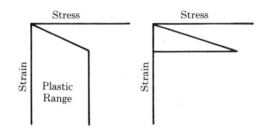

Figure 11.09c
Ideally Plastic   Ideally Brittle

rial exhibits elastic properties under stresses up to the peak stress. All strength is lost upon failure.

Most materials, including soils, fall somewhere between the two extremes described above.   For example, typical stress-strain curves for three types of soil are shown in Figure 11.09d.   All the soils exhibit nearly elastic behavior under small strains and nearly plastic behavior under large strains.   When the natural structure of the brittle clay is broken down by excessive strain, the strength becomes practically zero.   Freshly remolded clays have stress-strain curves closely resembling that for loose sand.   If a remolded clay is permitted to stand for an extended period of time, it regains a part of its strength because of thixotropy. After the thixotropic stiffening, the stress-strain curve resembles that for dense sand. Undisturbed clays exhibit all manner

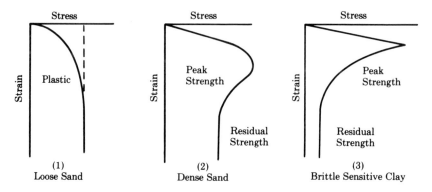

Figure 11.09d
Stress-Strain Curves for Different Types of Soil

of stress-strain relationships, varying between those shown in curves (2) and (3) of Figure 11.09d.

It is now appropriate to discuss the meaning of the term progressive failure which was mentioned in describing the direct shear test. In a progressive failure, the peak stress is not reached simultaneously everywhere along the failure plane because of the existence of a nonuniform strain condition. Failure occurs successively in elements along the failure plane as each in turn reaches the peak stress indicated by A in Figure 11.09e. The failure is accompanied by a reduction in strength, resulting in the transfer of additional stress to adjacent elements. The stress at any instant is computed by dividing the shearing force by the area of the shear plane. This gives the average shearing stress on the plane. The peak average stress, point B, can never be as great as the true strength of the soil, represented

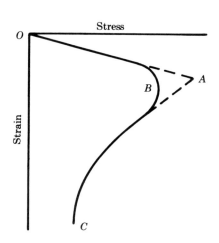

Figure 11.09e
Progressive Failure

by point A. The full curve, OBC, is the one obtained from a direct shear test. The peak strength, so determined, may be conservative by varying degrees.

In general, deformation of a cohesionless soil results in a change of its volume. Suppose that a drained test of a saturated cohesionless soil is carried out in a triaxial test apparatus. The specimen is first subjected to a confining pressure, represented by point A in Figure 11.09f. The axial load is increased to produce successively the deviator stresses p_1, p_2, p_3, etc. until finally the failure condition is reached.

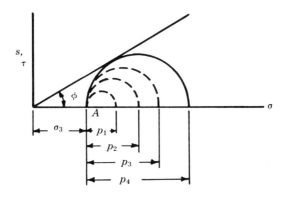

Figure 11.09f
States of Stress During Triaxial Test

The dashed circles in Figure 11.09f represent prefailure states of stress while the failure condition corresponds to the full circle. During the test the level of the water in the burette connected to the interior of the specimen should be observed frequently. If the specimen has been initially compacted to a dense state, it will be noticed that the water level in the burette falls, indicating the passage of water into the specimen. If, on the other hand, the specimen is initially in a loose condition, the water level in the burette rises, indicating the passage of water from the specimen into the burette. This indicates that dense cohesionless soils expand when undergoing deformation, while loose cohesionless soils decrease in volume. These phenomena can be easily demonstrated by means of two small rubber bulbs to which short lengths of transparent tubing are attached. Sand is thoroughly compacted in filling one bulb and the other loosely filled. In both bulbs the voids are filled with water to midheight of the attached tubes. Slight squeezing of the bulbs causes the water level to fall in the tube attached to the dense sand and to rise in the tube attached to the loose sand.

The volume changes produced by deformation represent changes in void ratio or porosity. Thus, the characteristics of the specimen are in a continuous state of change during the test. For this reason, the relationship between volume change and axial deformation is not linear. Sometimes, the curves showing the relationship between volume change and axial deformation may cross.

Typical relationships between deviator stress and strain and between volume change and strain are shown, in Figure 11.09g, for

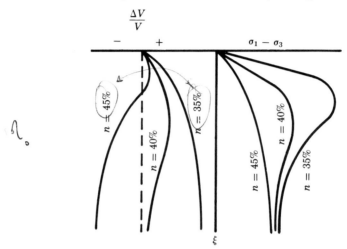

Figure 11.09g
Stress-Strain-Volume Change Relationships for Specimens of Different Initial Porosities

specimens of different initial porosities, varying from dense to intermediate to loose. The higher peak strength (maximum deviator stress) of the denser specimen is attributable to its lower initial porosity and consequent higher initial value of ϕ. It should be noted that the residual strengths of all specimens after large strains approach the same value. Therefore, the state of plastic equilibrium under constant deviator stress represents the attainment of constant and equal porosities in all specimens regardless of initial porosity. The preceding statement applies strictly to zones of failure only. Due to the non-uniform strain conditions previously discussed, some parts of the specimen, especially near the ends, may be little affected.

The illustration of the St. Peter sandstone, given previously, may now be completed by reference to Figure 11.09h. The remarkable characteristics of the undisturbed sandstone are believed to be due to the intimacy of contact between the grains, since the cohesion is practically zero.   The behavior of the sandstone is almost perfectly elastic up to the peak strength following which the behavior is almost identical to that of the dense sand.   In the case of the undisturbed sandstone, the peak stress is reached at about 1 per cent strain.   At this point, the disturbance of the natural structure is apparently severe enough to destroy the original intergranular contacts.   Although the dense

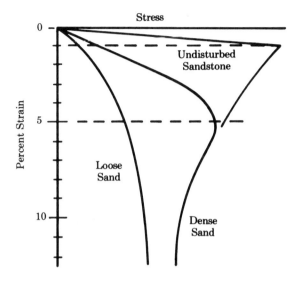

Figure 11.09h
St. Peter Sandstone

sand has the same void ratio as the sandstone from which it was derived, considerably greater strain (5 per cent) is required to develop its peak resistance.   Still larger strains, in excess of 10 per cent, are required to develop the maximum resistance of the loose sand.

d. Critical Void Ratio

Since deformation produces an increase in the volume of dense sands and a decrease in the volume of loose sands, it is logical to suppose that some intermediate density exists at which deformation of the sand is accompanied by no change in volume.   Drained shear tests on sand in both the loose and the dense states yield the approximate relationships shown in Figure 11.09*i*, in which the void ratio and the strain are plotted as functions of the stress.   The void ratio as plotted is intended to represent that of the failure zones.   At large strains, the void ratios of the two specimens approach a common constant value.   At this void ratio, shearing of the specimens may continue indefinitely without producing any further change in the volume. The void ratio corresponding to this condition is termed the critical void ratio and designated e_c in the figure.

If the same tests are performed at constant volume (i.e., an undrained test) pore pressures are developed within the specimens. In the dense specimen the pore water is stressed in tension and in the loose specimen in compression. The effective stress is the algebraic difference between the externally applied stress and the pore water stress. Thus, $\bar{\sigma}_1 = \sigma_1 - u$ and $\bar{\sigma}_3 = \sigma_3 - u$. The bar above the stress symbol σ is used to indicate effective or intergranular stress. Under these conditions it is possible for the pore water stress in the loose sand to become high enough to reduce the effective stresses to zero. This would result in the loss of all shear strength and consequent liquefaction of the soil mass. Such a phenomenon occurs in nature under proper circumstances. Both in fills and in natural deposits of fine grained, saturated, loose sands a sudden shock or shearing deformation may result in liquefaction and flowing of the material. The catastrophe which occurred during the construction of Fort Peck Dam is illustrative of this phenomenon.

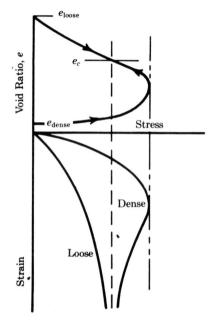

Figure 11.09i
Critical Void Ratio

Liquefaction may be demonstrated in the laboratory with a water tight box filled with fine sand and arranged so that water can be made to flow upward through the sand or allowed to drain downward through the sand. The sand can be loosened by causing the water to flow upward through the sand under a hydraulic gradient sufficient to lift the sand grains in suspension. When the flow is stopped and the elevation of the water maintained near the surface of the sand, a fairly heavy object placed gently on the surface of the sand will be supported by the sand. But, if the sand is deformed suddenly by striking the tank or the surface of the sand a sharp blow, the object will suddenly

and erratically sink into the sand.   Also, if the water is forced upward
through the loose sand supporting the object, the object will sink into
the sand.   When water is allowed to flow downward through the loose
sand, its ability to support loads is increased.   When the sand while
inundated is compacted to a dense state by vibration and tamping and
the water lowered to a considerable distance below the surface, its
ability to support loads is still further increased.   Vertical banks can
be cut into the surface of the sand without failure.   A fairly heavy
object can be placed on the surface near the edge of the vertical surface
without causing failure.   Sudden deformation caused by blows on the
tank do not cause failure of the dense sand containing the capillary
water, which increases the pressure between the grains as the dense
sand tries to expand when deformed.   This demonstration shows that
quicksand is not a specific material, but a condition which may exist in
any cohesionless soil.   The velocity of the upward flowing water
required to cause the soil to become quick is dependent upon the grain
size and to some extent the grain shape.   Fine sands of well rounded
grains are easier to make quick than coarser materials or those of
sharp, angular grains.

The first comprehensive investigation of the stress-deformation
characteristics of sand was performed by A. Casagrande, who reported
the results before the Boston Society of Civil Engineers in 1936.   His
laboratory tests were carried out using the direct shear apparatus.   On
the basis of these early tests, Casagrande concluded that there is for
each sand a critical initial e_c which marks the dividing line between the
safe and the potentially unsafe (i.e., subject to liquefaction) densities
for the sand.   Later tests using the triaxial apparatus have shown e_c
to be a function of the confining pressure σ_3, decreasing with increasing
confining pressure.

The critical void ratio corresponding to a given confining pressure
may be determined in the labora-
tory from a series of triaxial tests in
which the initial void ratio of the
specimens is varied.   If the volume
change corresponding to the failure
condition is plotted as a function of
the initial void ratio as in Figure
11.09j, the Casagrande critical void
ratio e_c can be determined from the
plot.   From several series of tests
made under different confining
pressures, the critical void ratio for
each different pressure σ_3 can be

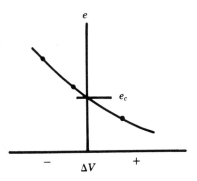

Figure 11.09j
Casagrande Critical Void Ratio

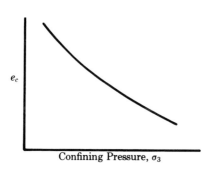

Confining Pressure, σ_3

Figure 11.09k
e_c–σ_3 Relationship

found. The critical void ratio may then be plotted as a function of the confining pressure. The resulting curve is similar to that shown in Figure 11.09k. This curve shows that the critical void ratio decreases with increasing confining pressure. Unfortunately, the physical conditions of the laboratory test are so dissimilar to those of the field that the laboratory determined critical void ratio is not directly applicable to field conditions. However, it is doubtful that liquefaction of a fill or deposit can occur when its void ratio is substantially below the critical value. On the other hand, void ratios higher than the critical do not unfailingly indicate inherent instability of the soil mass. While rapid deformation of masses of saturated, loose cohesionless soils will undoubtedly produce some rise in pore pressure with a consequent reduction of strength, there is no reason to suppose that the strength will always be reduced to zero. In the construction of fills and embankments it is not difficult to compact cohesionless soils to void ratios well below the critical, thus eliminating the danger of liquefaction.

e. Shear Strength

The shear strength of cohesionless soils has been shown to depend primarily upon the angle of internal friction ϕ, which is itself influenced by a variety of factors, and upon the normal pressure acting on the failure plane. In undrained compression tests the development of pore pressures is responsible for the test results yielding an apparent angle of internal friction ϕ_a (based on the total stresses), which differs from ϕ obtained from drained tests. In the drained test the pore pressure is zero so that the total applied pressure is carried by the soil grains, and the ϕ thus determined is sometimes called the true angle of internal friction. Under some conditions of loading in the field the drainage may occur slowly enough to permit the development of appreciable pore pressures. The engineer must decide upon the degree of similarity between laboratory and field conditions and use that strength data which, in his judgment, is most suitable. Measurement of pore pressures in the field enables the engineer to proceed with greater confidence, since an analysis can be made based on the more reliable drained (slow) test results.   The shear strength can then be estimated as $s = (p - u) \tan \phi$.

The drainage of clean, medium and coarse sands, and gravels generally occurs rapidly enough to prevent the building up of appreciable pore pressures during the field loading period.   Hence, values of the drained shear strength may be assumed to apply to these soils for almost any rate of loading; i.e., $u = 0$ and $s = p \tan \phi$.   Commonly, ϕ lies between 28° and 35° for loose sands and between 35° and 47° for dense sands, the higher values applying to sands having angular grains. Crushed rock and angular gravels have angles of internal friction comparable to the higher values for dense sand.

11.10   SHEARING RESISTANCE OF SILT

The behavior of silt possessing only feeble dry strength is governed by the same factors described for sands.   Because of its low permeability and sometimes high void ratio, silt is often regarded as a somewhat treacherous and unstable material of rather unpredictable properties.   The angles of internal friction for the fully drained condition for loose and dense silts usually fall within the lower range of values, given above, for loose and dense sands, tending to be somewhat lower. However, because of its low permeability, an undrained or partly drained condition is likely to govern the behavior of extensive deposits encountered in practice.   In this case, an analysis in terms of the apparent angle of internal friction ϕ_a, obtained from an undrained test, may be more nearly applicable.   If the pore pressure can be readily predicted or determined, much of the uncertainty can be eliminated by an analysis based on effective stresses.

Silts that have appreciable cohesion or dry strength may be investigated and their behavior analyzed by the methods used for clays.   The compressibility of silts, particularly those having a honeycomb structure, is likely to be quite high under either static or dynamic loading.

11.11   CEMENTED GRANULAR SOILS

When sand grains are cemented by either natural or artificial process, the bonds of cementation are likely to exist only at the very small areas of contact between the grains.   The voids rarely are completely occupied by the cementing substance.   The bonds are usually brittle in behavior, and consequently may be disrupted by comparatively small strains.   The strength of cemented granular soils is affected to some extent by the normal pressure.   The strength line has a positive intercept on the s axis as shown in Figure 11.11a and is inclined to the horizontal by an angle ϕ' which generally is smaller than the angle of internal friction ϕ exhibited by the same granular soil in an uncemented state.

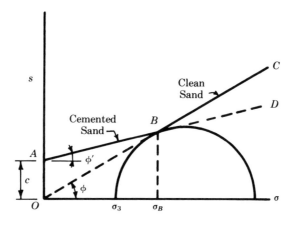

Figure 11.11a
Strength Line for Cemented Granular Soil

A cemented material undergoes two failures, one when the cohesive resistance of cementation is broken and again when the internal shearing resistance of the granular components is exceeded.   The strain required to develop the full shearing resistance of the granular components is much greater than that required to break the cohesive bonds. Line AD in Figure 11.11a represents the strength of the soil at that instant when the cohesive bonds are just on the verge of rupture. Very small strain is required to produce this rupture.   Line OC represents the shearing resistance which is available after much larger strains, when complete destruction of the bonds of cementation will have reduced the material to one having the properties of a clean granular soil.

The formula $s = c + \sigma \tan \phi$ cannot be blindly applied because the maximum strength is represented by the double segmented line ABC. The segment AB corresponds to $s = c + \sigma \tan \phi'$, while the segment BC corresponds to $s = \sigma \tan \phi$.   The two segments are applicable under mutually exclusive ranges of the confining pressure and for vastly different magnitudes of strain.

Figure 11.11a shows that there is one value of the normal pressure σ on the failure plane corresponding to point B for which the Mohr stress circle at failure is approximately tangent to both strength lines. An understanding of the behavior of the material may be gained from a study of the stress-strain curves which could be expected to result from a series of tests carried out at different confining pressures and terminated only after the development of large strains. Figure 11.11b(1)

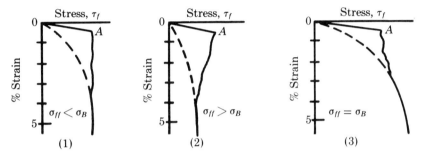

Figure 11.11b
Stress-Deformation Characteristics of Cemented Soils

shows the relationship expected for the particular state of stress illustrated by the Mohr circle in Figure 11.11a. For convenience, a loose structural arrangement of the grains has been assumed. The cohesive bonds are broken after a small amount of strain corresponding to point A. With continuing strain the resistance remains approximately constant while the grains readjust to the altered conditions. After a strain of perhaps 5 per cent the behavior is that of an ordinary granular material. Under low confining pressures, rupture of the cohesive bonds is followed by a period of readjustment during which the shearing resistance on the failure plane decreases as shown in Figure 11.11b(2). The decrease corresponds to the difference in strength represented by line AB and OB of Figure 11.11a. Similarly, under high confining pressures the readjustment following initial failure results in an increase of shearing resistance, which is in accord with the difference in length of ordinates to the lines BD and BC for a given magnitude of normal stress on the failure plane.

Data for plotting curves such as those in Figure 11.11b can be obtained only from tests in which the rate of strain is controlled. From these results it is clear that the cohesive strength and residual frictional resistance of cemented granular materials are not additive. This fact must not be overlooked in design.

11.12 SATURATED CLAYS

a. General Strength Relationships

A clay specimen may be consolidated in the laboratory triaxial apparatus under pressures which are the same in all directions (hydrostatic state of stress), or it may be consolidated under conditions in which σ_1 is made larger than σ_3. By varying the piston load the ratio of σ_1/σ_3 may be made equal to 1.5, 2.0, etc. However, there is a limit to the magnitude of this ratio because the shearing stresses on every potential failure plane increase as the ratio σ_1/σ_3 increases. The failure condition is defined by the maximum value of σ_1/σ_3 that can be imposed on the specimen. In the ordinary consolidation test, described in Chapter IX, the magnitude of σ_3 is unknown, but it is perhaps of the order of $0.7\sigma_1$ for normally consolidated clay. In the triaxial test both σ_1 and σ_3 are known.

In the discussion that follows it is always assumed, unless otherwise stated, that the soils are normally consolidated. More precisely, it should be assumed that the specimen has never been subjected to pressures greater than the lowest pressure for which the various relationships are plotted. This condition is approximately fulfilled by a soft remolded clay or by a very soft naturally deposited clay.

Figure 11.12a represents the results of a series of triaxial tests in which the ratio σ_1/σ_3 is controlled. The states of stress corresponding to the limiting values of σ_1/σ_3 are shown in part (1) of the figure. In part (2) the void ratio is plotted as a function of the minor principal stress σ_3. It is to be expected that for a given value of σ_3 the greater consolidation (or reduction of void ratio) occurs in the tests having the higher values of σ_1. The series of curves are limited by the hydrostatic stress condition (upper curve) and by the failure condition σ_1 max (lower curve). All curves for which σ_1/σ_3 is greater than 1 and less than σ_1 max$/\sigma_3$ will lie between these two curves. The curve for the ordinary consolidation test may correspond to one of the dashed curves in the figure.

The lower curve of Figure 11.12a(2) could be arrived at by following a somewhat different procedure than that which has been indicated. Suppose that the specimen is first consolidated under an all around pressure σ_3, as represented by specimen I and by the numeral I on the $e - \sigma_3$ curve. Then, with free drainage of the specimen permitted, the axial deviator stress may be slowly applied while σ_3 is held constant. The void ratio of the specimen will decrease, and at a given moment correspond to one of the intermediate dashed curves. As p increases the void ratio continues to decrease, finally reaching the point II on the σ_1 max curve at which time the specimen will fail. From the failure conditions the shear strength of the soil can be determined. Since no pore pressures have been allowed to develop, the

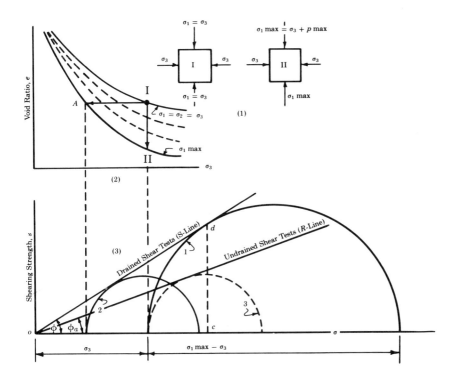

Figure 11.12a
Effect of Variation of the Principal Stress Ratio

strength for this condition is represented by the Mohr circle 1 in Figure 11.12a(3). A line drawn from the origin tangent to this circle, represents the strength of the soil in the drained condition. It should be noted that the shear strength corresponds to the ordinate cd, which is the shear stress on the plane of failure at failure. Theoretically, the same result would be found from a direct shear test in which the abscissa oc represents the normal (vertical) stress, and cd the shearing stress at failure.

If, on the other hand, the specimen is sheared from point I without any drainage being allowed, the pore pressure will increase and the effective stresses will decrease as indicated by the line IA. Again, failure would be expected to occur when $\bar{\sigma}_3$ has decreased to the value represented by point A on the σ_1 max curve.   The Mohr circle of effective stresses, circle 2, must be tangent to the strength line defined by the drained test because, in both cases, effective stresses have been plotted.   If total stresses, rather than effective stresses, are plotted there will be defined a strength line having a lesser slope than that for the drained tests.   This line will pass through the origin and will be tangent to the Mohr circle of total stresses for the failure condition represented by circle 3.

Total stress σ and effective stress $\bar{\sigma}$ are specifically related in terms of the excess pore pressure u, as $\bar{\sigma} = \sigma - u$.   The total stress is equal to the externally applied pressure divided by the cross sectional area of the specimen perpendicular to the direction of the stress.   Thus, while the terms effective stress and intergranular pressure are often used interchangeably, when so used the terms refer to an imaginary stress and not the actual (perhaps very high) stresses which exist at the small areas of the contact between individual soil particles.

In the drained tests that have been discussed, the significant fact to remember is that no pore pressures in excess of those due to the hydrostatic head of water are permitted to develop.   In practice it is impossible to avoid creating slight excess pore pressures, since otherwise no drainage could be induced.   However, if the tests are performed properly the excess pore pressure that exists at the instant of failure will be so small that it will have only negligible effect on the test results.

For clays of the type under consideration the envelope of the strength circles obtained from both drained (S) tests and consolidated undrained (R) tests are straight lines passing through the origin.   In other words, the strength may be assumed to be entirely dependent upon the frictional resistance, just as is the case with sand.   The envelope for the strength circles plotted from the results of S tests is referred to as the S-Line, and the one obtained from the R tests as the R-Line.   The slope angles of the two lines are designated, respectively, as ϕ and ϕ_a. Some other designations are also frequently used in soil mechanics literature.   For example, ϕ_d and ϕ_{cu} are often used in reference to the results of drained and consolidated undrained tests, respectively.

The slope of the S-Line for many clays is not very different from the value of ϕ for granular soils.   For silty clays ϕ is likely to be nearly 30°, while for very fat (highly plastic) clays ϕ may be around 15°. Natural soils containing a considerable amount of bentonite probably have smaller values of ϕ; however, these are difficult to determine satisfactorily.

In engineering practice, the values of ϕ from drained tests are rarely used.   The rate of loading during construction is usually rapid enough to cause substantial pore pressures to develop.   The use of ϕ_a may be more appropriate, although the proportion of the load carried by the pore water is not generally known.   However, the effective stresses may be closely determined if field piezometers are installed.

A relationship in addition to those already developed may be established if triaxial tests are made in which no drainage is permitted under either the all around pressure or the deviator stress.   Such tests

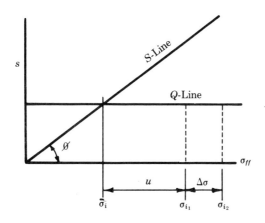

Figure 11.12b
Total Stress-Effective Stress Relationship in Q Test

are called undrained tests and should be carried out rather quickly in order to justify the assumption of no drainage.   Excess pore pressure will develop during both phases of the test.   If several of these tests are made using different confining pressures, a horizontal strength line will be obtained.   The water content of each specimen remains unchanged throughout the test.   These undrained tests are referred to as Q tests and the strength line is called the Q-Line.   It is emphasized that the shear strength depends upon the effective pressure. Since the strength in these tests is independent of the confining pressure, the normal pressure effective on the failure plane in all Q tests must correspond to the intersection of the Q-Line and the S-Line, because the latter is based on true effective stresses.   As shown in Figure 11.12b, any total normal pressure σ_i induces a pore pressure u of such magnitude that the effective normal pressure $\bar{\sigma}_i$ is always in the position shown.   Any increase in confining pressure shifts σ_i from position 1 to position 2 on the plot and gives rise to an equal increase in pore pressure; i.e., $\Delta u = \Delta \sigma$.   In the case of an unconfined compression test it is apparent that negative pore pressure (tension) must develop.

b. Total Stress—Effective Stress Relationships

The relationship between the strength circles of total and effective stresses should be considered in some detail.   On the two stressed elements shown in Figure 11.12c are shown the states of stress corresponding to the end of the consolidation phase and to an intermediate time during the shear phase of an R test.   The confining

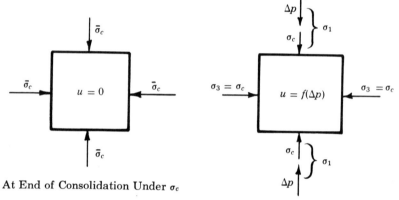

At End of Consolidation Under σ_c

Immediately After Application of Δp

Figure 11.12c
States of Stress Occurring During a R Test

pressure in the triaxial chamber is represented by σ_c, which is also the effective stress on every plane at completion of consolidation under σ_c. When the axial load is increased there results a change of pore pressure, Δu, the magnitude of which depends upon the axial pressure increase Δp and upon the stress-deformation characteristics of the soil.   The relationship between Δu and Δp is generally non-linear.   Of primary concern, are the relative magnitudes of these two quantities when the failure condition is reached.

Figure 11.12c illustrates that the total lateral stress σ_3 remains unchanged at σ_c when the total vertical stress σ_1 increases from σ_c to $\sigma_c + \Delta p$.   However, the effective stresses in all directions depend equally upon the change in pore pressure resulting from the increase in axial stress.

The manner of deformation of the clay specimen to a large extent controls the pore pressure that exists at failure.   The deformation characteristics depend primarily upon the soil structure.   In normally

consolidated, relatively soft clays of normal sensitivity the pore pressure at failure is likely to be about equal to the deviator stress.   In
this case, the relationship between the total and effective stress circles
is approximately that shown in Figure 11.12d(1) in which effective
stresses are represented by full circles and total stresses by dashed
circles.   In this case, application of the axial load does not cause any
change in the effective major principal stress, but only a reduction in
the effective minor principal stress.   The diameter of all strength

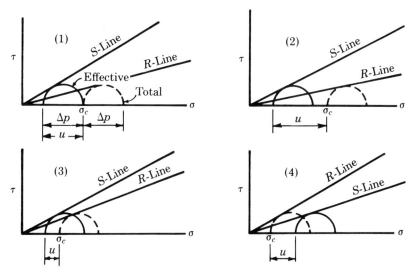

Figure 11.12d
Stress at Failure for Various Types of Clay

circles, both for total and effective stresses, must equal the deviator
stress at failure.

Brittle, highly sensitive clays are likely to undergo vast structural
alteration in the vicinity of the failure surface.   Such an occurrence
results in a rise of pore pressure of such magnitude that at failure the
pore pressure often exceeds the deviator stress responsible for the
failure.   In this event, the circles of total and effective stresses are
separated as indicated in Figure 11.12d(2).

The pore pressure rise in moderately overconsolidated clays is frequently rather small compared to the deviator stress at failure.   The
pore pressure is usually positive unless a low confining pressure is used.
The relationship between total and effective stresses for a moderately
overconsolidated clay is shown in Figure 11.12d(3).

Very dense clays such as glacial tills, compact sandy clays, and
heavily overconsolidated clays exhibit, similarly to dense sands, a

tendency to increase in volume when deformed. If this volume increase is prevented, as in an R test, negative pore pressures develop as the deviator stress is applied. As tension rises in the pore water, the effective confining pressure increases, being equal to the sum of σ_c and the pore water tension. The effective stress circle is then displaced toward the right from the total stress circle, as in Figure 11.12d(4). Under these circumstances the R-Line is more steeply inclined than is the S-Line. Whether such additional strength is actually available under field conditions must be judged by the engineer after consideration of the probable drainage condition and rate of deformation.

c. Interpretation of Laboratory Test Results

In a saturated sedimentary deposit of natural clay that has been consolidated only under the existing overburden, the vertical and horizontal stresses acting on an element of the soil at depth d beneath the surface are indicated on Figure 11.12e. The vertical stress is equal to the effective unit weight of the soil multiplied by the depth, and the horizontal stress is some fractional part K_0 (perhaps 0.7) of the vertical stress. As a result of consolidation under anisotropic stress conditions, the shear strength on a horizontal plane tends to be higher than that on a vertical plane. On the other hand, the horizontal stratification of sedimentary deposits tends to reduce the shear strength on a horizontal plane. Thus, there exist two more or less balancing factors which tend to equalize the strength along these orthogonal planes. Evidence indicates that for reasonably homogeneous clays the shear strength along vertical and horizontal planes is often about equal, and this tacit assumption is generally made.

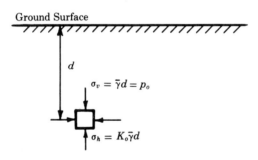

Ground Surface

d

$\sigma_v = \bar{\gamma}d = p_o$

$\sigma_h = K_o\bar{\gamma}d$

Figure 11.12e
Stresses on Element in Place

Specimens of a sample taken from the clay deposit described above may be tested in a number of different ways, as previously described. When the sample is removed from the ground, tension develops in the pore water as the soil attempts to expand under the reduced pressures. Only slight expansion is necessary to permit formation of full menisci in the pores, so that the restraint of the specimen may be only slightly less than the average stress to which the soil was subjected in the ground by the surrounding soil. However, an important difference exists in the two situations. The stresses in the ground are anisotropic whereas the capillary stresses are isotropic. The sudden alteration of the state of stress results in some deformation of the sample with somewhat the same effect as is produced by disturbance during sampling and trimming. The combined effects of the foregoing factors produce a significant reduction in the strength of the soil at unaltered water content. Voluminous data indicate that the strength of an ordinary good laboratory specimen is about 10 per cent less than the strength of the soil in place. The strength of badly disturbed specimens is much more affected.

If a specimen is subjected to a confining pressure with no drainage permitted, the rise in pore pressure is for practical purposes equal to the all around pressure. Since no deformation occurs, the load carried by the skeletal structure of the soil does not change. It was previously explained that the envelope of total stress circles obtained from a series of these undrained tests made under different confining pressures is a horizontal line—the Q-Line. It might therefore be concluded that the soil is a purely cohesive material, having an angle of internal friction of zero. The fallacy of this conclusion is made apparent by consideration of the effective stresses.

A deviator stress p_c sufficient to cause failure of the specimen in either a Q test or an unconfined compression test, causes a change in the pore pressure which is of such nature and magnitude that the results of any number of such tests can be expressed in terms of a single circle of effective stresses, as shown in Figure 11.12f. This circle is tangent to both the Q-Line and the S-Line. If the soil were a purely cohesive material, failure would occur on the plane of maximum shearing stress at 45° from the principal planes. Moreover, the strength (or size of the strength circles) would be independent of the effective stresses. Neither of these conditions is fulfilled by clay. Numerous test results have indicated that there is no statistical difference in the inclination of the failure planes obtained in U tests, Q tests, R tests, and S tests. The failure plane is inclined to the plane of major principal stress by an angle $45° + \dfrac{\phi}{2}$ in which ϕ generally lies between 10° and 30°. There is considerable evidence that the cohesion of clay is dependent upon the water content and, hence, the effective stresses. However, the increase of cohesion, with increasing effective stress, is insufficient to account for more than a modest fraction of the total increase in strength.

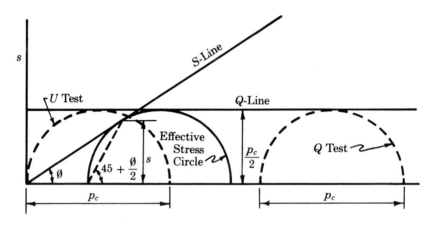

Figure 11.12f
Strength Obtained in Undrained Tests

The actual strength of the soil specimen, as shown in Figure 11.12*f*, must be $s = \dfrac{p_c}{2} \cos \phi$. However, it has already been explained that the strength circle obtained in the laboratory is of the order of 10 per cent smaller than one that represents the *in situ* strength of the soil. For $\phi = 30°$ and $\phi = 18°$, the difference in $p_c/2$ and $p_c/2 \cos \phi$ is 13 per cent and 5 per cent, respectively. It is therefore reasonable and practicable to use $p_c/2$ as the shear strength in order to compensate, roughly, for the loss in strength due to disturbance. This conclusion is supported by the fact that *in situ* shear tests using the shear vane have rather consistently shown somewhat higher strengths than those found by U or Q tests. A simple shear vane has been sketched in Figure 11.12*g*. The shearing resistance along a cylindrical surface is computed from a measurement of the torque required to twist the device and from the dimensions of the vane, which is held at constant elevation during performance of the test. Properly designed and used, placement of the vane into the soil at the bottom of a boring creates only slight disturbance of the soil. These instruments have in recent years come to be used

Figure 11.12g
Shear Vane

extensively for investigating shear strength in the field. For an excellent discussion of the shear vane and a special type known as the vane borer, the reader is referred to the papers by Cadling and Odenstad, and Bennett and Mecham listed at the end of this chapter.

Still considering the sample of saturated, normally consolidated clay which has been taken from the ground, its behavior is examined when subjected to triaxial R tests. The confining pressure of consolidation must be chosen for each specimen. It may be desired to attempt to reproduce the conditions of confinement existing in the ground. In

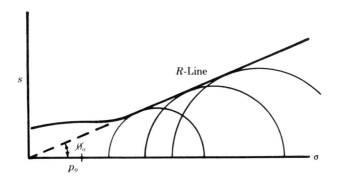

Figure 11.12h
Consolidated Undrained Tests
$\sigma_3 > p_0$

this case, the all around pressure during the consolidation phase of the test may be made equal to the computed pressure p_0 of the overburden at the elevation from which the sample was removed. Because sampling and trimming cause some disturbance of the specimen, and because the isotropic state of stress imposed in the test is unlike the state of stress which was present in the ground, the specimen undergoes consolidation and reduction of water content under the confining pressure. For this reason, the test specimen generally has a higher strength than in its natural position. For a specimen in good condition the difference is not great.

Tests also may be performed in which the confining pressure is either smaller or larger than the overburden pressure p_0. A series of tests in which the confining pressure $\sigma_c = \sigma_3 > p_0$ produces an array of stress circles for which the envelope (extended) is a straight line passing through the origin. This envelope is the R-Line indicated in Figure 11.12h. If, on the other hand, tests are performed in which $\sigma_3 < p_0$ the strength circles fall in the range of preconsolidation of the soil. In so far as their behavior in these tests is concerned, the soil specimens are overconsolidated. Under pressures lower than the preconsolidation pressure (p_0 in the case of a normally consolidated clay), the water content of the soil is always lower than that of specimens which are subjected for the first time to the same pressure (pressures lower than p_0). Consequently, the strengths of specimens tested in the range of preconsolidation are higher than indicated by a straight line through the origin. The envelope for the entire range of pressures is shown by the solid line in Figure 11.12h. This envelope shows that

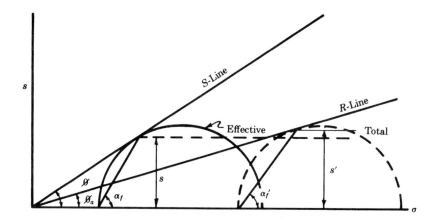

Figure 11.12i
Shear Strength Indicated by Consolidated Undrained Test

even for pressures equal to or slightly greater than p_0, the strengths are somewhat higher than those corresponding to the straight extension of the R-Line. This is due, as previously mentioned, to sample disturbance and the consequent reduction of void ratio during consolidation. For a better understanding of the effect of water content and void ratio on the strength of clay, the reader is admonished to remember the nature of clay minerals and their attached water.

If pore pressures are measured during the tests represented by the stress circles in Figure 11.12h, the corresponding circles for effective stress may be plotted. A relationship similar to that shown in Figure 11.12i is found to exist. It can be seen from the figure that the two methods of plotting the test results can not both correctly indicate the location of the failure plane since α_f is not equal to α_f'. The location of the failure plane is correctly indicated by α_f, and the shearing strength of the specimen is actually s rather than s' corresponding to the total stress circle. However, the difference between the actual strength s and the indicated strength s' is slight, so that the latter may be used in design when the construction procedures are appropriate for the use of undrained test results.

Frequently, ϕ_a is about equal to $\phi/2$. Therefore, if ϕ is known, a preliminary design may be made with a fair degree of safety on the assumption that $\phi_a = \dfrac{\phi}{2}$. For silty or lean normally consolidated clays, ϕ is likely to be around 28° and ϕ_a about 14°, in which case, $s = \sigma_c \tan 14° = 0.25\sigma_c$. Thus, the shear strength of normally consolidated clays is likely to be about one-fourth the effective stress on the soil skeleton in the ground. Overconsolidated clays have higher strengths. It seems that very plastic, fat clays should have strengths lower than $0.25\sigma_c$. However, it will be shown later that the rate of loading may have an important influence on the strength, particularly that part of the strength that is dependent upon cohesion, so that considerable experience and judgment is required in the prediction of *in situ* strengths of clays.

Drained (S) tests require so much time for their performance that they are not usually included in routine investigations of strength in the laboratory. The S-Line, similarly to the R-Line, is influenced by the preconsolidation pressure which has acted on the soil. The strength envelope for S tests of normally consolidated clays is curved for the range of confining pressures below p_0 as shown in Figure 11.12j. For confining pressures greater than p_0, the envelope is a straight line whose extension passes through the origin. Many so-called S tests are carried out so rapidly that the results are influenced by the existence of pore pressures of unknown magnitude. This results in a strength line which lies some-

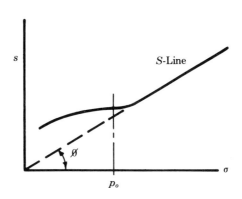

Figure 11.12j
Results of Drained Tests

where between the S-Line and the R-Line. In any case, the exact position of the curved portion of the line (below p_0) is likely to be highly dependent upon the rate of loading, approaching the linear extension (dashed line in Figure 11.12j) as the rate decreases.

It has been shown that the relationship between laboratory strength and confining pressure, in terms of total stress, is inextricably linked with the test method and procedures.   The relationship is further complicated by the influence of preconsolidation pressure.   The shear strength applicable to a particular field problem may be arrived at only after careful evaluation of the test results in the light of the most probable conditions which prevail in the field.   The following discussion serves as a brief illustration of the reasoning that may be employed.

Suppose that the. loading in the field is to occur so rapidly that no consolidation of the soil will occur during the loading period.   It is logical under these circumstances to use for the shear strength the

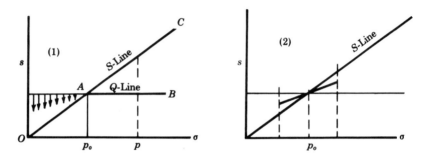

Figure 11.12k
Evaluation of Laboratory Test Results

constant value obtained from one or more U or Q tests.   During the period of loading the total stress on the soil will increase from p_0 to some higher pressure p, as in Figure 11.12k.   The least strength which the soil could have corresponds to zero per cent consolidation under the increased load and is represented by the ordinate erected at p to the Q-Line AB.   The maximum possible strength for this condition corresponds to 100 per cent consolidation and is represented by the ordinate to the S-Line OC.   If loading of the soil is to be preceded by excavation, the soil below and adjacent to the sides of the excavation will tend to swell under the reduced load.   There will be a consequent reduction in strength as indicated by the arrows in Figure 11.12k(1). The amount of reduction depends upon the length of time during which the excavation is open prior to addition of the construction loads. However, the strength cannot possibly decrease to values lower than those represented by ordinates to the S-Line OA.

It would be safe therefore to use the segmented line OAB for all conditions of loading.   Theoretically, it would be permissible to use a factor of safety of unity in connection with this strength line.   As a practical matter, however, uncertainties as to actual loads and manner of variation of the soil deposit make it desirable to use a factor of safety greater than unity. For field loading involving pressures lying within a fairly narrow range in the vicinity of p_0, it is reasonable to represent the strength by means of a straight line, as indicated in Figure 11.12k(2). The strength may then be expressed in the form $s = c + p \tan \phi$, with some appropriate factor of safety applied for practical applications.

d. Resumé of Strength Characteristics

The interrelationships among strength, normal pressure, and void ratio or water content may be conveniently represented as in Figure 11.12m. In the figure σ_n represents the pressure under which specimens taken from a natural clay deposit have been consolidated, and e_n and w_n represent the natural void ratio and the natural water content of these saturated specimens. For convenience of discussion only, let it be assumed that the results of direct shear tests are represented in Figure 11.12m so that the abscissa of any point represents the normal stress on the failure plane. The curve MNP is the normal consolidation curve for the soil.

(1) Confining Pressure Greater than In-Place Pressure

When a specimen of clay is subjected to a triaxial test in which the confining pressure σ_c is greater than σ_n, the behavior of the soil specimen varies according to the type of test performed.

(a) Undrained Test

The water content and, hence, the void ratio does not change during the entire test. Therefore, the water content and total stress on the failure plane correspond to point 1 in Figure 11.12m(2). The shear strength is represented by the ordinate s'' to the Q-Line in Figure 11.12m(1).

(b) Consolidated Undrained Test

Consolidation during the first phase of the test corresponds to a change from N to 3 along the consolidation curve MNP. During the second (shearing) phase the water content and total stress on the failure plane correspond to point 3 in Figure 11.12m(2). Since the water content has decreased, the strength s' is greater than the strength s'' obtained from the undrained test. In this case, the strength is represented by the ordinate s' to the R-Line.

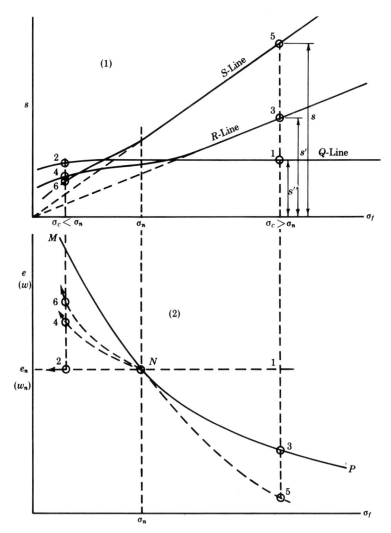

Figure 11.12m
Behavior of Saturated Clay During Q, R, and S Tests

(c) *Drained Test*

If the consolidation and shearing phases are slowly carried out simultaneously, the behavior of the specimen corresponds to the movement of a point from the point N along a path such as $N5$. No pore pressures develop, so the total and effective stresses are at all times equal. The failure condition is represented by points 5 in Figure

11.12m. The shearing stress found from this test is greater than that found in the other two type tests described above, because the water content was further reduced. The strength is given by the ordinate s to the S-Line.

(2) Confining Pressure Less than In-Place Pressure

The results obtained when the same tests are carried out under a confining pressure σ_c which is less than the in place pressure σ_n are described below.

(a) *Undrained Test*

The water content does not change from the initial condition. The shear strength, therefore, must be the same as found for the undrained test with σ_c greater than σ_n. It is implied that the effective normal stress on the failure plane is the same in both cases, and that the strength is independent of the total normal stress.

(b) *Consolidated Undrained Test*

During the consolidation phase the specimen swells since the confining pressure is lower than that to which it was subjected in the ground. The water content increases during the consolidation phase and then remains unchanged during the shearing phase. The water content at failure and the total normal stress correspond to point 4 in Figure 11.12m(2). The shear strength is given by the ordinate to point 4 on the R-Line. Because of its higher water content, the specimen has a lower strength than that found in the undrained test.

(c) *Drained Test*

If the specimen is permitted to swell under the low confining pressure as it is being sheared, the additional structural disturbance is likely to result in more swelling than occurred during the consolidated undrained test. At failure, the total (and effective) normal stress and the water content correspond to point 6 in Figure 11.12m(2). The shear strength of the soil is smaller than that found in either of the preceding tests, and is given by the ordinate to point 6 on the S-Line.

It may be generally concluded that the shearing resistance of saturated, normally consolidated clays is highly dependent upon the manner in which the load is applied and upon the magnitude of the load relative to the overburden pressure existing prior to the construction activities.

11.13 OVERCONSOLIDATED CLAYS

There exist vast areas on the earth where the soils at some time during their geological history have been consolidated under pressures greater than those of the existing overburden.   These higher pressures may have been produced by a thick layer of overburden which has since been removed by erosion; by the enormous continental ice sheet which advanced and receded several times over large areas of North America and Europe during the Pleistocene Epoch; or by desiccation during which the tensile stresses in the capillary water may reach surprisingly high values.   Overconsolidated clays can frequently be identified by their relatively low liquidity indexes, which may even be negative for heavily overconsolidated clays.

In its natural state, an overconsolidated clay has a water content which is low compared to that which it would have if normally consolidated under the same environment.   The shear strength of these clays depends upon the magnitude of the overconsolidation pressure which has permanently affected the water content of the soil.   If the soil is saturated, the Q-Line, representing the undrained strength, usually has the same appearance as that for a normally consolidated clay, except that it is displaced farther from the σ axis.   In the case of jointed or fissured clays, including most clays which have been overconsolidated by desiccation, the undrained strength when unconfined or under low confining pressure is usually lower than it is under high confining pressures.   For such clays the Q-Line resembles that shown in Figure 11.13a.   The clay immediately adjacent to the joints or fissures is often much softer than the clay in the interior of a block so that both the strength and the orientation of the failure surface are influenced.   Under higher pressures failure may be forced to occur on a plane at $45 + \frac{\phi}{2}$ degrees with the

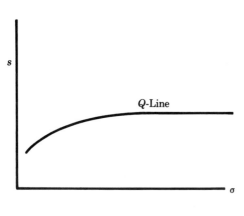

Figure 11.13a
Jointed and Fissured Clays

plane of the major principal stress, which generally does not coincide with the orientation of the joints or fissures.

If the clay is moderately to heavily overconsolidated, an entire series of R tests or S tests may be performed without the stress ever exceeding the preconsolidation pressure.   In this case the R-Line and the S-Line are likely to be curved over the entire plotted range, as indicated in

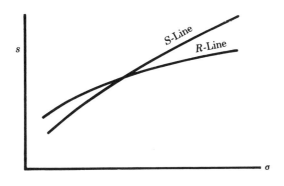

Figure 11.13b
Heavily Overconsolidated Clay

Figure 11.13b.   The curvature and position of these lines are similar to those for normally consolidated clays in the range of stresses less than the overburden pressure, because the clay is overconsolidated with respect to pressures less than those produced by the overburden. When these lines are used in design and analysis, that portion of the appropriate line which lies within the range of pressures under consideration in the problem is often approximated by a straight line, from which the constants c and ϕ in the expression $s = c + \sigma \tan \phi$ may be evaluated.   However, there is considerable evidence in the form of slope failures to indicate that the apparent cohesion c so obtained may be grossly in error on the unsafe side when applied to long term stability analyses.

11.14 PARTIALLY SATURATED CLAYS

Many natural clay deposits, particularly those that lie above the ground water table, have degrees of saturation less than 100 per cent. All compacted clays are initially only partly saturated.   The air, or other gas, which is in the voids is quite compressible compared to the water in the voids or even compared to the structural skeleton of the soil. Loads which are applied to the soil result in sudden compression of the air and a consequent transfer of stress to the soil. The stress in the liquid phase of the pore fluid also increases, but not so much as would be the case if the soil were saturated. Furthermore, the stresses carried by the soil structure and by the pore water are to some extent dependent upon time, because the air entrapped under pressure requires a short time to enter into solution in the water until final equilibrium is established in accordance with Henry's Law. J. W. Hilf

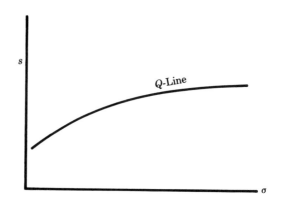

Figure 11.14a
Undrained Strength of Compacted Soil

has given a thorough discussion of the pore pressures in compacted cohesive soils in *Technical Memorandum 654* published by the U.S. Bureau of Reclamation.

As a result of the increase of effective stresses when external pressures are applied, compacted soils exhibit in undrained tests strengths which are dependent upon the applied pressure. Figure 11.14a indicates that the relationship between s and σ is not linear. As the external pressure increases, more and more air enters into solution. If high enough pressure is exerted, all the air will enter into solution, provided that the initial degree of saturation is not too low. Under still higher pressures the soil behaves as a saturated soil with all additional stress taken by the pore water. In this higher pressure range the Q-Line becomes parallel to the σ axis indicating a strength which is, in this pressure range, independent of the confining pressure.

During drained or consolidated undrained triaxial tests, soils which are not initially saturated are likely to become very nearly saturated so that their behavior during these tests is nearly the same as that discussed previously for saturated clays.   The shear strength of compacted cohesive soils is dependent upon the water content used during compaction.   The "as molded" strength is greatest for soils compacted at water contents well below optimum, decreasing with increasing water content.   However, if soil specimens compacted at different moisture contents are then saturated prior to testing, it is often found that the highest strengths in the "soaked" condition correspond to intermediate values of the molding water content.   Usually, the soaked strength deserves the greater consideration in field applications of the laboratory test results.

11.15   PRESTRESS INDUCED DURING THE R TEST

A. Casagrande and S. D. Wilson first pointed out in 1953 that triaxial consolidated undrained tests actually determine the shearing strength of a prestressed clay, even though the soil may not have been prestressed in nature.   This is explained by the fact that the confining pressure (effective stress) during the consolidation phase of an R test is almost always greater than the effective minor principal stress at failure.   This is always true for those soils in which positive pore pressures develop during application of the deviator stress.   Moreover, the positive pore pressures are commonly high enough to reduce the effective normal stress on the failure plane at failure to a value lower than the stress under which the soil was consolidated.   As a consequence, the results of R tests over a rather large range of confining pressures will represent the behavior of a prestressed clay.   The strength line for a series of R tests expressed in terms of total stress or effective stress is likely to be curved throughout the entire range of test pressures.

The relationships shown in Figure 11.15a are typical of the results of an R test carried out on a soft saturated clay.   The effective stress conditions at the conclusion of the consolidation phase of the test are represented by a point circle located on the σ axis at σ_c, the confining pressure.   Consolidation under the hydrostatic stress σ_c has occurred on every plane passing through any element of the soil mass.   During the axial loading undrained phase of the test, positive pore pressures are induced resulting in a separation of the circles of total and effective stresses.   At failure, the effective stress circle is shifted to the left through a distance u, relative to the total stress circle.   The effective stress σ_{ff} on the plane of failure at failure is smaller than the stress σ_c

under which consolidation occurred. For this reason, the effective stress circle does not lie tangent to the *S*-Line passing through the origin, but tangent to some line such as the dashed line in Figure 11.15a, which represents a prestressed soil.

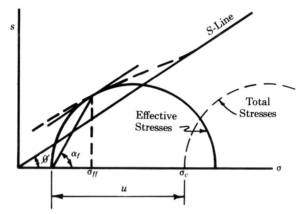

Figure 11.15a
Effect of Prestress During *R* Test

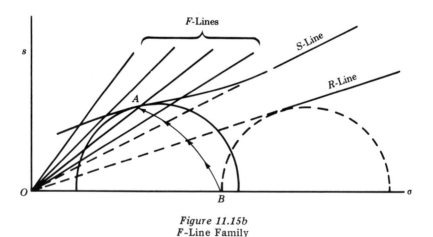

Figure 11.15b
F-Line Family

Casagrande and Wilson have designated a straight line from the origin passing through that point corresponding to the failure plane on the strength circle as the *F*-Line. It has been shown that the inclination of the *F*-Line is dependent upon the prestress ratio $R_p = \dfrac{\sigma_c}{\sigma_{ff}}$ or $\dfrac{p_p}{\sigma_{ff}}$, whichever is greater, becoming steeper with increasing value of R_p.

If, in a series of tests, the prestress ratio is made to vary from test to test, a corresponding family of F-Lines may be drawn as shown in Figure 11.15b. Only one pair of stress circles has been drawn in the illustration. It should be observed that the F-Line is not tangent to the effective stress circle, but is the vector sum $\overline{OA}$ of the normal and shearing stresses on the failure plane. The strength line in the range of prestress passes through the point A of all vectors $\overline{OA}$ represented by the F-Lines. The position of the S-Line (or length of vectors $\overline{OA}$) depends upon the magnitude of σ_c in the prestress ratio σ_c/σ_{ff}, rising higher with increasing σ_c.

If pore pressures are measured more or less continuously throughout the test, it is often informative to plot the vector curve BA. A line drawn from the origin to any point on the curve BA represents the effective resultant stress on the failure plane at some stage of the test, but only the vector $\overline{OA}$ corresponds to the failure condition.

11.16   OTHER FACTORS AFFECTING THE SHEAR STRENGTH

It has been pointed out in the preceding articles that the strength of a cohesive soil is influenced by: sample disturbance, anisotropic consolidation, the stress history of the soil (including induced prestress), the soil structure, and the type of test equipment and test methods employed. There are some factors, in addition to those listed above, which may have an important effect on the strength of the soil.

a. Rate of Loading

A number of experimental investigations have shown that the strength of a soil tested in the laboratory depends to a remarkable extent upon the rate or duration of loading employed in the test. This fact was recognized more than 100 years ago by A. Collin, a French engineer, but was for many years neglected as a field of study. Much of the recent work in this field was pioneered by A. Casagrande. Under field conditions, a strength-load duration relationship seems to exist, as evidenced by the failure of slopes after long periods of stability. However, there is some doubt that delayed failure in the laboratory and in the field are due to the same basic causes.

In engineering practice, soils may be required to withstand loads which may vary in duration from a fraction of a second to many years.

The stress which a soil can resist without failure under transient load may be of the order of twice that which causes failure after being sustained for one minute.   On the other hand, there is evidence that the load which can be indefinitely carried by some soils may be no more than 25 per cent of the load which causes failure in one minute.   The variation of strength with rate of loading is vastly different for different types of soils so that it is impossible to state rules which are generally applicable.   A relationship between strength and time to failure for

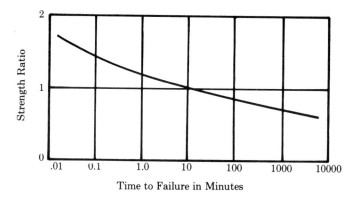

Figure 11.16a
Effect of Time of Loading

undrained tests of saturated clay may be expected to have somewhat the form illustrated in Figure 11.16a.   The strength ratio is determined on the basis of the strength corresponding to the usual Q test in which the total loading period is approximately ten minutes.   In investigating the strength under loads of long duration, both slow loading and creep tests have been used.   In the former the load is slowly increased at a constant rate until failure occurs.   In creep tests, loads of various magnitudes are applied quickly and thereafter maintained constant until failure occurs or until it appears that failure will never occur under the load.   The greatest stress which can be sustained indefinitely is referred to as the creep strength of the soil.

b. Repeated Loading

Most investigations of the effect of repeated loading have dealt with compacted (therefore, unsaturated) clays.   To some extent the effects of stress history and soil structure are basically involved in the test results.   H. B. Seed and co-workers at the University of California, in recent years, have studied the effects of repeated loading of compacted

clays. Their work has shown that for clays tested at constant moisture content the shear strength is increased by a large number of
repetitions of the stress. If the stress intensities are too high, the
cumulative deformation results in failure after too few repetitions to
have a beneficial effect on the strength. However, a large number of
repetitions of relatively low stress result in an improvement of
strength such that the stress intensity may be stepped up with further
beneficial effect. The num-
ber of repetitions required
to produce an appreciable
increase of strength is not
known, but appears to be
in excess of 1000. The
overall effect of repeated
loading can perhaps best
be represented as in Figure
11.16b, which shows typical
curves for two initially
identical specimens. One
specimen has been sub-
jected to several thousand
cycles of repeated stress
before being loaded to

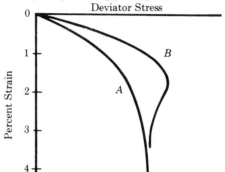

Spec. A not previously loaded.
Spec. B previously subjected to
repeated stress applications.

Figure 11.16b
Effect of Repeated Loading

failure in a normal undrained test, while the other has been brought
to failure with no pre-treatment. Evidently, not only is the strength
affected but the stress-deformation characteristics as well. The practical implications of the effects of repeated loading of soils are of considerable importance in highway engineering.

c. Intermediate Principal Stress

In the usual laboratory triaxial test the intermediate principal stress
is equal to the minor principal stress. However, it is possible to perform the test (axial extension test) in such a way that the intermediate
principal stress is equal to the major principal stress. It is possible to
evaluate the effect of variations of the intermediate principal stress by
comparing the results of the usual triaxial test with those of the axial
extension test. A knowledge of this influence may be of practical use
in evaluating the shear strength of soils in cases where the intermediate
principal stress differs from both the major and minor principal stresses,
such as in the soil adjacent to slopes and retaining walls.

Investigations of the type implied above have been carried out by
Hirschfeld, Clough, and others, utilizing constant-volume (undrained)

tests. From the results of these investigations it appears that the strength of saturated clay is about 15 to 25 per cent smaller when $\sigma_2 = \sigma_1$ than when $\sigma_2 = \sigma_3$. The difference in strength in the two cases is generally ascribed to the greater pore pressure rise accompanying shear in the tests in which $\sigma_2 = \sigma_1$. The effective stress and, therefore, the strength on the failure plane is directly related to the pore pressure. It seems reasonable to conclude that truly intermediate values of σ_2 have some proportional effect on the strength.

d. Dimensions of Test Specimen

Compression test specimens are generally, for convenience, made in cylindrical form. The trimming may be accomplished using a device in which the soil can be held and rotated to different positions. For soft clays a thin wire drawn along a vertical guide is used to carve away the soil lying outside the desired cylinder as the sample is rotated. When the trimming is done freely by hand, it is often easier to carve a specimen which is square in cross section. Such square specimens are suitable for unconfined compression tests provided the cross section is uniform throughout the length of the specimen; but because of the nature of the equipment used they are not suitable for triaxial tests. A sharpened cylinder with a rolled-in edge may be used for soils that cannot be trimmed by other methods.

The loading head and base of compression testing devices offer some frictional restraint against lateral deformation of the test specimens near their ends. As a result, the state of stress in conical masses of the soil adjacent to the head and base is unknown, and differs from the state of stress elsewhere in the specimen. If the failure surface intersects either the loading head or base, the test results will be affected, indicating a strength which may be unreliable. Some tests have been performed using segmented loading caps that offer little, if any, restraint to lateral deformation. However, the apparatus has not been generally adopted because of the additional complications introduced into the testing procedure.

In compression tests of isotropic specimens, the plane of failure is rarely inclined to the horizontal by an angle greater than about 61°. Consequently, a specimen having a length 1.8 times its diameter may be barely long enough to accommodate a failure surface which intersects neither end. In practice, the ratio of length to diameter is usually made to fall within the range of $2:1$ to $2\frac{1}{2}:1$. In some of the earlier compression tests of clay, cubes of soil were used, with the result that the strength of the soil was overestimated.

e. Properties of Membrane

The measured strength of a specimen enclosed in a thin elastic membrane depends to a limited extent upon the physical properties of the membrane. In routine soil testing, the effect of the membrane is usually neglected.

Gilbert and Henkel, at the Imperial College of Science and Technology in London, performed careful experimental investigations to evaluate the effect of rubber membranes used in compression tests. The effects of the membrane were found to be somewhat greater for triaxial compression tests, in which the necessary corrections agree well with the compression shell theory, than in unconfined compression tests where the corrections are more nearly in agreement with the hoop tension theory. In any case, the strength contributed by the membrane at failure of the soil specimens was found to depend upon the thickness of the membrane and the strain at failure; but to be independent of the confining pressure and the strength of the specimen. Specifically, Gilbert and Henkel reported the effects to be as indicated in Table 11.16a when failure corresponded to 15 per cent axial strain.

Table 11.16a
Corrections in psi to be Deducted from Compressive Strength
After Gilbert and Henkel

Test	Thickness of Rubber Membrane		
	Thick (0.5 mm)	Standard (0.2 mm)	Thin (0.1 mm)
Triaxial	1.4	0.6	0.3
U/C	0.7	0.25	0.1

It was also indicated, but not well substantiated, that the corrections should be about 50 per cent of the tabulated values for failure at 5 per cent strain, and about 25 per cent for failure at $2\frac{1}{2}$ per cent strain. These corrections may constitute a significant part of the strength of soft clays, and even of medium stiff clays if thick membranes are used.

Rubber latex membranes permit some air to migrate from the chamber fluid into the specimen, complicating particularly the interpretation of the results of triaxial tests of long duration. This problem may be largely overcome by using deaired water in the triaxial chamber. Water, also, will pass through the membrane by osmosis when there exists a difference in the concentration of salts in the pore water and in the water used in the compression chamber. Osmotic flow may become a serious problem in tests of more than a few hours duration if the difference in salinity is great.   In such cases, it is sometimes helpful to use two membranes which are separated by a layer of silicone oil or grease.   Passage of either air or water into the specimen complicates the measurement of pore pressures and the interpretation of test results.

11.17   THE MEASUREMENT OF PORE PRESSURES

After the relationship among total and effective stresses and pore water stress was set forth by Terzaghi during the third decade of this century, it became apparent that a knowledge of the magnitude of the pore pressure would permit a closer understanding of the compressibility and strength characteristics of soils.   By 1936, Rendulic had succeeded in measuring the pore pressures which develop during undrained triaxial tests.   Rendulic reported the results of these tests at the First International Conference on Soil Mechanics, held at Harvard University in 1936.   Since that time, many different types of apparatus have been developed for the measurement of pore pressures, both in the laboratory and in the field.

The purpose of all pore pressure measurement devices is to permit the measurement of stress in the water which is contained in the soil voids without, at the same time, affecting the quantity being measured.   It seems unlikely that this has been accomplished.   However, the most satisfactory types of apparatus in use today nearly do attain this goal.   In principle, all of these devices establish an egress for the escape of pore water from the specimen, and then provide a means for the immediate application and measurement of that pressure just necessary to prevent any flow of the pore water.   In practice, the procedure is complicated by the presence of small quantities of air or gas in the specimen or in the measuring system, by the volumetric changes which occur within the system as the pressures vary, and by variation of the temperature during long time tests.

The pore pressures measured are those near the ends of the specimen, when porous stones provide the egress, or those in the immediate vicinity of a porous needle-like probe inserted through the membrane into the interior of the specimen.   There is always a question as to how nearly the measured pressures represent the pore pressures in the usually restricted zone of failure.   If slow rates of strain are used in the test, it appears that the pore pressure is practically uniform

throughout the specimen, at least up to the moment of failure.   At the
time of failure, the comparatively large deformations in the failure
zone so affect the soil structure that it is logical to expect the development of an appreciable hydraulic gradient within the specimen.
Whether the pore pressures existing in the failure zone, at the instant
of failure, should be counted the cause or the consequence of failure is a
question which probably has not been fully answered.   Yet, the
general reliability and practical value of the results achieved with the
instruments now available has been established beyond doubt.

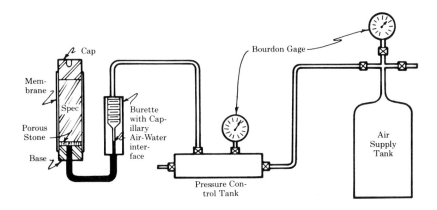

Figure 11.17a
Pore Pressure Measurement Apparatus

All pore pressure measuring systems contain some device (often
called the null apparatus) by means of which the internal pore pressures
can be balanced or nullified by an externally applied pressure.   This
may take the form of an interface inside a capillary tube, between
water and air or between water and mercury; it may consist of some
type of pressure cell or diaphragm which separates the internal and
external systems, or an expandable bellows to separate the two systems.   As an illustration, a system of the first mentioned type is
sketched in Figure 11.17a.   An elaborate explanation of the device is
not necessary.   The soil specimen encased in a membrane is positioned
inside a pressure cell, together with cap and base.   The water in the
voids is a part of a saturated system which includes the porous stone
and the tube leading to the capillary.   With the air-water interface
initially stabilized opposite a reference mark on the capillary tube, any

change in cell pressure or axial load produces a dislocation of the interface. To the maximum extent possible, the water surface is maintained stationary at the reference mark by adjusting the air pressure in the pressure control tank. The pressure required to maintain equilibrium is indicated on the Bourdon gage and is equal to the pressure which exists in the pore water. In practice, it may be necessary to add make-up water to the system, or to provide a movable reference mark, to compensate for compression of entrapped air and expansion of the pressure line between the specimen and the capillary tube.

11.18   PORE PRESSURE COEFFICIENTS

The pore pressure changes which accompany changes in confining pressure or in deviator stress were discussed qualitatively in Article 11.12. In 1954, A. W. Skempton of Imperial College suggested that a general quantitative expression for the change of pore pressure may be written in the form

$$\Delta u = B[\Delta\sigma_3 + A(\Delta\sigma_1 - \Delta\sigma_3)] \qquad \text{(Eq 11.18}a\text{)}$$

in which A and B are empirical parameters which have been termed pore pressure coefficients. The coefficient B depends upon the porosity of the soil and upon the relative compressibilities of the fluid and the soil structure. The coefficient A is influenced principally by the type of soil structure and by the stress history and sensitivity of the soil.

For practical purposes B is equal to unity for saturated soils, to zero for dry soils, and is intermediate for partially saturated soils. The coefficient A generally falls within a range of about -0.6 for heavily overconsolidated clays to $+1.5$ or 2.0 for extrasensitive clays. The coefficients may be determined experimentally during laboratory tests in which pore pressures are measured. When so determined and utilized in preliminary design, the coefficients should be checked by means of pore pressure measurements taken during construction. Expressions for the shear strength of clay, in terms of the coefficients A and B and the slope and position of the strength line based on effective stresses, are given by Skempton and Bishop in the reference cited at the end of this chapter.

There is a growing tendency to employ effective stress principles and pore pressure coefficients in the design of earth structures.

11.19   TRUE VALUES OF c AND ϕ

The design of earth structures and foundations can be accomplished quite satisfactorily with no knowledge of the true magnitude of the individual components, cohesion and friction, which combine to give cohesive soil its strength.   The parameters c and ϕ used in the preceding articles are experimentally determined mathematical constants which may have no relation to the true physical properties of the soil. However, a considerable amount of effort by many of those engaged in soil mechanics research has been directed toward separating and measuring these elusive properties.   Yet, certainly, the efforts have not met with unqualified success.

As an illustration, suppose that the true angle of friction is to be determined, as has been done, by inference from the orientation of the failure planes in test specimens.   From mechanics it is known that the true angle of internal friction of an isotropic material is given by the expression $\phi = 2\alpha_f - 90°$, in which α_f is the angle between the failure plane and the plane of major principal stress.   The failure plane must be identified and the angle measured before excessive deformation alters the orientation.   When this procedure is carried out, ϕ is usually presented as the mean value obtained from a number of tests of presumably identical specimens.   The deviation of individual values is generally important enough to foster some skepticism concerning the reliability of the mean as a true physical property.

If the true cohesion is assumed to be entirely dependent upon the water content, as seems reasonable, a different procedure may be used to determine the true angle of internal friction for a given soil.   In Figure 11.19a are shown typical results of strength tests for a clay initially remolded at a water content near the liquid limit so that its cohesion is practically zero at the start.   The S-Line may be established by the normal procedure.   Suppose that a specimen is consolidated under some pressure p_p, and then permitted to swell under a reduced pressure σ_B.   Point B lies on the branch of the S-Line which represents a prestressed soil.   The position of point B may be found by loading the specimen to failure in a drained test.   There exists some point A, below p_p on the normal S-Line for which the water content is exactly equal to that of the specimen whose strength corresponds to B. Since the water contents are equal, the cohesion in the two cases is the same, and the difference in strength between B and A must be due entirely to a difference in the frictional component of the shearing resistance. Thus, the slope angle ϕ of the line BA must be the true angle of internal friction, and the difference in strength between B and A is given by $\Delta s = (\sigma_A - \sigma_B) \tan \phi$. The total

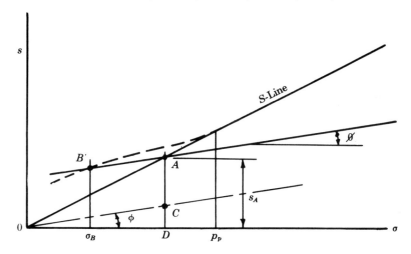

Figure 11.19a
Determination of "True" Value of ϕ

strength s_A is now seen to comprise a frictional component DC and a cohesive component CA.

Unfortunately, the positions of point B and, to a lesser extent, point A are likely to be quite dependent upon the rate at which the tests are performed. Consistent procedures usually produce consistent results for c and ϕ obtained in this manner. But, who is to say which procedure should be used to obtain "true" results?

In summary, it can probably be said that some true value for the angle of internal friction does exist, whether it can be determined or not. On the other hand, it may be questioned whether there even exists a soil property that deserves to be called true cohesion. The cohesive resistance is at best a rather complicated function of the mineral properties and the water content of the soil. Indeed, the cohesion of clay has been found to depend not only upon the water content at failure, but also upon the water content at which the clay was remolded during preparation of the specimen. Possibly, the argument is entirely one of semantics.

11.20 TEST PROCEDURE

A description of apparatus and instructions for conducting shear tests on cohesionless and cohesive soils is given in the corresponding section of Chapter XII, Laboratory Testing. Examples of test data and the resulting curves for several types of tests are given in that chapter.

REFERENCES

Bennett, G. B., and J. G. Mecham, "Use of the Vane Borer on Foundation Investigation of Fill," *Proc. 32nd Annual Meeting Highway Research Board*, 1953.

Bishop, A. W., and D. J. Henkel, *The Measurement of Soil Properties in the Triaxial Test*, London: E. Arnold, 1957.

Bjerrum, L., "Theoretical and Experimental Investigations on the Shear Strength of Soils," *Norwegian Geotechnical Institute*, Pub. No. 5, 1954.

Cadling, L., and S. Odenstad, "The Vane Borer," *Proc. No. 2, Royal Swedish Geotechnical Institute*, Stockholm, 1950.

Casagrande, A., and W. L. Shannon, "Research on Stress-Deformation and Strength Characteristics of Soils and Soft Rocks Under Transient Loading," *Harvard Soil Mechanics Series*, No. 31, 1948.

Casagrande, A., and S. D. Wilson, "Prestress Induced in Consolidated-Quick Triaxial Tests," *Harvard Soil Mechanics Series*, No. 42, 1953.

Casagrande, A., and S. D. Wilson, "Effect of Stress History on the Strength of Clays," *Harvard Soil Mechanics Series*, No. 43, 1953.

Casagrande, A., and S. D. Wilson, "Effect of Rate of Loading on the Strength of Clays and Shales at Constant Water Content," *Geotechnique*, Vol. 11, No. 3, (1951).

Collin, A., *Landslides in Clay*, A translation by W. R. Schriever of Collin's work published in 1846 in Paris. University of Toronto Press, 1956.

Henkel, D. J., and G. D. Gilbert, "The Effect of the Rubber Membrane on the Measured Triaxial Compression Strength of Clay Samples," *Geotechnique*, Vol. 111, No. 1, (1952).

Hilf, J. W., "An Investigation of Pore Water Pressure in Compacted Cohesive Soils," *Tech. Memo. 654*. U.S. Dept. of Interior, Bur. of Reclamation, Denver, 1956.

Hirschfeld, R. C., "Factors Influencing the Constant Volume Strength of Clays," Ph. D. Thesis, Harvard University, May, 1958.

Seed, H. B., R. L. McNeill, and J. de Guenin, "Increased Resistance to Deformation of Clay Caused by Repeated Loading," *Journal Soil Mechanics and Foundations Division*, Proc. Am. Soc. C. E., Vol. 84, No. SM2, (May, 1958).

Skemptom, A. W., "The Pore Pressure Coefficients A and B," *Geotechnique*, Vol. 1V, No. 4, (1954).

Skempton, A. W., and A. W. Bishop, "Soils," *Building Materials—Their Elasticity and Inelasticity*, Editor, M. Reiner, New York: Interscience Publishers, Chapter X.

PROBLEMS

11.1 Name three factors which have an important influence on the angle of internal friction of cohesionless soils. What is the probable physical explanation for differences in the angle of internal friction which occur with variations of these factors?

11.2 What is meant by the term *critical void ratio*? Explain its practical significance and discuss the suitability of laboratory determined values of the critical void ratio for use in field applications.

11.3 Under what conditions will there be danger of a flow slide in an earthen embankment?

11.4 For a soil whose shear strength is given by the expression $s = \sigma \tan \phi$ derive an expression for

(*a*) the angle α_f between the failure plane and the plane of the major principal stress, in terms of the angle of internal friction ϕ, and

(*b*) the relationship among the principal stresses, σ_1 and σ_3, and ϕ.

11.5 In a bar of steel the major principal stress is 10,000 pounds per sq in. compression and the minor principal stress is 3000 pounds per sq in. tension. Find the normal and shearing stresses on a plane making an angle of 10° with the plane on which the minor principal stress is acting.

11.6 A sample of cohesionless sand is subjected to a vertical pressure of 3 kg per sq cm in a direct shearing apparatus and fails after a shearing force of 2 kg per sq cm has been applied. Determine by means of Mohr's diagram the magnitude and direction of the principal stresses acting on an element in the *region* of failure.

11.7 The following stress conditions are known at a point in a mass of dense cohesionless sand:

Normal stress on horizontal plane = 3.5 kg cm^{-2}
Normal stress on vertical plane = 2.2 kg cm^{-2}
Shearing stress on each plane = 1.0 kg cm^{-2}

Determine by means of Mohr's diagram the magnitudes and directions of the principal stresses. Is this state of stress safe against failure?

11.8 A cylindrical sample of cohesionless sand is subjected to a triaxial compression test. At the start of the test both principal stresses were 3.2 kg cm^{-2}. During the test the lateral pressure was kept constant at 3.2 kg cm^{-2}, while the axial pressure was increased to a total of

11.5 kg cm^{-2} whereupon failure occurred. *Compute* the angle of internal friction of this sand.

11.9 Failure during a triaxial test on a dense sand occurred under a confining (chamber) pressure of 1.0 kg/sq cm and a total axial pressure of 4.0 kg/sq cm.

(a) Assuming that the relationship between normal stress and shearing strength is linear, compute the angle of internal friction and the angle which the failure planes make *with the axis* of the sample.

(b) If the same sample had been exposed at the start to an all around hydrostatic stress of 6.0 kg/sq cm, how much would one have to reduce the axial stress to cause failure? What then would be the angle of the failure planes with the axis of the sample?

11.10 During a triaxial S test, a cylindrical specimen of fine sand failed when the *deviator stress* reached a value of 5 kg/sq cm. The failure plane was well defined, making an angle of 28° with the axis of the specimen. The confining pressure in the chamber is unknown because of a broken pressure gage. Estimate the value of the confining pressure.

11.11 In a drained triaxial compression test in which a chamber pressure of 2 kg/sq cm was maintained, a sand specimen failed when the axial *deviator stress* reached a magnitude of 3.32 kg/sq cm. Determine the angle of internal friction of the sand. State whether you believe the sand specimen to have been in a loose or a dense state, giving reasons for your belief.

11.12 An unconfined compression test was performed on a cylindrical specimen of clean silt at the shrinkage limit. Failure occurred under a vertical load when the stress reached a magnitude of 4 kg/sq cm. A well defined shear plane was formed at an angle of 70° with the horizontal. Estimate the height to which capillary water will rise above the water table in this soil.

11.13 Briefly explain why the strength of an intact, saturated clay specimen is independent of the confining pressure which is used during a Q test.

11.14 Explain why the effective stresses determined on the basis of pore pressure measurements during a series of R tests may not define a strength line which is the same as that found from a series of S tests on the same soil.

11.15 In a triaxial S test on a normally consolidated clay, at failure, the lateral pressure was 3 kg/sq cm and the total vertical pressure 9 kg/sq cm. Locate the failure plane and determine the stresses on this plane at failure.

11.16 A specimen of soft clay is consolidated under a confining pressure of 2 kg/sq cm in a triaxial test apparatus and then loaded to failure in an S test. The deviator stress at failure was 4 kg/sq cm. Determine ϕ.

An identical specimen of the soft clay was consolidated as above and loaded to failure in an R test. The specimen failed under a deviator stress of 2.5 kg/sq cm. What was the pore pressure u at failure?

11.17 An undisturbed sample of a normally consolidated, homogeneous clay is obtained from a depth of 10 feet below the surface, and two triaxial test specimens are prepared from the sample.

(a) A consolidated-drained (S) test is performed, in which failure occurs under an all-around pressure of 2 kg/sq cm and a total axial stress of 6 kg/sq cm. Sketch Mohr's strength circle and compute the corresponding angle of internal friction.

(b) A consolidated-undrained (R) test is performed, in which failure occurs under an all-around pressure of 2 kg/sq cm and a total axial stress of 4.1 kg/sq cm. Sketch Mohr's strength circle for total stresses and compute the apparent angle of internal friction.

(c) Show on a sketch the position of the strength circle for the R test if effective stresses rather than total stresses are plotted on the Mohr diagram. Compute the neutral (pore water) stress in the specimen at the time of failure in the R test.

11.18 A sample of clay at the liquid limit was loaded in a direct shearing machine and consolidated under a vertical pressure of 3.0 kg/sq cm. The corresponding shearing resistance was found to be 1.6 kg/sq cm. The pressure was then increased to 5.0 kg/sq cm and after a time, insufficient for complete consolidation, the shearing resistance was found to be 2.1 kg/sq cm. What was the degree of consolidation under the increase in pressure at the time the second shearing test was made?

CHAPTER XII

Laboratory Testing

12.01 GENERAL

Intelligent testing to determine physical properties of materials cannot be carried on without a thorough knowledge of the effect which the manner of testing has upon the results of the tests. For this reason, it is important that all testing for determining the properties of soils be done under the direct supervision of the engineer responsible for the design of the structure.

The test procedures listed below are not presented as standard methods for determining the physical properties of soils. Standard tests for some of the physical properties have been adopted by A.S.T.M. It is important that some tests, such as those for determining liquid and plastic limits, be made following a standardized procedure so that results of the tests are comparable with those for other soils and with those from other laboratories.

Since the results of tests are of use only as aids to the engineer's judgment, the engineer should feel free to change the procedure of some of the tests and to improvise other tests in an effort to provide information of maximum usefulness.

The equipment described, except for a few standardized items, such as the liquid limit device, is made in a variety of forms and may vary from laboratory to laboratory. This is especially true of loading devices. Some of the equipment, such as that described for the constant head and falling head permeability tests, is of a simple type which can be assembled from easily procurable parts. Equipment to perform the same function but of somewhat different design is available from manufacturers of test equipment. In general, this commercially available equipment does not perform its function better than that described here but, in some cases, is more convenient to use.

12.02   IDENTIFICATION FOR REFERENCE

In order to preserve continuity and to show conformity of the test procedures with the rational development of theoretical considerations, the same identification numbers (but with the addition of the letter L) are used in this section as were used for the same subject in the earlier text. For example: In Chapter II, Specific Gravity of Solids, Section 2.04, Test Procedure, the reader is referred to the chapter on Laboratory Testing for detailed instructions for performing the test for determining the specific gravity of the soil solids in a sample of soil. In this chapter on Laboratory Testing, the detailed instructions are designated II-L, Specific Gravity of Solids, Section 2.04L, Test Procedure.

Specific Gravity of Solids

2.04L TEST PROCEDURE

(Using 500 cc volumetric flask)

a. Cohesionless Material

1. Place somewhat more soil than required for the sample in an evaporating dish, dry in an oven at 105° C and cool in a desiccator.
2. Weigh to nearest 0.1 g and record as Wt Sample dry + Tare.
3. Pour through a funnel, being careful not to lose any material, about 80 to 100 g of the material into a clean 500 cc volumetric flask in which there has already been placed about 150 cc of water.
4. Weigh dish and remainder of material and record as Tare.
5. Fill flask with distilled water at room temperature to within about ½ inch of the neck.
6. Apply vacuum to flask for about 10 minutes, gently rolling the flask to remove entrapped air.

7. Fill flask with water until the bottom of the meniscus is even with the calibration mark and remove all excess water from inside the neck and outside the flask.
8. Weigh and record as W_{bws}.
9. Determine temperature of water in flask to nearest 0.2° C and record.
10. Determine W_{bw} as follows, or take from calibration curve for flask used.

> Fill same flask as used for determining W_{bws} with distilled water to slightly below the calibration mark and subject to vacuum for a few minutes to remove dissolved air.
>
> Add water until bottom of meniscus is even with calibration mark and remove all water from inside neck and outside flask.
>
> Weigh to nearest 0.1 g and record as W_{bw}.
>
> Determine temperature of water in flask to nearest 0.2° C and record. This temperature should not differ by more than 0.5° C from the temperature for W_{bws}.

11. Compute G_s from data.

b. Cohesive Material

1. Mix with a spatula in an evaporating dish about 100 to 125 g of the sample with enough water to form a uniform paste or slurry.
2. Place sample in cup of mechanical mixer, add water to make about 250 cc of suspension and mix for about 15 minutes or until there is complete dispersion of the soil particles in the water.
3. Pour and rinse sample into a clean volumetric flask and add distilled water at room temperature, keeping the water level well below the neck of the flask.
4. Apply vacuum to flask for about 15 minutes, gently rolling flask to remove entrapped air.   Add water to about $\frac{1}{2}$ inch below the calibration mark and apply vacuum for 2 to 3 minutes.
5. Add water until bottom of meniscus is even with calibration mark and remove all excess water from inside the neck and from outside the flask.   If this water is added gently, it will remain on top of the muddy water so that the meniscus can be clearly seen.
6. Weigh and record as W_{bws}.
7. Determine temperature to nearest 0.2° C and record.

8. Pour and rinse all water and sample into a 1000 cc evaporating dish, place in a drying oven at 105° C and leave until all water is driven off. Place in desiccator to cool.
9. Weigh dish and dry sample and record as Wt Dry + Tare.
10. Determine W_{bw} as described for Cohesionless Material.
11. Compute value of G_s.

c. Example Data

On the following page is given the data from a test for determining the specific gravity of solids in a soil.

2.05L CALIBRATION OF FLASK

The following procedure may be followed for calibrating flasks:

1. Determine W_{bw} at room temperature in the same manner as given in **a.** 10 under Test Procedure and record both W_{bw_1} and T_1 on Data and Computation Sheet for Calibration of Volumetric Flask.
2. Place the flask filled with deaired distilled water in a water bath at a constant temperature approximately 10° C above room temperature and leave until the water in the flask is the same temperature throughout.
3. Take temperature and see that bottom of meniscus is at the calibration mark before removing flask from bath. Record temperature as T_2.
4. Remove flask from bath, wipe off excess water from inside of neck and from outside of flask. Weigh and record as W_{bw_2}.
5. Determine α_v from the relationship

$$\alpha_v = \frac{(W_{bw_1} - W_{bw_2}) - (\gamma_{T_1} - \gamma_{T_2})V_b}{(T_1 - T_2)V_b\gamma_T}.$$

6. Repeat determination for several trials, and use consistent results or average value of α_v.
7. Using the value of α_v as determined above, compute values of ΔW_{bw} between room temperature and temperature of 10°, 20°, 25°, 30°, and 35° C from the relationship

$$\Delta W_{bw} = [(\gamma_T - \gamma_{T_1}) + \alpha_v(T - T_1)]V_b.$$

SOIL MECHANICS LABORATORY
SPECIFIC GRAVITY DETERMINATION
Data and Computation Sheet

Tested by ___Parcher___ Sample No. ___IC-1___

Date ___Oct. 8, 1954___ Test No. ___1___

Tested for ___—___ Sheet No. ___1___

Location of Sample ___Portland, Maine___

Position of Sample _____

Description of Sample ___Gray, medium plastic clay___

Trial No.	1		
Flask No.	C4		
Method of Air Removal	Vac.		
W_{bws}	706.74		
Temperature, T° C	24.5		
W_{bw}	655.67		
Evap. Dish No.	G4		
Wt. Sample Dry + Tare	472.67		
Tare	392.52		
W_s	80.12		
G_s	2.76		

W_{bws} = Weight of Flask + Water + Sample at T° C

W_{bw} = Weight of Flask + Water at T° C

W_s = Weight of Dry Soil

G_s = Specific Gravity of Solids = $\dfrac{W_s}{W_s + W_{bw} - W_{bws}}$

8. Add these differences to W_{bw_1} at room temperature to obtain values of W_{bw} corresponding to these temperatures.

9. Plot curve showing the relationship between W_{bw} and T to such a scale that W_{bw} can be read to the closest 0.1 g for a difference of 0.2° C.

Because the values used in these determinations are differences between numbers of small differences, it is necessary to make all weighings accurate to the nearest 0.01 g and temperature readings accurate to the nearest 0.1° C. This will be readily apparent from a study of the example which follows.

The curve may also be determined experimentally by making duplicate weighings at not less than 3 different temperatures and plotting the relationship between W_{bw} and T. In using this experimental method, it is desirable to have rooms of three different temperatures: a cool room, a normal room, and a warm room. Temperatures of the rooms should differ by about 10° C.

Data and computation taken and used in the calibration of a 500 cc volumetric flask are given on the following pages.

SOIL MECHANICS LABORATORY
CALIBRATION OF VOLUMETRIC FLASK
Data and Computation Sheet

Calibrated by___*Parcher*___   Flask No.___*5*___
Date___*Oct. 6, 1954*___   Cal. Temperature___*20°C.*___
                                   Cal. Vol., V_{bc}___*500* cc___

Relationship between Weight of Flask + Water, W_{bw}, and Temperature, T.

$$W_{bw} - W_{bw_1} = (\gamma_T - \gamma_{T_1})V_{b_1} + \alpha_v(T - T_1)\gamma_T = \Delta W_{bw}$$

For practical purposes use $V_{b_1} = V_{b_1}, \gamma_T = V_{bc}$

$$W_{bw} - W_{bw_1} = [(\gamma_T - \gamma_T) + \alpha_v(T - T_1)]V_{bc}$$

Determination of Coefficient of Volume Expansion, α_v

$$\alpha_v = \frac{\Delta W_{bw} - V_{bc}\Delta\gamma_T}{V\Delta_{bc}\Delta T}$$

Computation of Weight of Flask + Water, W_{bw}, corresponding to Temperature, T.

$$\Delta W_{bw} = (\Delta\gamma_T + \alpha_v\Delta T)V_{bc}$$
$$W_{bw} = W_{bw_1} + \Delta W_{bw}$$

Trial No.	1	2	
W_{bw_1}	646.85	646.80	
$(-)W_{bw_2}$	645.78	645.75	
ΔW_{bw}	1.07	1.05	
T_1	19.2	19.6	
$(-)T_2$	28.5	28.5	
ΔT	-9.3	-8.9	
γ_{T_1}	.998365	.998284	
$(-)\gamma_{T_2}$	.996088	.996088	
$\Delta\gamma_T$	.002277	.002196	
$V_{bc}\Delta\gamma_T$	1.1385	1.0980	
$\Delta W_{bw} - V_{bc}\Delta\gamma_T$	-0.0685	-0.0480	
$V_{bc}\Delta T$	-4650	-4450	
$\alpha_v \times 10^4$	0.147	0.108	
$\alpha_v(av)$	0.1275 ×10⁻⁴		

T	10° C	15° C	20° C	25° C	30° C	35° C
$(-)T_1$	19.2	19.2	19.2	19.2	19.2	19.2
ΔT	-9.2	-4.2	0.8	5.8	10.8	15.8
$\alpha_v\Delta T \times 10^{-4}$	-1.1730	-0.5353	0.1020	0.7395	1.3770	2.0145
γ_T	.99970	.99910	.99820	.99704	.99565	.99403
$(-)\gamma_{T_1}$	.99836	.99836	.99836	.99836	.99836	.99836
$\Delta\gamma_T$	.00134	.00064	-.00016	-.00132	-.00271	-.00433
$\alpha_v\Delta T + \Delta\gamma_T$	.00117	.00059	-.00015	-.00124	-.00252	-.00413
ΔW_{bw}	0.585	0.295	-0.075	-0.62	-1.26	-2.065
W_{bw_1}	646.85	646.85	646.85	646.85	646.85	646.85
W_{bw}	647.43	647.14	646.78	646.23	645.59	644.79

$(-)$ means subtract pos. value from value above
W_{bw_1} = Wt. Flask + Water at room temp., T_1
W_{bw} = Wt. Flask + Water at temperature T
T_1 = Room temperature
T = Temp. corresponding to W_{bw}
γ_{T_1} = Unit wt. water at room temp.
γ_T = Unit wt. water at temperature T

Weigh to nearest 0.01 gm.
Take temperature to nearest 0.1° C

Remarks. _____

SOIL MECHANICS LABORATORY
CALIBRATED CURVE FOR VOLUMETRIC FLASK
Showing

Relationship between Weight of Flask + Water and the corresponding Temperature.

Calibrated by___Parcher_____   Flask No._____5_____
Date____Oct. 6, 1954_____   Cal. Temp._____20° C_____
⊙ Computed points   Cal. Vol., V_{bc}____500 cc_____
⊛ Measured points

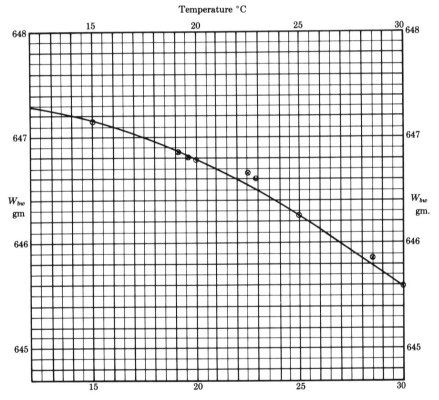

Temperature °C

Temperature °C

Grain Size Distribution

3.05L CALIBRATION OF HYDROMETER AND CYLINDER

The distance the grains of diameter D' have settled in time t in terms of the hydrometer reading may be determined by the following procedure.

1. Partially fill lucite tube with water and allow to run through the connection to the burette to eliminate the air in the connection. Close stop, empty burette and allow water to run into burette until bottom of meniscus is at 0 mark. Fill cylinder to hook gauge.
2. Insert hydrometer into cylinder and allow to float free.
3. Open cock and allow water to run into burette until surface of water is again at hook gauge. Record burette reading and take difference in readings as V_b.
4. Raise burette until exactly half the water has run back into the cylinder.
5. Tape end of stem of hydrometer to meter stick and immerse hydrometer until surface of water is at hook gauge. Surface of

382

water is at center of immersion.   Read meter stick at index and record for zero H'.

6. Lower stick and hydrometer until bottom mark on hydrometer stem is approximately even with the hook gauge.   Lower water in cylinder until surface is at hook gauge.   Read hydrometer at surface of water and record as R_H.   Read index and record.

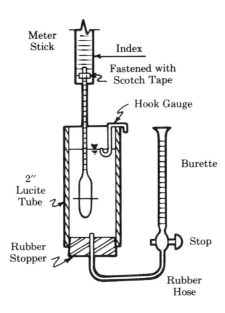

Figure 3.05a
Hydrometer Calibration

7. Lower hydrometer to next major division on stem and lower water to hook gauge.   Read and record R_H and index.
8. Repeat until R_H and index readings have been determined for all major divisions on hydrometer.
9. Difference between index reading for any R_H and the initial index reading is the value of H' for that R_H.

The correction to be subtracted from the measured distances H' found above in order to determine the depth of settlement when the hydrometer is out of the suspension, H, can be determined as follows:

1. Put about 250 cc of water into the cylinder that is to contain the suspension, and mark the height of the water surface on the outside of the cylinder.   A wet piece of paper is a good mark.

2. Add a carefully measured volume of water (about 500 cc) to the water in the cylinder and again mark the surface of the water in the cylinder.
3. Carefully measure the distance between the two marks by means of a vernier caliper or by other accurate means.
4. Determine the cross-sectional area of the cylinder by dividing the volume of added water by the height of rise in the cylinder.
5. Determine the correction by dividing the volume of the hydrometer bulb, V_b, by twice the area of the cylinder.

The volume of the bulb can be determined by the procedure listed above, by immersion in a small diameter cylinder, or by weight. Because the hydrometer measures specific gravity, its volume in cubic centimeters is its weight in grams.

SOIL MECHANICS LABORATORY
CALIBRATION OF HYDROMETER AND CYLINDER
Data Sheet

Calibrated by __Parcher__   Date __Oct. 1954__

Hydrometer No. __6895__   Cylinder No. __E 2__

Meniscus Correction __0.2__

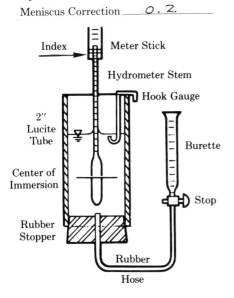

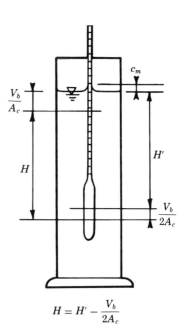

All readings taken with water surface at hook gauge. R_H read at surface of water.

	Bur. Read.	
Hydrometer out	73.20	cm³
Hydrometer in	25.96	cm³
Difference = V_b	47.24	cm³
For $V_b/2$	49.58	cm³

$$H = H' - \frac{V_b}{2A_c}$$

Vol. of Water added __500__ cm³

Hgt. Rise in Cylinder __17.3__ cm

Area of Cylinder __28.90__ cm²

$$\frac{V_b}{2A_c} = 0.817 \text{ cm}$$

R_H	Index cm	H' cm	H cm
C.I.	1.00	0	-0.82
40	9.11	8.11	7.29
30	11.93	10.93	10.11
20	14.86	13.86	13.04
10	17.74	16.74	15.92
0	20.74	19.74	18.92

SOIL MECHANICS LABORATORY
CALIBRATION OF HYDROMETER AND CYLINDER

$R_H - H$ Curve

Calibrated by __Parcher__   Date __Oct. 1954__

Hydrometer No. __6895__   Cylinder No. __E 2__

Meniscus Correction __0.2__

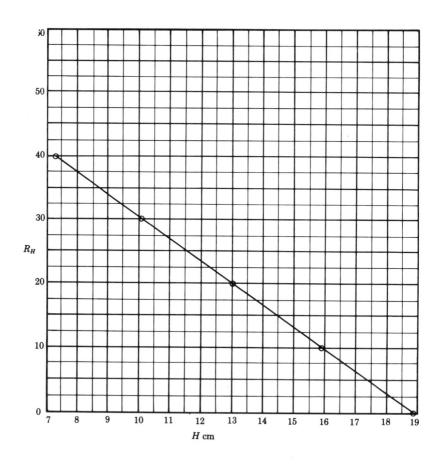

H cm

3.08L   TEST PROCEDURE FOR HYDROMETER ANALYSIS

1g. Calibrate hydrometer and glass cylinder as described in the preceding section, and mark values of R_H on the H scale of nomographic chart for solution of Stokes' Law.

2g. Determine meniscus correction c_m and record on data sheet. This correction can be determined by reading the top and the bottom of the meniscus around the stem when the hydrometer is inserted in clear water.   The difference in these readings is the meniscus correction. The stem of the hydrometer must be clean for a fully developed meniscus.   The hydrometer is calibrated to read the correct specific gravity only when the meniscus is fully developed around the stem.

a. Cohesionless Soil

1. Separate sample into coarse and fine fractions by washing through a 100 mesh sieve.
2. Oven dry and determine dry weight of two fractions.
3. Weigh out to nearest 0.01 g between 50 and 100 g of the oven dry fine fraction.   Record as W_s.
4. Add this sample to about 800 cc of distilled water in the calibrated glass cylinder and fill cylinder to the 1000 cc mark.   (If colloidal material is present, the addition of a dispersing agent may be necessary in order to prevent flocculation.   See Cohesive Material.)
5. Determine temperature of suspension to nearest 0.1° C.   Record on data sheet opposite first hour readings.
6. Place palm of hand over top of cylinder so that water will not leak when cylinder is turned upside down, and then reverse positions of top and bottom of cylinder about once every 2 seconds for about 1 minute to obtain a uniform suspension.
7. Set cylinder upright and immediately start stop watch   Record time of start.
8. Insert hydrometer into the suspension.   Push hydrometer slightly below floating position to wet stem so that meniscus will be fully formed.
9. Read hydrometer at the top of the meniscus as soon as possible. Record elapsed time and hydrometer reading R_H on the data and computation sheet.
10. Read hydrometer at $\frac{1}{2}$ min, 1 min, and 2 min from start of sedimentation and record these elapsed times and readings.

11. Immediately after the 2 min reading, withdraw hydrometer very slowly from suspension. Wipe with dry, clean cloth and lay on table until next reading period. Do not hold hydrometer by the stem, except in a vertical position.

12. Take readings at elapsed times of 4 min, 8 min, 15 min, 30 min, 1 hr, 2 hr, etc. after start of sedimentation, doubling elapsed time between readings. After 8-hr reading, read once each day. One should start inserting the hydrometer about 20 to 30 seconds before reading is taken. Remove and dry hydrometer after each reading. It is not important that the readings be taken at exactly the intervals stated above, but the exact time at which the readings are taken should be recorded. This doubling of time intervals between readings spaces points fairly evenly on the grain size distribution curve when the diameter is plotted to a logarithmic scale.

13. After the first hour, take and record temperature immediately before or after each reading.

14. Continue to take readings until the readings remain constant, indicating that all particles larger than colloidal sizes have settled out.

15. Compute W_D per cent from relationship developed in Section 3.04 and determine corresponding diameters from solution of Stokes' Law. The nomographic chart may be used for determining diameters from Stokes' Law.

16. Draw grain size distribution curve on semilog paper as described in Section 3.09, using points found as result of sieve and hydrometer analyses.

b. Cohesive Soil

A great many clays flocculate when placed in suspension in distilled water. For these clays, it is necessary to add a dispersing agent to the mixing water. The same agent will not work for all clays, but sodium silicate can be used successfully with many clays and is quite commonly used. Sodium hexametaphosphate is frequently used as a dispersing agent. The solution of sodium silicate or other dispersing agent should be used for all mixing of the soil as well as for the suspending fluid.

The amount of dispersing agent required to prevent flocculation varies with different clays. An amount sufficient to increase the specific gravity of the suspending water 0.0002 to 0.0005 is satisfactory for most clays, although considerably more is required for some clays containing considerable salts. About $\frac{1}{2}$ to 1 cc of 40° Baumé solution of sodium silicate is sufficient to increase the specific gravity of 1000 cc of water 0.0002 to 0.0005.

1. Place 1000 cc of distilled water at room temperature in a glass cylinder and check its specific gravity.
2. Add to the water in the glass cylinder an amount of dispersing agent which has been found adequate to prevent flocculation of a sample of the clay to be analyzed for grain size distribution. Determine the specific gravity of the solution.
3. Determine the difference in specific gravity of the distilled water and the solution, and record as c_d.
4. To a small amount of water in an evaporating dish, add a sample of soil large enough to weigh 30 to 50 g after drying, using a spatula to form a smooth paste.
5. Place the sample with about 300 cc of water in the cup of a mechanical mixer and mix for about 15 minutes in order to break the sample into individual particles. The water used for mixing should be the solution of dispersing agent which has been checked for increase in specific gravity. If the increase in specific gravity produced by the addition of a specific amount of the concentrated dispersing agent solution is known, steps 1, 2, and 3 may be omitted, and the total amount of concentrated solution added to the sample in the mixing bowl and distilled water used for mixing.
6. Rinse the dispersed slurry into the calibrated cylinder and add water with the dispersing agent to form 1000 cc of suspension. If the total amount of dispersing agent and distilled water were used for mixing, add distilled water to form 1000 cc of suspension.
7. Carry out steps 5 to 14 as described for cohesionless soils.
8. After all readings have been taken, pour and rinse suspension into an evaporating dish and evaporate the water in a drying oven at 105° C. Hydrochloric acid added to the suspension will cause the colloidal particles to flocculate and settle out in two or three hours. The clear water can be siphoned off leaving considerably less water to be evaporated in the oven.
9. Weigh dried sample and record as W_s.
10. Compute W_D per cent and D' for all readings.
11. Plot grain size distribution curve on semilog paper.

c. Example Data

Data and computation for a hydrometer analysis follow. This soil did not require a dispersing agent to prevent flocculation.

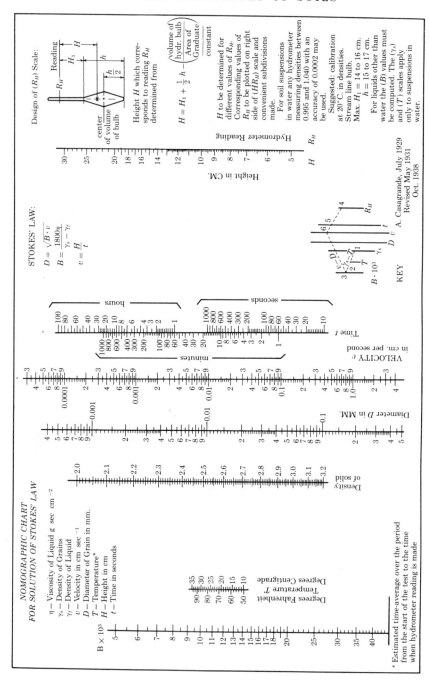

NOMOGRAPHIC CHART
FOR SOLUTION OF STOKES' LAW

η – Viscosity of Liquid g sec cm^{-2}
γ_s – Density of Grains
γ_l – Density of Liquid
v – Velocity in cm sec^{-1}
D – Diameter of Grain in mm.
T – Temperature*
H – Height in cm
t – Time in seconds

STOKES' LAW:

$$D = \sqrt{B \cdot v}$$
$$B = \frac{1800\eta}{\gamma_s - \gamma_l}$$
$$v = \frac{H}{t}$$

A. Casagrande, July 1929
Revised May 1931
Oct. 1938

Design of (R_H) Scale:

$H = H_1 + \frac{1}{2}h - \left(\frac{\text{volume of hydr. bulb}}{\text{Area of Graduate}}\right)$ constant

H to be determined for different values of R_H. Corresponding values of R_H to be plotted on right side of (HR_H) scale and convenient subdivisions made.

For soil suspensions in water any hydrometer measuring densities between 0.995 and 1.040 with an accuracy of 0.0002 may be used.

Suggested: calibration at 20° C. in densities. Stream line bulb.
Max. H_1 = 14 to 16 cm.
h = 15 to 17 cm.
For liquids other than water the (B) values must be computed. The (γ_s) and (T) scales apply only to suspensions in water.

Height H which corresponds to reading R_H determined from

Hydrometer Reading R_H

Height in CM.

Time t

hours

seconds

minutes

VELOCITY v
in cm. per second

Diameter D in MM.

Density of solid

Temperature T

Degrees Centigrade

Degrees Fahrenheit

*Estimated time-average over the period from the start of the test to the time when hydrometer reading is made

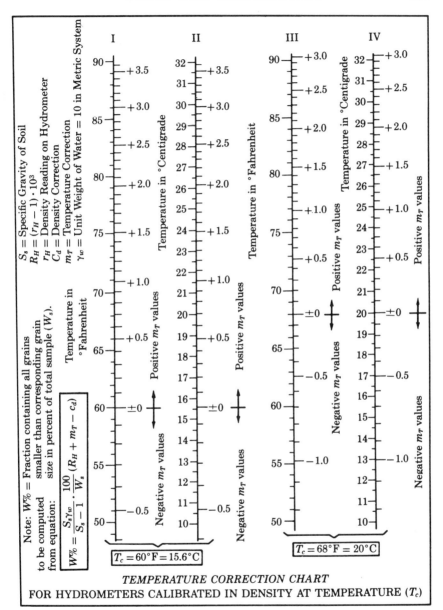

Note: $W\%$ = Fraction containing all grains smaller than corresponding grain size in percent of total sample (W_s).

to be computed from equation:

$$W\% = \frac{S_s \gamma_w}{S_s - 1} \cdot \frac{100}{W_s} (R_H + m_T - c_d)$$

S_s = Specific Gravity of Soil
$R_H = (r_H - 1) \cdot 10^3$
r_H = Density Reading on Hydrometer
C_d = Density Correction
m_T = Temperature Correction
γ_w = Unit Weight of Water = 10 in Metric System

$T_c = 60°F = 15.6°C$

$T_c = 68°F = 20°C$

TEMPERATURE CORRECTION CHART
FOR HYDROMETERS CALIBRATED IN DENSITY AT TEMPERATURE (T_c)

Properties of Water
0° to 50° C

Temp. °C	Volume cm³	Spec. Gravity	Density g cm⁻³	Viscosity × 10⁴		
				Absolute		Kine-matic cm sec
				dyne sec/cm²	g sec/cm²	
0	1.00013	0.99987	0.999841	179.21	0.1827	0.1827
1	1.00007	0.99993	0.999900	173.13	0.1765	0.1765
2	1.00003	0.99997	0.999941	167.28	0.1706	0.1706
3	1.00001	0.99999	0.999965	161.86	0.1652	0.1652
4	1.00000	1.00000	0.999973	156.74	0.1598	0.1598
5	1.00001	0.99999	0.999965	151.88	0.1549	0.1549
6	1.00003	0.99997	0.999941	147.28	0.1502	0.1502
7	1.00007	0.99993	0.999902	142.84	0.1457	0.1457
8	1.00012	0.99988	0.999849	138.60	0.1413	0.1413
9	1.00019	0.99981	0.999781	134.62	0.1373	0.1373
10	1.00027	0.99973	0.999700	130.77	0.1333	0.1333
11	1.00037	0.99963	0.999605	127.13	0.1296	0.1296
12	1.00048	0.99952	0.999498	123.63	0.1261	0.1262
13	1.00060	0.99940	0.999377	120.28	0.1226	0.1227
14	1.00073	0.99927	0.999244	117.09	0.1194	0.1195
15	1.00087	0.99913	0.999099	114.04	0.1163	0.1164
16	1.00103	0.99897	0.998943	111.11	0.1133	0.1134
17	1.00120	0.99880	0.998774	108.28	0.1104	0.1105
18	1.00138	0.99862	0.998595	105.59	0.1077	0.1079
19	1.00157	0.99843	0.998405	102.99	0.1050	0.1052
20	1.00177	0.99823	0.998203	100.50	0.1025	0.1027
21	1.00198	0.99802	0.997992	98.10	0.1000	0.1002
22	1.00221	0.99780	0.997770	95.79	0.0977	0.0979
23	1.00244	0.99756	0.997538	93.58	0.0954	0.0956
24	1.00268	0.99732	0.997296	91.40	0.0932	0.0935
25	1.00294	0.99707	0.997044	89.37	0.0911	0.0914
26	1.00320	0.99681	0.996783	87.37	0.0891	0.0894
27	1.00347	0.99654	0.996512	85.45	0.0871	0.0874
28	1.00375	0.99626	0.996232	83.60	0.0853	0.0856
29	1.00405	0.99597	0.995944	81.80	0.0834	0.0837
30	1.00435	0.99567	0.995646	80.07	0.0816	0.0820
31	1.00466	0.99537	0.995343	78.40	0.0799	0.0803
32	1.00497	0.99505	0.995023	76.79	0.0783	0.0787
33	1.00530	0.99473	0.994703	75.23	0.0767	0.0771
34	1.00563	0.99440	0.994373	73.71	0.0752	0.0756
35	1.00598	0.99406	0.994033	72.25	0.0737	0.0741
36	1.00633	0.99371	0.993683	70.85	0.0722	0.0727
37	1.00669	0.99336	0.993333	69.47	0.0708	0.0713
38	1.00706	0.99299	0.992963	68.14	0.0695	0.0700
39	1.00743	0.99262	0.992593	66.85	0.0682	0.0687
40	1.00782	0.99224	0.992213	65.60	0.0669	0.0674
41	1.00821	0.99186	0.991833	64.39	0.0656	0.0661
42	1.00861	0.99147	0.991443	63.21	0.0644	0.0650
43	1.00901	0.99107	0.991043	62.07	0.0633	0.0639
44	1.00943	0.99066	0.990633	60.97	0.0622	0.0628
45	1.00985	0.99025	0.990223	59.88	0.0611	0.0617
46	1.01028	0.98982	0.989793	58.83	0.0600	0.0606
47	1.01072	0.98940	0.989373	57.82	0.0589	0.0595
48	1.01116	0.98896	0.988933	56.83	0.0579	0.0585
49	1.01162	0.98852	0.988493	55.88	0.0570	0.0577
50	1.01207	0.98807	0.988043	54.94	0.0560	0.0567

SOIL MECHANICS LABORATORY
HYDROMETER ANALYSIS

Tested by __Parcher__ Sample No. __H-211-B__

Date __Oct. 15, 1954__ Test No. __1__

Tested for _____ Sheet No. __1__

Location of Sample __Portland, Maine__

Position of Sample _____

Description of Sample __Gray, medium plastic clay__

Hydrometer No. __6895__ Evaporating Dish No. __FP__

Cylinder No. __E2__ Wt. Sample Dry + Dish __433.43 g__

$$W_D\% = \frac{100}{W_s} \cdot \frac{G_s}{G_s - 1} (R_H + m_T - c_d)$$

Tare __387.60__

$$W_D\% = (3.42)(R_H + m_T - c_d)$$

$\dot{W}_s$ __45.83__

Meniscus Correction, $c_m =$ __0.2__

Spec. Gr. Solids, G_s __2.76__ Est. Det.

$c_d =$ __0__ Correction for increase in density of liquid due to addition of dispersing agent.

Date	Temp. °C	Time	Elapsed Time	R'_H	$R_H = R'_H + c_m$	D mm	m_T	$R_H + m_T - c_d$	$W_D\%$
10/15	28.7	1611	0	—	—	—	—	—	—
			30 sec	1.0257	25.9	0.0600	1.95	27.85	95.2
		1612	1 min	1.0255	25.7	0.0420		27.65	94.5
		1613	2	1.0252	25.4	0.0300		27.35	93.5
		1616	5	1.0252	25.4	0.0190		27.35	93.5
		1619	8	1.0242	24.4	0.0150		26.35	90.1
	28.5	1626	15	1.0237	23.9	0.0110	1.91	25.81	88.2
	28.2	1641	30	1.0224	22.6	0.0080	1.81	24.41	83.5
	27.7	1711	60	1.0218	22.0	0.0057	1.69	23.69	81.0
	27.2	1811	120	1.0199	20.1	0.0042	1.47	21.67	74.1
	26.3	2011	240	1.0190	19.2	0.0030	1.35	20.55	70.3
10/16	25.8	0811	16 hrs.	1.0149	15.1	0.0016	1.22	16.32	55.8
10/18	20.5	0811	64	1.0130	13.2	0.00086	0.09	13.29	45.4
10/19	20.8	0911	89	1.0119	12.1	0.00072	0.15	12.25	41.9
10/20	24.2	1011	114	1.0109	11.1	0.00062	0.84	11.94	40.8
10/22	24.9	1421	164.2	1.0091	9.3	0.00052	1.01	10.31	35.3

SOIL MECHANICS LABORATORY
GRAIN SIZE DISTRIBUTION CURVE

Tested by___Porcher_____   Sample No.___H-211-B____

Date_____   Test No.___1____

Tested for_____   Sheet No.___2____

Location of Sample___Portland, Maine_____

Position of Sample_____

Description of Sample___Gray Medium Plastic Clay___

Dia. at 60%, $D_{60} =$ ___0.0017___ mm   Spec. Gravity, G_s___2.76___

Effective Dia., $D_e = D_{10}$___.001___ mm   Est.   Det.

Uniformity Coefficient, $C_u = \dfrac{D_{60}}{D_{10}} =$ ___1.7_____

Coefficient of Permeability by Hazen's Formula, $k = 100 D^2{}_{10}$_____ cm sec^{-1}

(Valid only for very coarse clean sands and gravels) (Note: D_{10} in cm.)

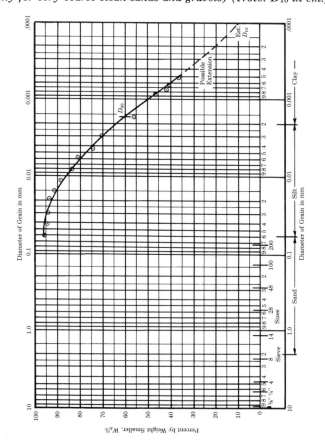

Plasticity

4.07L PROCEDURE FOR LIQUID LIMIT TEST

The determination of the liquid limit can be carried out as follows (preferably in a humid room to prevent excessive evaporation):

1. Mix in an evaporating dish about 500 g of soil with enough water to form a stiff paste somewhat below the liquid limit. Work thoroughly with a spatula to a uniform smooth paste. Remove any pebbles or other lumps that would interfere with cutting a smooth groove.
2. Check the fall of the cup in the liquid limit device and adjust, if necessary, so that the height of fall is exactly 1 cm. It is important that this measurement be made between the base and the point on the cup which comes in contact with the base.
3. Place enough of the paste in the cup to cover about the front three-fourths of the cup to a depth sufficient that the grooving tool will cut a full depth groove through the middle portion of the sample. With a spatula, work the sample in the cup to a uniform consistency, being careful to work out any entrapped air.

Smooth out the surface with the broad side of the spatula run along the edge of the cup.

4. Cut a groove in the sample by holding the grooving tool perpendicular to the inside surface of the cup with the rounded edge to the front, and pulling it through the sample down the middle from back to front in line with the hinge at the back. Scrape the material off the tool onto the edge of the cup.

5. Be sure that the bottom of the cup and the top of the base are clean. Turn the crank about 2 revolutions per second until the groove comes together for about $\frac{1}{2}$ inch along the bottom, keeping a record of the number of blows required to close the groove. Rework the sample, cut a new groove, and check the number of blows to close the groove. When the number of blows checks for two repetitions, record the number of blows and immediately remove about 5 g of the sample along the groove with a clean spatula and place in one of two matched watch glasses or in a small aluminum moisture can. Cover the sample immediately with the other matched watch glass or with the moisture can lid to prevent evaporation. Weigh the container with the sample to the nearest milligram, as soon as possible, and record as Sample + Tare. Open the container and place in a drying oven at 105° C.

   If the sample is so stiff in the beginning that more than 35 blows are required to close the groove, a small amount of water should be worked with the sample so that not more than 35 blows are required. If the mixture is too stiff, the time required to close the groove may be so great as to make it difficult to check the number of blows because of excessive evaporation. There is also the possibility that the point may lie below the linear portion of the flow curve thereby introducing an error.

6. Add a small amount of water to the sample in the cup to reduce the number of blows required to close the groove by 3 to 5. Work to a uniform consistency and repeat step 5.

7. Add enough water to reduce the stiffness such that about 15 to 17 blows will close the groove and repeat step 5.

8. Add water to reduce the number of blows by 3 or 4 and repeat step 5.

9. After the samples taken from the groove have thoroughly dried, remove the containers from the oven, replace the covers, and place in a desiccator to cool. Be sure that the parts of the containers are properly matched.

10. Weigh the containers and covers with the dry samples to the nearest milligram and record as Sample Dry + Tare.

11. By subtraction determine the weight of the water and the weight of the solids for each trial, and compute the water content corresponding to each number of blows.

12. Plot these 4 to 6 points on semilogarithmic paper with the water content w to an arithmetic scale and the number of blows to the logarithmic scale. Draw the straight line best fitting these 4 to 6 points. Find on this flow curve the water content at 25 blows, which is the liquid limit w_L. Determine the difference in water content over one cycle of the logarithmic scale and record as the flow index I_F.

4.09L PROCEDURE FOR PLASTIC LIMIT TEST

The plastic limit may be determined as follows:

1. Roll out a small pellet of soil of uniform consistency and somewhat above the plastic limit on a dry smooth glass, paper towel, or other smooth surface, into a thread about $\frac{1}{8}$ inch in diameter.

2. Gather up the thread, knead and roll into a pellet and again roll into a $\frac{1}{8}$ inch thread.

3. Repeat step 2 until the thread breaks or crumbles into parts $\frac{1}{8}$ to $\frac{1}{2}$ inch long.

4. Place the crumbles of the thread into a covered container, weigh, dry out, reweigh, and compute the water content as described for liquid limit. Two or three determinations should be made for a check.

SOIL MECHANICS LABORATORY
WATER CONTENT AND LIMIT TESTS

Tested by __W. H. Hall__   Sample No. __C - 10__

Date __11/8/50__   Test No. __1__

Tested for __O.S.U. Library__   Sheet No. __1__

Location of Sample __Library site, OSU Campus, Stillwater, Ok.__

Position of Sample __Bottom Elevator Pit NW corner__

Description of Sample __Dark reddish brown silty clay__

Liquid Limit, w_L

Trial No.	1	2	3	4		
Container No.	4	7	11	21		
No. of Blows	40	31	16	13		
Wt. Sample Wet + Tare	31.5102	33.2511	32.0819	35.8427		
Wt. Sample Dry + Tare	30.3332	31.7154	30.4286	34.2895		
Wt. Water	1.1770	1.5357	1.6533	1.5532		
Tare	26.9178	27.3623	26.0930	30.2683		
Wt. Dry Soil, W_s	3.4154	4.3531	4.3356	4.0212		
Water Content, w_L%	34.7	35.3	38.1	38.7		

Plastic Limit, w_P

Trial No.	1	
Container No.	19	
Wt. Sample Wet + Tare	31.7504	
Wt. Sample Dry + Tare	30.9590	
Wt. Water	0.7914	
Tare	26.7690	
Wt. Dry Soil, W_s	4.1900	
Water Content, w_P%	18.9	

Natural Water Content, w_n

Trial No.	1	
Container No.	E-5	
Wt. Wet + Tare	275.97	
Wt. Dry + Tare	251.81	
Wt. Water	24.16	
Tare	138.37	
Wt. Dry Soil, W_s	113.44	
Water Content, w_n	21.2	

Shrinkage Limit, w_{SL}

State		
Container No.		
Vol. of Container, V_1		
Wt. Sample Wet + Tare		
Tare		
Wt. Sample Wet, W_1		
Wt. Dry Soil, W_s		
Wt. Dish + Hg.		
Wt. Dish		
Wt. Hg.		
Vol. Dry Soil, V_{SL}		
Shrinkage Limit, w_{SL}%		

Spec. Grav., $G_s = \dfrac{2.74}{\text{Est.} \qquad \text{Det.}}$

Spec. Grav. of Mercury, $G_q = 13.6$

$$w_{SL} = \frac{(V_{SL} - W_s/G_s\gamma_0)\gamma_w}{W_s}$$

$$w_{SL} = \frac{W_1 - (V_1 - V_{SL})\gamma_w - W_s}{W_s}$$

$$G_s = \frac{W_s}{V_1\bar\gamma_0 - W_1 + W_s}$$

SOIL MECHANICS LABORATORY
WATER CONTENT AND LIMIT TESTS
Flow Curve and Summary

Tested by __W. H. Hall__   Sample No. __C-10__

Date __11/8/50__   Test No. __1__

Tested for __Library site__   Sheet No. __2__

Data from Preceding Form

Flow Curve

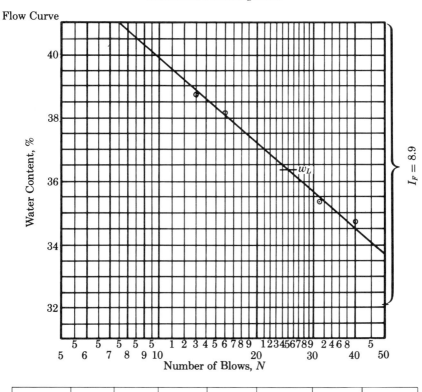

Nat. Water Con. w_n	Liquid Lim. w_L	Plastic Lim. w_P	Plastic Index, I_P	Flow Index, I_F	Toughness Index, I_T	Liquidity Index, I_L	Shrinkage Limit, w_{SL}
21.2	36.4	18.9	17.5	8.9	1.98	0.131	

Structure

5.11L TEST PROCEDURE FOR STANDARD COMPACTION TEST

The standard compaction can be made by following the procedure listed below.

1. Air-dry sample of soil to 5 to 10 per cent moisture content (well below optimum).
2. Screen out particles larger than those passing the No. 4 sieve.
3. Thoroughly pulverize and mix about 3000 g of the soil, preferably in a flat square baking pan using a trowel with the point cut off.
4. Place in the mold enough soil to compact to about one third the volume of the compaction cylinder.
5. Place tamper on top of soil, lift to top of travel (1 foot) and allow to fall free to compact the soil. Move hammer, lift and let fall. Repeat until 25 blows have been struck and the entire mass has been compacted to the same degree.
6. Add to the mold enough soil that will compact to fill the middle third of the compaction cylinder and compact same as first third.

7. Add enough soil to compact to slightly above the top of the lower portion of the mold, and compact same as for bottom and middle thirds.

8. Carefully remove the extension without disturbing the compacted soil in the lower part of the mold, remove excess soil with a butcher knife and form a smooth flat surface even with top of the compaction cylinder with a straight edge.

9. Remove cylinder with sample from base. Weigh the cylinder and sample and record as Wt. Cylinder + Soil.

10. Remove sample from cylinder into mixing pan saving a generous sample from the center of the sample in a moisture can. Determine water content of sample removed from center of sample and record.

11. Add with a sprinkler a small amount of water to the entire sample in the mixing pan and thoroughly mix to a homogeneous state.

12. Repeat steps 4 to 10 inclusive using this slightly wetter mixture.

13. Add another small amount of water and repeat for another water content and wet density determination.

14. Continue this procedure until the weight of the compacted sample is less than that of the previous trial. This lower weight indicates that the density of the compacted soil is decreasing with increased water content and that the optimum moisture content has been exceeded.

15. Compute wet and dry densities and void ratios for all trials and plot unit dry weight and void ratio against water content and estimate the optimum water content from the curve.

16. Plot zero air voids curve on the water content-dry density graph.

Test data, computations and curves for two compactive efforts are given on the following pages.

SOIL MECHANICS LABORATORY
COMPACTION TEST FOR OPTIMUM MOISTURE CONTENT
Data and Computation Sheet

Tested by __W. H. Hall__   Sample No. __3__

Date __7/9/47__   Test No. __2__

Tested for __Cities Ser. Gas Co.__   Sheet No. __1__

Location of Sample __Compressor Site - Grant Co., Kan.__

Position of Sample __Test Pit - 12 ft. below surface__

Description of Sample __Lqt. gray silt (loess) w/white lumps $CaCO_3$__

Compaction Cylinder No. __1__   Dia. of Compaction Cyl. __10.16 cm.__

Vol. of Compaction Cyl. __946 cm^3__   Hgt. of Compaction Cyl. __11.67 cm.__

Method of Compaction __Standard Proctor - 3 lifts - 25 blows__

Test No.	1	2	3	4	5	
Wt. of Cylinder + Soil	3698	3796	3905	3975	3979	
Wt. of Cylinder	1966	1966	1966	1966	1966	
Wt. of Compacted Soil	1732	1830	1939	2009	2013	
Wet Density $gm\ cm^{-3}$	1.83	1.935	2.05	2.12	2.12	
Tare No.	P-1	P-2	P-3	P-4	P-5	
Wt. Sample Wet + Tare	65.0	77.1	69.3	94.2	99.9	
Wt. Sample Dry + Tare	63.2	74.3	66.1	88.7	91.7	
Wt. of Water	1.6	2.8	3.2	5.5	8.2	
Tare	28.2	30.2	27.6	32.8	27.8	
Wt. of Dry Soil	35.2	44.3	38.5	55.9	63.9	
Water Content, w	4.55	6.34	8.32	9.85	12.84	
Dry Density, $\gamma\ dry\ {}^{gm\ cm^{-3}}_{lb\ ft^{-3}}$	1.75 / 109	1.82 / 113	1.89 / 118	1.92 / 120	1.88 / 117	
Void Ratio, e	0.530	0.471	0.417	0.395	0.425	

Wet Density $= \dfrac{\text{Wt. of Compacted Soil}}{\text{Volume of Compaction Cylinder}}$

Dry Density $= \dfrac{\text{Wet Density}}{1 + \text{Water Content}}$

$G_s =$ __2.68__

      Est.   Det.

Void Ratio, $e = \dfrac{G_s \gamma_0}{\gamma_{dry}} - 1$

Remarks _____

SOIL MECHANICS LABORATORY
OPTIMUM MOISTURE CONTENT

Tested by __W. H. Hall__   Sample No. __3__

Date __7/9/47__   Test No. __2 & 3__

Tested for __Cit. ser. Gas Co.__   Sheet No. __3__

Location of Sample __Compressor Sta. Site - Grant Co., Kan.__

Position of Sample __Test Pit - 12 ft. below surface__

Description of Sample __Lgt. gray silt (loess) w/white CaCO₃__

Method of Compaction __Standard Proctor - 3 lifts__

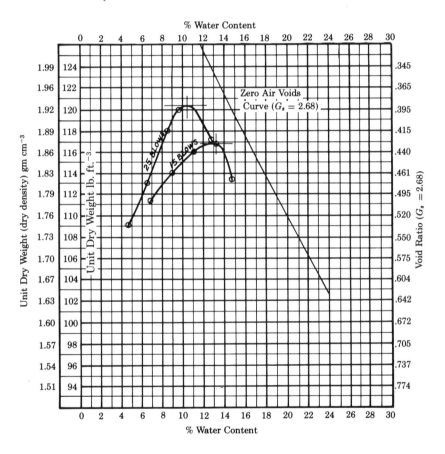

5.16L   DENSITY OF SOIL IN PLACE

a. Sand Cone

Instructions for calibration and use.

1. Fill fruit jar with clean sand, weigh, and record weight.
2. Screw cone onto jar.
3. Place density plate over container of known volume, such as the $\frac{1}{30}$ cu ft mold used in the Proctor compaction test.
4. Place cone with jar of sand over opening in plate.
5. Open valve and allow sand to fill container, cone and plate.
6. Close valve, lift cone off plate, turn upside down, and remove cone from fruit jar.
7. Weigh jar and remaining sand.   Difference in two weighings is weight of sand required to fill container, cone, and plate.
8. Strike off surface of sand in container, being careful not to vibrate to a denser state.
9. Weigh sand in container and determine its unit weight by dividing weight of sand by volume of container.
10. Determine weight of sand required to fill cone and plate from difference in weight of sand emptied from fruit jar and weight of sand used to fill the container.

In using the sand cone apparatus, the following procedure may be followed.

1. Smooth off surface of soil and lay plate on the smooth surface.
2. Excavate from inside the opening in the plate a pit about 6 inches deep, removing all loose particles and saving all soil removed in an airtight container for future drying out to determine W_s.
3. Place cone and weighed jar of calibrated sand over the opening.
4. Open valve and allow sand to fill cone and pit.
5. Close valve, turn jar and cone upside down, remove cone, and weigh jar and remaining sand.
6. Determine weight of sand required to fill pit from difference in weighings of jar and sand minus weight of sand required to fill cone and hole in plate as determined by calibration.
7. Divide weight of sand required to fill pit by unit weight of sand to determine volume of pit, which is total in place volume of the sample removed.

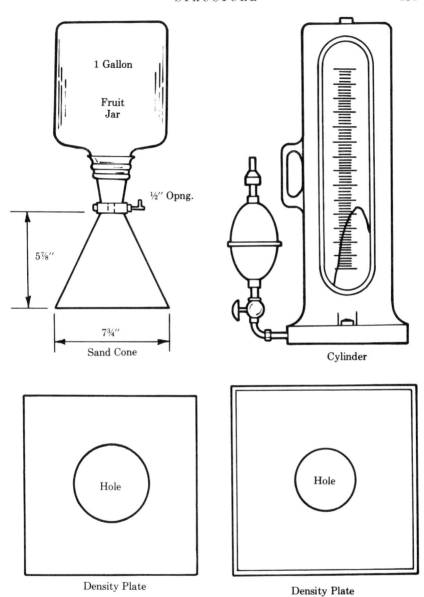

1 Gallon

Fruit
Jar

½" Opng.

5⅞"

7¾"

Sand Cone

Cylinder

Hole

Hole

Density Plate

Density Plate

Figure 5.16b
Sand Cone Density Apparatus

Figure 5.16c
**Rubber Balloon Density
Apparatus**

In using the sand cone method, the density of the sand as used must be the same as that used in the calibration. If the sand in the cone and pit is vibrated more or less than in the calibration, the density will not be the same and the volume of the excavated soil will not be determined accurately.

b. Rubber Balloon

The volume of a pit from which a sample of soil has been removed can be determined with the rubber balloon apparatus by following the instructions listed below.

1. Smooth off surface of soil and lay plate on this surface.
2. Place cylinder over plate, open air valve, and pump air into the cylinder until balloon is completely deflated against the surface of the soil in the opening, and read volume of water in cylinder.
3. Remove cylinder from plate.
4. Excavate pit through hole in plate, removing all loose particles and saving all soil in an airtight container for future drying out to determine W_s.
5. Place cylinder over opening in plate, open air valve and pump air into cylinder forcing balloon into excavation until cylinder is raised off plate.
6. Push cylinder down on to plate and read volume of water in cylinder.
7. Determine volume in place of soil sample removed from the excavated pit.

In using this method, care should be exercised to have the inside of the pit smooth with rounded corners so that the balloon will completely fill the excavated hole and not bridge over small pits and corners in the surface.

Permeability

8.08L CONSTANT HEAD PERMEABILITY TEST

A test for determining the coefficient of permeability of a cohesionless material can be performed in the permeameter described in Section 8.08 by following the procedure listed below.

1. Dry in an oven at 105°C a quantity of sand to be tested, and store in a desiccator to cool.
2. With a vernier caliper, carefully measure the diameter of the lucite tube, and record on the data sheet.
3. Clamp the screen over the bottom end of the lucite tube and place a filter of sand or gravel coarse enough to contribute negligible resistance to the flow of water through the specimen and the filter. The filter material should be fine enough to prevent the entrance of the finer particles of the specimen into the pores of the filter. Place copper screen disk on filter.
4. Weigh the dried sand and dish and record as Wt Dry + Tare.
5. Being careful not to spill any sample, pour into the permeameter (through a funnel) enough sample to form a specimen of the

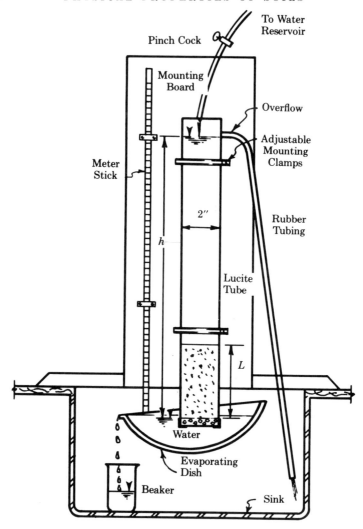

Figure 8.08a
Constant Head Permeameter

desired length.   The specimen can be left loose or compacted to
a medium or dense state by vibrating and rodding the sand as it
is placed in layers.   Weigh the dish and remaining sand and
record as Tare.

6. Support an evaporating dish under the lucite tube containing
   the specimen at such an elevation that the surface of water in
   the filled dish is at or slightly above the bottom of the specimen
   (above the top of the filter).   Tilt the dish so that it will overflow at the lip.

It is important that all the water that flows through the specimen in time t is collected for accurate measurement. Usually, all the water can be made to flow into the collecting vessel by laying a bent nail on the lip of the evaporating dish as shown in Figure 8.08c. Spreading a

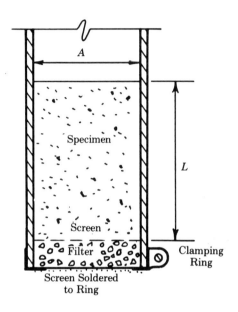

Figure 8.08b
Detail of Sample

film of oil or grease on the outside of the dish at the lip also helps to prevent water from flowing down the outside of the dish.

7. Carefully measure the length of the specimen, L, and record on the data sheet.

8. Allow water to run into the lucite tube to fill the tube to over-flowing and adjust the flow so that the overflow tube will carry off the excess water and maintain a constant surface elevation of the water in the permeameter.

9. Carefully measure the distance between the surfaces of water in the dish and in the lucite tube and record as h.

10. Determine temperature of water and record on data sheet.

11. After the water has flowed through the specimen long enough to establish steady flow, start a stop watch and insert a beaker under the lip of the evaporating dish to collect the water that flows through the specimen.

12. After enough water for an accurate measurement (50 to 100 cc) has been collected, stop the watch and remove the beaker with the collected water.
13. Record the time in seconds required to collect the water.
14. Weigh or measure the quantity of water collected to the nearest 0.1 cc and record as Q.
15. Compute the coefficient of permeability from Darcy's Law
$$k = \frac{QL}{hAt}.$$
16. Compute void ratio of specimen.
17. Compute k for 20° C for comparison with results of other tests made at other temperatures.

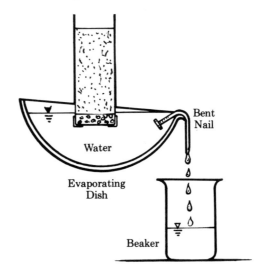

Figure 8.08c
Directing Water Into Beaker

The constant head permeability test described above is suitable for cohesionless soils having coefficients of permeability of 10^2 to 10^{-3} cm sec^{-1}. For soils having coefficients less than 10^{-3} cm sec^{-1}, the falling head permeability test is more suitable than the constant head.

Data from a test performed as described is given on the following page.

SOIL MECHANICS LABORATORY
CONSTANT HEAD PERMEABILITY TEST
Data Sheet

Tested by __Parcher__   Sample No. __A-1__

Date _____   Test No. __1__

Tested for _____   Sheet No. _____

Location of Sample __Arkansas River, Ralston, Okla.__

Position of Sample __Surface__

Description of Sample __Well graded silica-feldspar sand__

Permeameter No. __1__   Spec. Gr. Solids, G_s __2.68__

Diameter of Specimen __4.58__ cm   Est.   Det.

     Tare No. __6__

Length of Specimen, L __10.2__ cm   Wt. Sample Dry + Tare __1110__ gm

Area of Specimen, A __16.5__ cm^2 Tare __818__ gm

Volume of Specimen, V __168__ cm^3 Wt. of Dry Sample, W_s __292__ gm

$$\text{Void Ratio, } e = \frac{VG_s - W_s}{W_s} = 0.541$$

Temp. of Water $T°$ C	Time t sec	Head h cm	Hydraulic Gradient h/L	Quantity Q cm^3	k_T cm sec^{-1}	k_{20} cm sec^{-1}
22	63	44.7	4.38	178.4	392×10^{-4}	372×10^{-4}
22	120	44.7	4.38	312.0	359×10^{-4}	341×10^{-4}
22	120	44.7	4.38	300.9	346×10^{-4}	328×10^{-4}

$$k_T = \frac{QL}{hAt}$$

$$k_{20} = k_T \frac{\nu_T}{\nu_{20}}$$

8.09L FALLING HEAD PERMEABILITY TEST

The procedure for performing a falling head permeability test depends to some extent upon the type of soil being tested. For coarse grained cohesionless materials the constant head method is usually preferable to a falling head. However, for medium to fine sands a falling head permeameter with a standpipe of the same area as the specimen can be used. The procedure listed below is for a simple permeameter which is suitable for fine sands and silts having coefficients of permeability between 10^{-1} cm sec^{-1} and 10^{-6} cm sec^{-1}, using a large standpipe for the more permeable materials.

a. Cohesionless Material

A falling head permeability test may be made on cohesionless soils by following the procedure given below.

1. Oven dry at 105° C a quantity of cohesionless material to be tested and store in a desiccator.
2. Carefully measure the diameter of the lucite tube with a vernier caliper and record on data sheet.
3. Clamp the screen over the bottom of the lucite tube and place a filter of sand coarse enough to contribute negligible resistance to the flow of water through the specimen and filter. The filter material should be fine enough to prevent the entrance of the finer particles of the specimen into the pores of the filter. Place a copper screen disk on the filter.
4. Weigh the dried material and dish and record as Wt Dry + Tare.
5. Being careful not to spill any sample, pour into the permeameter (through a funnel) enough sample to form a specimen of the desired length. The specimen can be left loose or compacted to any density desired by vibrating and rodding the material as it is placed in layers. Weigh dish and remaining sand and record as Tare.
6. Place a copper screen on top of the specimen in the lucite tube and place a filter as described in 3 leaving room for the insertion of the rubber stopper in the top of the tube.
7. Determine the area of the standpipe, a. This can be done by weighing the dry glass tube forming the standpipe. The tube is then filled almost full of water, fingers placed over the ends,

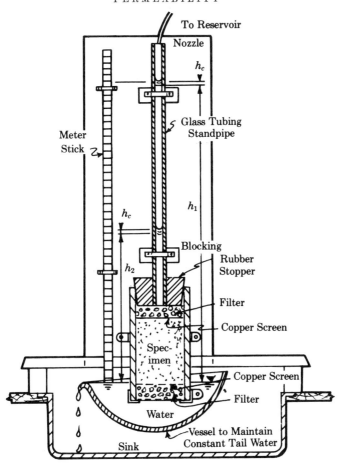

Figure 8.09b
Falling Head Permeameter

and the tube laid horizontally on a balance. When the fingers are removed from the ends of the tube, the water will be held in the tube by capillary forces. The tube and water are weighed and the distance between menisci measured. From the difference in weights the volume of water can be determined and, knowing the length of the column of water in the tube, the area of the tube can be determined. Be sure all water is wiped off the outside of the standpipe before weighing tube and water. If trouble is experienced in keeping the water in the tube with both ends open, one end may be stopped with a small plug of plasticene before the first weighing.

8. Compute the capillary rise in the standpipe from the relationship, $h_c = \dfrac{0.3}{d}$.

9. Place the tube with the prepared specimen on the mounting board and insert the rubber stopper with the standpipe into the top of the lucite tube. The stopper must fit snugly inside the tube or the head of water in the standpipe will force the stopper out of the tube allowing leakage around the outside of the stopper.

10. Carefully measure the length of the specimen, L, and record on the data sheet.

11. Support an evaporating dish or other vessel under the tube containing the specimen at such an elevation that the surface of the water in the filled vessel is at or slightly above the bottom of the specimen.

12. Allow water to run into the top of the standpipe and through the specimen long enough to fill the tail water vessel and remove as much of the entrapped air as practicable. Preferably, the water should be allowed to run into the top of the standpipe so that the specimen is under the pressure of the head in the standpipe only. If a rubber tubing is pushed over the standpipe, the specimen will be under the full head from the reservoir which may force the stopper out of the lucite tube, and which may cause the surface of the water in the standpipe to bounce when the tubing is removed. When water is fed into the standpipe through a nozzle, it is practically impossible to maintain a head without allowing the standpipe to run over, in which case one cannot tell whether or not water is leaking from the standpipe between the stopper and lucite tube or between the standpipe and the stopper. During filling and before starting the test, the rubber stopper may be loosened to allow accumulated air at the top of the specimen to escape. Before testing, the stopper should be tightened leak tight.

13. Determine the temperature of the water and record on the data sheet.

14. Cut off the flow of water into the standpipe to allow the water to fall to a predetermined head, $h_1 + h_c$, and start a stop watch when the water in the standpipe reaches this elevation. When the water has fallen to a predetermined head, $h_2 + h_c$, stop the watch and record the elapsed time t in seconds and the head in cm. Readings should be made to bottom of meniscus.

15. Determine h_1 and h_2 by subtracting h_c from standpipe readings.

16. Compute void ratio of specimen.

17. Compute coefficient of permeability from the relationship
$$k = \frac{La}{tA} 2.303 \log \frac{(h_1 - h_c)}{(h_2 - h_c)}$$ and record as k_T.

18. Compute the coefficient of permeability at 20° C and record as k_{20}.

Data from a test performed as described above are given on a following page.

b. Undisturbed Samples

A falling head permeability test on an undisturbed sample can be performed following the procedure listed below.

1. Form a specimen of the undisturbed sample with relatively smooth surfaces of a size that will fit inside the permeameter, leaving about $\frac{3}{8}$ inch clearance on the sides.

2. Measure the dimensions of the undisturbed specimen, compute the cross sectional area and record A and L. If the natural water content is desired, the specimen should be weighed.

3. Place the specimen in the center of the copper screen attached to the clamp, and tamp lightly into place a thin layer of warm paraffin or wax of very soft consistency between the specimen and the clamp.

4. Place the lucite tube in the clamping ring, push down into the soft paraffin, and clamp the screen containing the specimen to the lucite tube.

5. Mix bentonite and water to a consistency that will flow through a funnel and fill through a funnel the space between the specimen and the wall of the tube, being careful not to allow the bentonite to cover any part of the top surface of the specimen.

6. Allow sufficient time for the bentonite to gel, then cover the top of the gel with fine sand applied through a funnel. Cover the entire arrangement with a layer of coarse sand.

7. Set up the permeameter and run the test as described for cohesionless soils.

8. Compute k_T and k_{20}.

9. Remove specimen, carefully clean off the paraffin and gel, being careful not to lose any of the specimen, and dry in an oven at 105° C. Cool in desiccator and weigh.

10. Compute void ratio from dimensions and weight.

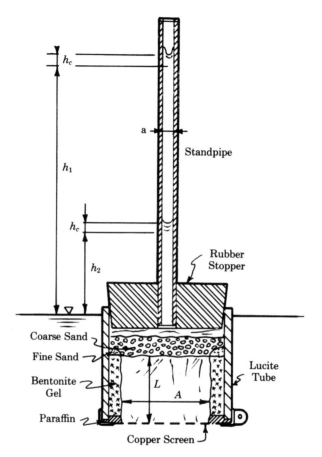

Figure 8.09c
Permeability Test for Undisturbed Sample

SOIL MECHANICS LABORATORY
FALLING HEAD PERMEABILITY TEST
Data Sheet

Tested by __Parcher__   Sample No. __1-P__

Date_____   Test No. __2__

Tested for_____   Sheet No. __1__

Location of Sample __Franklin Falls, N.H.__

Position of Sample __River Bed__

Description of Sample __Clean light tan silty sand__

Permeameter No. __3__   Spec. Gr. of Solids, G_s __2.70__

Diameter of Specimen __4.29__ cm   Tare No. __F.__   Est.   Det.

Length of Specimen, L __10.00__ cm   Wt. Sample Dry + Tare __364.35__ gm

Area of Specimen, A __14.46__ cm^2 Tare __142.72__ gm

Volume of Specimen, V __144.6__ cm^3 Wt. of Dry Sample, W_s __221.63__ gm

Standpipe No. __3-R__   Area of Standpipe, a __0.261__ cm^2

Hgt. of Capillary Rise, h_c __0.52__ cm $h_c = \dfrac{0.3}{d}$ (Dimensions in cm)

Void Ratio, $e = \dfrac{VG_s - W_s}{W_s} = $ __0.76__ $k_T = 2.303 \dfrac{aL}{At} \log \dfrac{(h_1 - h_c)}{(h_2 - h_c)}$

Temp. of Water $T°$ C	Time t Sec.	Head at Start, h_1 cm	Head at End, h_2 cm	k_T cm sec^{-1}	k_{20} cm sec^{-1}	
26	66.6	125.3	107.6	4.11×10^{-4}	3.57×10^{-4}	$k_{20} = k_T \dfrac{\nu_T}{\nu_{20}}$
	153		87.6	4.22×10^{-4}	3.66×10^{-4}	$= k_T (0.868)$
	264		67.6	4.25×10^{-4}	3.69×10^{-4}	

8.11L HORIZONTAL CAPILLARITY TEST

A horizontal capillarity test may be carried out in the following manner.

1. Prepare a specimen in a length of 2 in. diameter lucite tube in the same manner as for the falling head permeability test, except that a short length of glass tubing with a short piece of rubber tubing should be used in place of the long standpipe. See Figure 8.11c.
2. Measure and record dimensions and weights as described for the falling head test.
3. Partially fill a shallow pan with water to a depth barely to cover the lucite tube when immersed.
4. Take temperature of water and record.
5. Immerse the tube with the specimen as shown in Figure 8.11c, keeping the free end of the vent tube exposed to the atmosphere. Start stop watch at the instant of immersion.

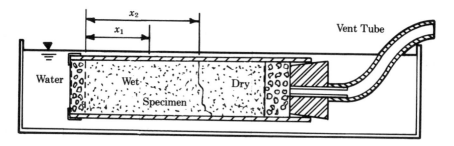

Figure 8.11c
Horizontal Capillarity Test

6. Rotate tube with specimen to keep advance line of wetting approximately vertical.
7. After line of wetting has progressed a short distance, mark its position and record elapsed time from beginning of test.
8. Allow line of wetting to progress another short distance, mark its position and record elapsed time from beginning of test.
9. Repeat step 8 for several values of x and t.
10. Remove tube with specimen from the water and measure the distances from the wet end of the specimen (not including the

filter) to the grease pencil marks and record these distances as x with their corresponding elapsed times t.

11. Plot values of x^2 and t on arithmetic graph paper. These plotted points should fall approximately on a straight line. Draw a straight line best fitting the plotted points and compute m_T as the tangent of the angle between the base and the line showing the rate of advancement of the seepage.

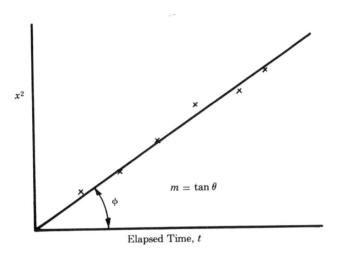

$$m = \tan \theta$$

Elapsed Time, t

Figure 8.11f
Seepage Rate in Capillarity Test

12. Compute m_{20} for 20° C.
13. If value of coefficient of permeability k is available from a direct permeability determination, estimate the value of k for the void ratio of the specimen in the capillarity test from the relationship

$$k_1 = k_2 \frac{e_1^2}{e_2^2}.$$

14. Compute value of z_{20} from the relationship

$$z_{20} = \frac{m_{20}^2}{k_{20}}.$$

Test data from a horizontal capillarity test for a medium sand are given on the following pages.

SOIL MECHANICS LABORATORY
HORIZONTAL CAPILLARITY TEST

Data Sheet

Tested by___Parcher_____   Sample No. ____A____

Date____Nov. 15, 1954_____   Test No. ____101____

Tested for_____   Sheet No. _____

Location of Sample____Franklin Falls, N.H.____

Position of Sample_____River Bed_____

Description of Sample___Clean, light tan silty sand___

Cylinder No. ____3____   Spec. Gr. Solids, G_s __2.70__

Diameter of Specimen __4.29__ cm

Tare No. _____F_____   Est.   Det.

Length of Specimen ___13.70__ cm

Wt. Sample + Tare __540.56__ gm

Area of Specimen, A __14.46__ cm^2

Tare _____228.92_____ gm

Vol. of Specimen, V __198.0__ cm^3

Wt. of Dry Specimen, W_s __311.64__ gm

Temp. of Water, $T°$ C __21__

Void Ratio, $e = \dfrac{VG_s - W_s}{W_s} = 0.72$

Time t min	Distance x cm	x^2
½	3	9.0
1	4	16.0
1½	4.7	22.1
2	5.2	27.0
2½	5.8	33.6
3	6.3	39.7
4	7.3	53.3
5	8.2	67.3
6	8.8	77.5
7	9.7	94.1
8	10.3	106.0

From Plot, $m_T = \dfrac{x^2}{t} = 13.33 \text{ cm}^2\text{min}^{-1}$

$m_{20} = m_T \dfrac{\nu_T}{\nu_{20}} = 13.33(0.977) = 13.02$.

If z is known for soil tested,

$k_{20} = \dfrac{m_{20}^2}{z_{20}} = \underline{\hspace{3cm}}$

If k_{20} is known for soil tested,

$z_{20} = \dfrac{m_{20}^2}{k_{20}} = \dfrac{\overline{13.02^2}}{3.65} = 46.5$

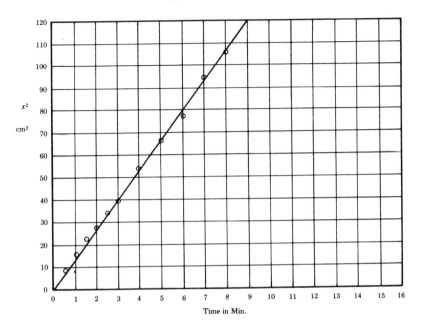

Figure 8.11h
Data from Horizontal Capillarity Test Plotted to Arithmetic Scales

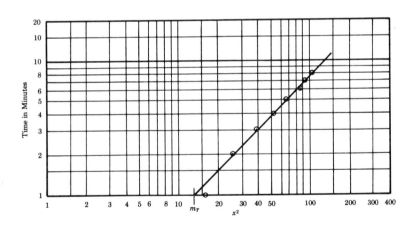

Figure 8.11i
Data from Horizontal Capillarity Test Plotted to Logarithmic Scales

8.16L   WATER SUPPLY FOR PERMEABILITY TESTS

For constant and falling head permeability tests using ordinary tap and distilled water, the test water can be stored in a 5 gallon carboy supported on a shelf above the permeameter.   A rubber stopper in the bottle can be fitted with a glass tube long enough to extend to about 1 inch from the bottom of the bottle and a short glass tube to act as a vent.

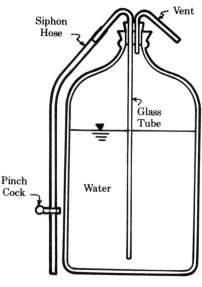

Figure 8.16a
Water Reservoir

A length of rubber tubing can be used to conduct water to the permeameter.   Once this hose is filled with water, the bottle can be emptied by simply opening the pinch cock.   A length of hose can be provided on the vent tube and used for filling the bottle from a water faucet without removing the bottle from the shelf, provided the stopper is loosened.   Also the siphon hose can be used to fill the bottle or a third opening through the stopper can be provided for that purpose.

Deaired water is required for accurate measurement of the coefficient of permeability of soils in the 10^{-4} cm sec^{-1} range.   A method of removing the dissolved air from water was devised by Professor G. M. Fair while Head of the Department of Sanitary Engineering at Harvard University.   It was found that the deaired water standing in an open bottle or carboy slowly reabsorbed air until after 24 hours the oxygen content rose to approximately 5 cc per liter.   Modifications for deairing and storing the deaired water for use in permeability testing were made by G. E. Bertram while a graduate student of Soil Mechanics at Harvard University.   By Bertram's method an oxygen content as low as 0.5 cc per liter could be produced.

A description of the method of deairing and storing water to prevent reabsorption of air as developed by Bertram follows.

In order to facilitate the removal of air, the water to be deaired is heated to about 40° C.   The heating can be done conveniently in a water bath. A suction tube extends from the bottom of the storage reservoir to a spray nozzle at the top of a 5 ft long 2 in. diameter lucite tube. After falling through the deairing tube, which is subjected to a vacuum, the deaired water flows through a connection leading from the bottom of the deairing lucite tube into the storage carboy through one of 4 openings in the stopper.

A vacuum is applied through 2 of the 4 openings in the stopper. Air is prevented from being reabsorbed by the water by a thin rubber balloon of the type used for weather observations. The balloon is of such size that it will completely fill the storage carboy when fully distended. The opening in the balloon is secured to one of the two vacuum lines at the stopper so that a vacuum can be applied to the interior of the balloon. Through the other line a vacuum can be applied to the carboy outside the balloon and through the fill opening to the deairing tube.

A rubber hose is attached to the fourth opening through the stopper and extended to the bottom of the storage carboy on the outside of the balloon.

Vacuum is applied by a vacuum pump pulling on a mercury manometer provided with a trap to prevent mercury being pulled into the pump. A trap is also provided in the vacuum line leading into the carboy to catch water that might be pulled into the vacuum line. Lines through the rubber stoppers and junctions of rubber tubing are made with short lengths of glass tubing. Connections between units are made with thick walled rubber tubing that will not collapse under atmospheric pressure. All joints must be made air tight. Pinch cocks can be used as stops in the lines.

The arrangement of the deairing and storing apparatus is shown in Figure 8.16b.

Water can be deaired by following the procedure given below.

1. Preheat the water to be deaired in a water bath or other method to about 40° C.
2. Close the discharge line from the storage carboy and the suction line from the water supply.
3. Apply a vacuum of about 70 cm mercury to the entire system. The vacuum should be applied to the inside of the balloon to keep it deflated.
4. Open the suction line to allow a small amount of water to be pulled into the deairing tube. The rate of flow should be such as to fill a 5 gallon carboy in about 1 hour.
5. When the storage carboy is full and some water has been allowed to flow into the water trap, close the flow line from the deairing

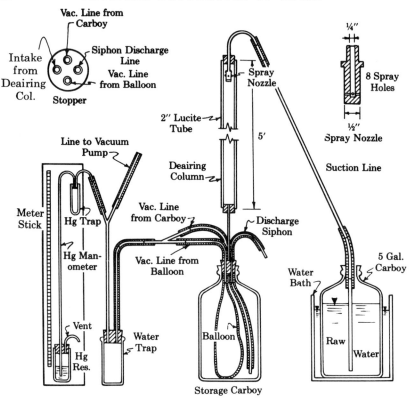

Figure 8.16b
Apparatus for Deairing and Storing Water

column to the storage carboy. Then with all lines leading into the carboy closed, open the vacuum line to the inside of the balloon to admit atmospheric pressure, which will inflate the balloon against the sides of the carboy and the surface of the water sealing against absorption of air.

Water may be siphoned from the storage carboy by opening the discharge line. As water is withdrawn from the carboy, atmospheric pressure keeps the balloon inflated to preserve the air seal.

REFERENCES

Bertram, G. E., "An Experimental Investigation of Protective Filters," *Soil Mechanics Series No. 7*, Graduate School of Engineering, Harvard University.

Compressibility and

Consolidation

9.19L BRIEF INSTRUCTIONS FOR CONSOLIDATION TEST PROCEDURE

The steps to be taken in carrying out a consolidation test are given below. Detailed instructions for carrying out the different steps are given in Chapter IX, Section 9.18.

1. Measure height and diameter of ring to nearest 0.01 cm and record. Weigh ring to nearest 0.01 g and record.
2. Fit specimen into ring.
3. Weigh specimen and ring to nearest 0.01 g and record.
4. Moisten porous stones or leave dry according to conditions of test and place on specimen for floating ring or fixed ring test as desired.
5. Place ring with specimen and stones in vessel.

6. Place vessel with specimen in loading frame, bring loading head in contact with stone, fix dial indicator, and set indicator to initial reading.

7. Provide for prevention of evaporation of moisture from specimen with moist paper towel or by flooding.

8. Apply first increment of load and read dial at 6 sec, 15 sec, 30 sec, 1 min, etc. doubling the elapsed time for successive readings to 8 hrs. Thereafter, read after each 8-hr interval. Record load, dial reading, and elapsed time for each reading. Plot dial reading-time curve as consolidation progresses. Leave this load unchanged until well after 100 per cent consolidation. Take and record temperature of flooding water.

9. Add second increment of load equal to first increment making the total load double the preceding one. Read and record as for first load. Allow load to remain unchanged for same time as for first load.

10. Continue to add load increments, doubling previous load each time, until the total load is the desired maximum. Read and record for each increment same as described above.

11. After consolidation under maximum load, remove the last two increments. Leave remaining load on specimen for same time as for consolidation. Plot dial reading-time curves for swelling.

12. Continue to remove two increments each time until all load has been removed and allow specimen to swell under zero load.

13. Record final dial reading and remove dial and setup from loading frame.

14. Remove porous stones from specimen. Place any spoil from specimen back on specimen and remove excess water from ring and specimen.

15. Weigh specimen and ring to nearest 0.01 g and record.

16. Place ring and specimen in drying oven at 105°C until dry. Cool in desiccator.

17. Weigh dry specimen and ring to nearest 0.01 g and record.

18. Place same setup without specimen using same porous stones as used in test in the loading frame and determine deformation of machine and other parts for same loads in same order as used in test. Record on form for that purpose.

19. Record and compute unit pressure on Summary and Computation Sheet. At same elapsed time after 100 per cent consolidation find on dial reading-time curves the dial reading corresponding to this elapsed time and record. Compute dial change for each load increment. Subtract correction for deformation of

machine parts and porous stones from total dial change to find net change in height of specimen.   Compute and record change in void ratio Δe as ratio of change in height of specimen to equivalent height of solids.   Subtract Δe from original void ratio e_0 to determine void ratio corresponding to unit pressure p.

20. Plot void ratio-pressure curve on 3-cycle semilog paper.

21. Compute and record on sheet with e-p curve and consolidation-time curve the properties of the soil, C_c, C_s, a_v, c_v, and k.

9.20L   TEST DATA

The following data are from a consolidation test made on a sample of red silty clay taken by pushing a sharpened 3 in. diameter thin wall tube into the soil at the bottom of an augered hole.   The test was made for the purpose of estimating settlement produced by added load.

It is obvious from a study of the $e-\log p$ curve that this clay is overconsolidated, probably by desiccation.   The large elastic recovery indicates that the clay is expansive and can be expected to shrink and swell a great deal with change in water content.   A great deal of damage has occurred to buildings with foundations on this clay because of volume change caused by change in water content.

Considerable information about the history of the clay can be obtained from the shape of the virgin $e-\log p$ curve.   The virgin curves for clays of low sensitivity are straight or nearly so.   The virgin curves for highly sensitive clays are curved concave upward.

SOIL MECHANICS LABORATORY
CONSOLIDATION TEST
Water Content & Computation Sheet

Tested by ___J.V.P. & R.F.___ Sample No. ___2___

Date ___6/17/61___ Test No. ___1___

Tested for ___Sinclair Pipe Line Co.___ Sheet No. ___1___

Location of Sample ___Pauls Valley, Okla.___

Position of Sample ___9'-11" to 10'-0"___

Description of Sample ___Red silty clay___

W. C. DETERMINATION	AT START OF TEST		AT END OF TEST	
Ring No.	B-5	*M-22	B-5	
Wt. Specimen Wet + Tare	154.06	49.78	153.66	
Wt. Specimen Dry + Tare	127.22	42.32	127.22	
Wt. of Water	26.84	7.46	26.44	
Tare (Ring)	28.55	15.49	28.55	
Wt. of Dry Soil, W_s	98.67	26.83	98.67	
Water Content, w	w_1 27.2 %	w_1 29.35	w_2 26.8	w_2
Average w	w_1 27.2 %		w_2 26.8 %	

No. of Test Machine ___6___

Dia. of Test Ring ___6.43___ cm

Area of Test Ring, A ___32.5___ cm^2

Dial Reading at Start ___2.0000___ cm

Dial Reading at End ___1.9423___ cm

Diff. in Dial Readings ___0.0577___ cm

Mach. Defor. 0 to 0 ___0.0034___ cm

Change in Ht. Specimen, ΔH .054 cm

Ht. of Specimen at End of Test
$H_2 = H_1 - \Delta H =$ ___1.896___ cm

No. of Test Ring ___B-5___

Spec. Gr. of Solids, G_s ___2.70___
 Det. Est.

Ht. of Ring = Ht. of Specimen at
 Start of Test, $H_1 =$ ___1.95___ cm

Ht. of Solids, $H_s = W_s/AG_s =$ 1.124 cm

Ht. of Voids, $H_{v1} = H_1 - H_s =$ .826 cm

Init. Void Ratio, $e_0 = H_{v1}/H_s =$ 0.735

Ht. Water at Start of Test
$H_{w1} = w_1 H_s G_s =$ ___0.826___ cm

Ht. Water at End of Test
$H_{w2} = w_2 H_s G_s =$ ___0.813___ cm

Degree of Saturation at Start of Test $S_1 = \dfrac{H_{w1}}{H_1 - H_s} \dfrac{100}{} =$ ___100___ %

Degree of Saturation at End of Test $S_2 = \dfrac{H_{w2}}{H_2 - H_s} \dfrac{100}{} =$ ___105.2___ %

* W.C. check

FOR COMPUTATION OF VOID RATIOS USE THE FOLLOWING VALUES:

$e_0 =$ ___0.735___ $H_s =$ ___1.124___ $H_1 =$ ___1.950___

SOIL MECHANICS LABORATORY
MACHINE CALIBRATION

Machine No. _____ 6 _____

Place same setup without specimen using same porous stones as used in test in the machine and determine deformation of machine and other parts for same loads in same order as used in test.

Load kg	Loading		Unloading	
	Dial Reading	Dial Change	Dial Reading	Dial Change
0	1.0000	0	0.9966	0.0034
6	0.9994	0.0006		
12	0.9992	0.0008	0.9956	0.0044
24	0.9988	0.0012		
48	0.9986	0.0014	0.9941	0.0059
96	0.9977	0.0023		
192	0.9958	0.0042	0.9917	0.0083
384	0.9917	0.0083		
768	0.9863	0.0137		

PHYSICAL PROPERTIES OF SOILS

SOIL MECHANICS LABORATORY
CONSOLIDATION TEST
Data Sheet

Tested by __J.V.P. & R.F.__   Sample No. __2__
Date __6/17/61__   Test No. __1__
Tested for __Sinclair Pipe Line Co.__   Sheet No. __3__
Note: Temp. refers to specimen temperature.   Machine No. __6__

Date	Time	Temp. C°	Load kg	Elapsed Time	Dial Reading	Date	Time	Temp. C°	Load kg	Elapsed Time	Dial Reading
6/22	0958	26	192	0	1.9488	6/24	0955	26	768	0	1.8299
				6 sec.	—					6 sec	1.8180
				15	1.9410					15	1.8158
				30	1.9393					30	1.8132
	0959			1 min	1.9377		0956			1 min.	1.8100
	1003			5	1.9299		0957			2	1.8059
	1006			8	1.9268		0959			4	1.7997
	1013			15	1.9219		1003			8	1.7913
	1028			30	1.9169		1010			15	1.7823
	1058			60	1.9131		1025			30	1.7729
	1203			125	1.9103		1055			60	1.7637
	1403			245	1.9083		1155			120	1.7568
	1758	28		480	1.9065		1355			240	1.7523
	2338			820	1.9055		1755			480	1.7492
6/23	0928	26		1410	1.9049	6/25	0025			870	1.7473
							0940			1425	1.7461
6/23	0930	26	384	0	1.9049						
				6 sec	—	6/25	0942		192	0	1.7461
				15	—					6 sec	1.7558
				30	1.8810					15	1.7570
	0931			1 min	1.8786					30	1.7582
	0932			2	1.8752		0943			1 min	1.7596
	0934			4	1.8704		0944			2	1.7613
	0938			8	1.8639		0947			5	1.7649
	0945			15	1.8567		0953			11	1.7684
	1000			30	1.8487		1005			23	1.7731
	1042			72	1.8413		1029			47	1.7775
	1200			150	1.8377		1052			70	1.7793
	1510			340	1.8348		1137			115	1.7813
	2250			800	1.8322		1314			212	1.7828
6/24	0953	26		1463	1.8299		1746			484	1.7836

CONSOLIDATION TEST
DIAL READING-TIME-CONSOLIDATION CURVE

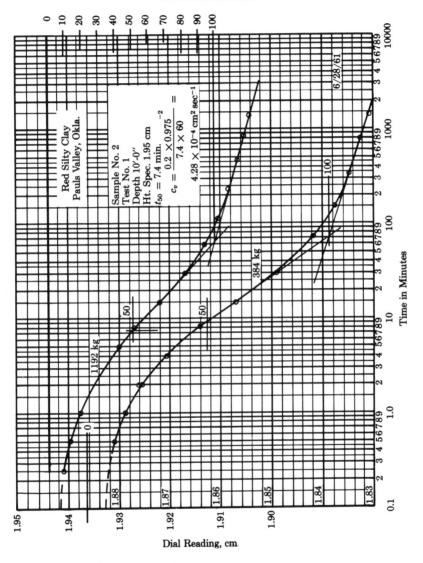

Red Silty Clay
Pauls Valley, Okla.

Sample No. 2
Test No. 1
Depth 10'-0"
Ht. Spec. 1.95 cm

$$t_{50} = 7.4 \text{ min.}$$

$$c_v = \frac{0.2 \times 0.975^{-2}}{7.4 \times 60} = $$

$$4.28 \times 10^{-4} \text{ cm}^2 \text{ sec}^{-1}$$

6/28/'61

Percent Consolidation

Time in Minutes

Dial Reading, cm.

SOIL MECHANICS LABORATORY
CONSOLIDATION TEST
Summary & Computation Sheet

Tested by ___J.V.P. & R.F.___   Sample No. ___2___

Date ___7/4/61___   Test No. ___1___

Tested for ___Sinclair Pipe Line Co.___   Sheet No. ___5___

Duration of Test ___13 Days___   Machine No. ___6___

Elapsed Time for Computation of Void Ratio ___1440___

For Computations (from Sheet No. _1_): $H_s = 1.124$ cm , $H_1 = 1.950$ cm , $e_0 = .735$

Date Increment Applied	Elapsed Time for Increm.	Load p kg	Pressure p kg cm^{-2}	Dial Reading cm	Dial Change cm R_D'	Correction cm	Correct Dial Change R_D	Δe $\frac{R_D}{H_s}$	Void Ratio e
6/17	–	0	0	2.0000	0	0	0	0	0.735
6/17	633	6	0.185	2.0059	.0059	.0006	.0065	.006	0.741
6/18	1391	12	0.369	2.0028	.0028	.0008	.0036	.003	0.738
6/19	1496	24	0.738	1.9931	.0069	.0012	.0057	.005	0.730
6/20	1402	48	1.476	1.9790	.0210	.0014	.0196	.017	0.718
6/21	1470	96	2.95	1.9488	.0512	.0023	.0489	.045	0.690
6/22	1410	192	5.90	1.9049	.0951	.0042	.0909	.081	0.654
6/23	1463	384	11.81	1.8299	.1701	.0083	.1618	.144	0.591
6/24	1425	768	23.63	1.7461	.2539	.0137	.2402	.214	0.521
6/25	1423	192	5.90	1.7843	.2157	.0083	.2074	.184	0.551
6/26	1435	48	1.476	1.8393	.1607	.0059	.1548	.134	0.597
6/27	1453	12	0.369	1.8890	.1110	.0044	.1066	.095	0.640
6/28	3390	0	0	1.9423	.0577	.0034	.0543	.048	0.687

CONSOLIDATION TEST

VOID RATIO-PRESSURE CURVE

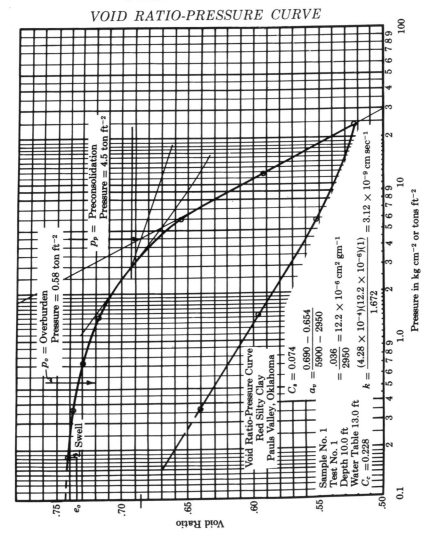

p_o = Overburden
Pressure = 0.58 ton ft^{-2}

p_p = Preconsolidation
Pressure = 4.5 ton ft^{-2}

Swell

Void Ratio-Pressure Curve
Red Silty Clay
Pauls Valley, Oklahoma

Sample No. 1
Test No. 1
Depth 10.0 ft
Water Table 13.0 ft
$C_c = 0.228$

$C_s = 0.074$

$a_v = \dfrac{0.690 - 0.654}{5900 - 2950}$

$= \dfrac{.036}{2950} = 12.2 \times 10^{-6} \text{ cm}^2 \text{ gm}^{-1}$

$k = \dfrac{(4.28 \times 10^{-4})(12.2 \times 10^{-6})(1)}{1.672} = 3.12 \times 10^{-9} \text{ cm sec}^{-1}$

Void Ratio

Pressure in kg cm^{-2} or tons ft^{-2}

Stress-Deformation and Strength Characteristics

11.20L TEST PROCEDURE

a. Cohesionless Soils

The most common strength tests for cohesionless soils are the drained triaxial test and the direct shear test. Two procedures are rather common in performing the triaxial test. In one procedure the confining pressure is supplied by the atmosphere when a vacuum is pulled on the inside of the membrane covering the dry soil specimen. In the second procedure for which detailed instructions are given below, the test is performed on a saturated specimen and the confining pressure is supplied by air or water under pressure in a triaxial cell pressure chamber. The volume changes can be more easily and accurately measured when the second procedure is used.

434

(1) Test Apparatus

A simple triaxial apparatus that is suitable for the testing of cohesionless soils is shown in Figure 11.20a. This apparatus can also be used for cohesive soils, but no provision is made for the measurement of pore pressures.

The apparatus consists of three parts: an arrangement for saturating the specimen and measuring volume change, a pressure chamber in which the specimen is enclosed with provision for applying confining pressure and axial loads to the specimen and for measuring vertical deformation of the specimen, and a pressure reservoir for supplying a liquid under pressure to the pressure chamber containing the specimen.

The mercury manometer can be made with a small, large-mouth bottle, a glass tubing standpipe about 1 meter long, and a meter stick. This assembly can be mounted on a board with provision for adjusting the meter stick vertically to the surface of the mercury in the reservoir.

The water reservoir can be a 2 liter aspirator bottle. The water trap is a wide mouth bottle of about 1 pint capacity. The pipette should be one to which air tight connections can be made with rubber tubing at each end, having a capacity of 10 ml and graduated in 0.1 ml. The graduation marks should be far enough apart that volumes can be estimated to 0.01 ml.

The triaxial pressure chamber consists of a base, a top, a loading head, a piston, and an enclosing lucite cylinder with 3 hold down bolts. The base is provided with a cylindrical pedestal $1\frac{1}{2}$ inches high, having the same diameter as the specimen. A porous stone slightly smaller in diameter than the specimen is recessed into the pedestal and connected to the exterior of the base through drilled holes. A common diameter of specimens is 1.4 in. or 3.56 cm. Another hole is drilled for connection between the loading cap and the exterior of the base. The hole drilled in the loading cap is provided with a small brass union for connection to a length of plastic tube. The exits for both these openings are provided with connections for rubber tubing. An opening is also provided through the base for allowing fluid to flow from the pressure reservoir into the triaxial pressure chamber. Provision is made on the outside of the base for connection to the pressure hose from the pressure reservoir. This connection can be made with one-half of a $\frac{3}{8}$ in. brass union.

The other half of the union together with a ground key stop is attached to the pressure hose from the pressure reservoir. Three tapped holes are provided in the base for reception of the hold down bolts. The top of the chamber is a disk with a polished hole in the center for the piston, three holes for the hold down bolts, and a petcock

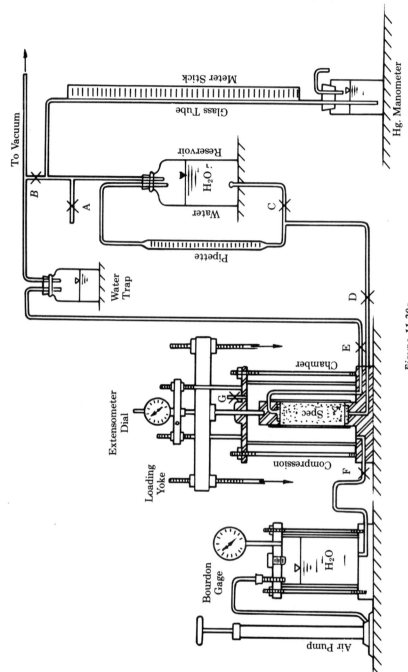

Figure 11.20a

Triaxial Set-Up for Cohesionless Soil

to allow air to escape while filling the chamber. The loading head is the same diameter as the base pedestal and is fitted in the same manner with a porous stone connected to the exterior with a drilled opening. Provision is made for attaching the small plastic tubing which leads to the base inside the chamber. A semi-spherical or conical depression is provided in the center of the head for the reception of the piston. The piston should be ground to a fit that will allow it to fall slowly through the hole in the chamber top under its own weight. The ends of the piston are ground to fit the depressions in the loading head and the loading bar. Provision should be made for the confining fluid to enter the space between the piston and the loading head so that the confining pressure is applied to the entire upper surface of the loading head and can thereby provide an equal all around pressure to the specimen. The small plastic tubing connecting the top of the specimen to the opening in the base is usually wound spirally around the specimen. The wall of the pressure chamber is a thick walled lucite tube capable of withstanding at least 100 psi internal pressure. It can be from 4 to 6 in. in diameter and about 10 in. long. Rubber gaskets placed between the ends of the cylinder and the base and top provide leak tight joints.

The pressure reservoir consists of a lucite tube, similar to that used for the pressure chamber, between a base and a top as shown in the sketch. The base is provided with an opening for connection to a pressure hose. The top is provided with a filler plug, a Schraeder tire valve, and a Bourdon gage registering to 100 psi. A tire pump can be used for producing the desired confining pressure.

Connections between the parts described above can be made with thick wall rubber tubing or plastic tubing capable of withstanding a vacuum without collapse. Transparent tubing is advisable for the connection between the bottom of the specimen and the pipette so that air bubbles can be seen and worked out.

Stops in the connecting lines can be glass stops or screw type pinch cocks depending upon the type of tubing used for the connections.

(2) Test Procedures

(a) *Drained Triaxial Test*

Directions for carrying out a drained triaxial test on cohesionless material are given below.

1. Oven dry at 105° C in an evaporating dish a quantity of sand somewhat greater than that required to form the specimen. Cool in a desiccator.

2. With the pressure chamber disassembled, place the end of a rubber membrane having the same inside diameter as the pedestal, and a length about 2 in. greater than the form for the specimen, over the pedestal, lapping about 1 in.   Break a rubber band and wrap tightly around the overlapped end of the specimen to form a tight connection between membrane and pedestal. Fasten end of band under a tight turn to maintain the tension. Keep rubber band at least $\frac{1}{4}$ in. below top of pedestal.

3. Measure thickness of membrane and record.

4. Clamp two halves of a form around pedestal and membrane. Pull membrane up taut and break over top of form to lap down about $\frac{3}{4}$ in. to hold membrane open to receive specimen.

   The form can consist of a square block of aluminum split lengthwise into two halves with a cylindrical hole having a diameter equal to that of the pedestal plus twice the thickness of the membrane and a height about 3 times the diameter plus the height of the pedestal on the base.   As clearance for the rubber band, the bottom $\frac{3}{4}$ inch of the cylindrical hole should be increased by about $\frac{1}{4}$ inch.   About 1 inch of the top of the form should be turned down on the outside to form a cylinder having walls $\frac{1}{8}$ inch thick to receive the turned down end of the membrane.

5. Weigh the evaporating dish and the sample of sand to the nearest 0.01 g and record as Wt Specimen $+$ Tare.

6. For a loose specimen, pour through a funnel enough of the dry sample to fill the membrane to within about $\frac{1}{4}$ in. of the top of the form.   For a dense specimen, place the sand in layers, tamping and vibrating each layer as it is placed.   For specimens of intermediate density, compact according to the density desired.

7. Place the loading head in contact with the top of the specimen and pull the membrane over the head.   Wrap with a rubber band in the same manner as for the bottom joint.

8. Fasten the connecting lines to the top and bottom of the specimen and with all stops open, except A, apply the available vacuum.   An excellent source of vacuum is a water pressure filter pump on a water faucet.

9. Remove form from specimen.

10. Close stop B and crack stop A to allow the mercury to fall in the manometer enough to create a differential between top and bottom of the specimen to pull water up through the specimen into the trap.   The flow through the specimen should be regulated to catch a trap full in about 15 minutes.   Close stop A

and continue pulling air out and water into the specimen until completely saturated.

11. Close stop E and measure the diameter of the specimen at top, middle, and bottom with a vernier caliper and record. Measure and record the height of the specimen. The height can be determined by measuring the distance from the top of the base to the top of the head and subtracting the combined height of the pedestal and head.

12. Enclose the specimen in the pressure chamber by screwing the hold-down bolts into the base, placing the lucite tube over the specimen with rubber gaskets top and bottom, placing the top over the hold down bolts, tightening the nuts on the hold-down bolts to produce tight joints, and inserting the piston through the hole in the top to a firm contact with the recess in the loading head.

13. Place the pressure chamber, with specimen, in a loading frame and bring the loading bar into firm contact with the piston with no load on the bar.

14. Close stop C and read pipette. Record as V_0. Read vacuum pressure and record as p_1.

15. Open stop F and pump up pressure in the pressure reservoir to force liquid into the pressure chamber. When liquid flows out of stop G, close stop F. Pump pressure in reservoir to the confining pressure desired. See that piston is in contact with loading head and secure indicator dial in place.

16. Close stop G, hold loading bar so that piston is in contact with loading head, and open stop F. Pressure on the bottom of the piston will push the piston upward against the loading bar. Apply just enough load to compensate for this upward pressure, being sure that the piston is in contact with the loading head. Do not count this compensating load as part of the applied piston load or deviator stress.

Record confining pressure.

Read pipette and record as V_1.

17. Open stop A to apply atmospheric pressure to interior of specimen.

Read pipette and record as V_2.

Set and read dial and record on second Data and Computation Sheet.

18. Estimate total load P required to produce failure by estimating ϕ and computing from $\sigma_1 = \sigma_3 \tan^2 \left(45 + \dfrac{\phi}{2} \right)$. Apply load increments of $0.1P$ to loading bar at 1 minute intervals.

Read dial and pipette at 55 second interval after application of each load increment. Record each load, dial reading, and pipette reading.

Continue until failure, noting type of failure, whether by bulging or shear plane. Draw a sketch of the specimen after failure before removal of loads.

19. Remove load and confining pressure and disassemble the pressure chamber and remove the collapsed specimen and membrane.

20. Calibrate rubber tubing for volume change due to pressure change from vacuum pressure p_1 to atmospheric pressure. Close stops A, C, D, and E and subject the system to vacuum. Close stop B and crack stop A until the mercury drops to the vacuum pressure p_1 at which the confining pressure was applied. Read the pipette with this pressure on the tubing.

Open stop A and allow atmospheric pressure to enter the tubing. Read pipette under atmospheric pressure conditions.

Compute the difference in pipette readings for the above change in pressure and record as V_3.

21. Carry out the computations called for on the Data and Computation Sheets and plot Per Cent Volume Change-Stress-Strain curves.

22. Determine failure stresses and compute value of ϕ.

(b) Undrained Triaxial Test

Under some conditions it might be desirable to conduct an undrained test on fine grained cohesionless soils. The test may be carried out as follows:

1. Steps 1 to 17 inclusive, same as for drained test.
18. Close stop D and apply loads in increments. Read and record dial readings.
19. Plot Stress-Strain curve.
20. Determine failure conditions and draw stress circle at failure.

(3) Example Data

On the following pages are shown laboratory forms which are suitable for recording the data taken during a triaxial test in which the foregoing procedure is used. To illustrate the use of the forms, actual data from a test on Arkansas River sand is recorded on the sheets. The stress-strain and stress-volume change curves are on the same graph. It may be seen that the sand was in a fairly dense state when tested, resulting in a general increase in the volume of the specimen when deformed. The volume change of the specimen was relatively insignificant for strains less than about one per cent.

SOIL MECHANICS LABORATORY
TRIAXIAL COMPRESSION TEST
(Cohesionless Material)

Data and Computation Sheet

Name __Parcher__ Sample No. __1__ Test No. __1__

Date __Jan. 30, 1947__ Tested for _____ Sheet No. __1__

Location of Sample __Rocky Mts.__ Position of Sample __Stream Bed__

Description of Sample __Arkansas River sand passing No. 28__

Initial Wt. Specimen + Tare __571.30__ gms.   Specific Gr. of Solids, G_s __2.68__

Final Wt. Specimens + Tare __412.70__ gms.   Vol. of Solids in Specimen

Dry Wt. of Soil in Specimen __158.60__ gms. $V_s = \dfrac{W_s}{G_s} = $ ___59.2___ cm³

Initial Conditions (Vacuum inside specimen by no external pressure)

Thickness of Rubber Membrane = __0.085__ cm
(Deduct twice to obtain net dia.)

	Gross Dia.	Net Dia.	Area cm²
Top	3.71	3.54	$A_t = 9.83$
Mid.	3.67	3.50	$A_c = 9.60$
Bot.	3.61	3.44	$A_b = 9.27$

$A_{av} = \dfrac{A_t + A_c + A_b}{3} \dots\dots\dots = $ __9.58__ cm²

Height of Specimen. . . . . = __9.60__ cm

Volume of Specimen, V. . . . . = __91.9__ cm³

Vol. of Voids = $V_v' = V - V_s$ . . . = __32.7__ cm³ ←

Void Ratio, $e_o = \dfrac{V_v}{V_s} = $ __0.552__ Pipette $V_o = $ __4.0__ cm³

Vacuum Pressure = __39.7__ cm Hg. × 0.01355 = __0.538__ kg cm⁻²

Conditions under Pressure (Vacuum plus liquid pressure)

Liquid Pressure = __30__ lb in⁻² × 0.0703 = __2.109__ kg cm⁻² = $\sigma_1 = \sigma_2 = \sigma_3$

Pipette $V_1 = $ __3.72__ cm³ Vol. of Voids $V_v'' = V_v' - (V_o - V_1) = $ __32.42__ cm³

Void Ratio under Pressure $e_p = \dfrac{V_v''}{V_s} = $ __0.548__

Determination of Residual Air

Atmospheric Pressure, p_2, assume equal to. . . . . __76__ cm Hg

Vacuum Pressure, p_1. . . . . __39.7__ cm Hg

Absolute Pressure = $(76 - p_1)$. . . . . __36.3__ cm Hg

Pipette Reading V_1 corresponding to p_1. __3.72__ cm³

Pipette Reading V_2 corresponding to p_2. __7.10__ cm³

Volume Change of Tubing $(p_1$ to $p_2) = V_3$ from calibration. . . . . . __2.908__ cm³

Total Volume Change $(p_1$ to $p_2) = \Delta V = V_2 - V_1 - V_3$. . . . . __0.472__ cm³

Volume of Residual Air at Atmos. Pres. = $V_a = \dfrac{76 - p_1}{p_1} \Delta V$. . . . . __0.432__ cm³

Percentage of Residual Air at Atmos. Pres. $V_a\% = \dfrac{V_a}{V_v''} \cdot 100$. . . . . . __1.33__ %

Degree of Saturation = $100 - V_a\%$. . . . . __98.67__ %

Remarks _____

SOIL MECHANICS LABORATORY
TRIAXIAL COMPRESSION TEST
(Cohesionless Material)

Data and Computation Sheet

Name __Parcher__ Sample No. __1__ Test No. __1__

Date __Jan. 30, 1947__ Tested for __R E M__ Sheet No. __2__

Elapsed Time Min.	Piston Load kg	Dial Reading cm	Dial Change cm	Pipette Reading cm³	Volume Change cm³	Void Change %	Strain ε	Correct Area cm²	Piston Load kg cm⁻²	Vertical Stress kg cm⁻²
0	0	0		7.11	0	0	0	9.575	0	2.11
1	5	0.0284		7.21	+0.10	0.306	0.00294	9.600	0.52	2.63
2	10	0.0298		7.20	0.09	0.275	0.00311	9.603	1.04	3.15
3	15	0.0315		7.20	0.09	0.275	0.00328	9.607	1.56	3.67
4	20	0.0350		7.19	0.08	0.245	0.00365	9.609	2.08	4.19
5	25	0.0378		7.18	0.07	0.214	0.00394	9.610	2.60	4.71
6	30	0.0431		7.17	0.06	0.184	0.00449	9.617	3.12	5.23
7	35	0.0495		7.17	0.06	0.184	0.00516	9.635	3.63	5.74
8	40	0.0590		7.16	0.05	0.153	0.00614	9.642	4.15	6.26
9	45	0.0730		7.20	0.09	0.275	0.00760	9.650	4.66	6.77
10	50	0.0852		7.24	0.13	0.398	0.00888	9.660	5.17	7.28
11	55	0.1064		7.28	0.17	0.520	0.01110	9.695	5.67	7.78
12	60	0.1247		7.34	0.23	0.704	0.01299	9.700	6.12	8.23
13	65	0.1461		7.42	0.31	0.948	0.01523	9.733	6.68	8.79
14	70	0.1689		7.53	0.42	1.286	0.01760	9.750	7.18	9.29
15	75	0.2010		7.74	0.63	1.928	0.02095	9.790	7.66	9.77
16	80	0.2190		7.88	0.77	2.355	0.0228	9.805	8.15	10.26
17	85	0.2699		8.25	1.14	3.49	0.0281	9.850	8.62	10.73
18	90	0.3190		8.63	1.52	4.65	0.0332	9.900	9.09	11.20
19	95	0.4186		—	—	—	0.0436	10.01	9.49	11.60

(handwritten note running vertically in Dial Change column: "reading on dial same as")

Initial Conditions

$\sigma_1 = \sigma_2 = \sigma_3 = 0.538$ kg cm⁻²

$V_v = \underline{\quad 32.7 \quad}$ cm³

$V_s = \underline{\quad 59.2 \quad}$ cm³

$e_0 = \underline{\quad 0.552 \quad}$

$D_d' = \underline{\qquad}$ %

Degree of Density $D_d = \dfrac{e_L - e}{e_L - e_D} \, 100$

Corrected Area $= \dfrac{V - \Delta V}{H - \Delta H}$

At Start of Test

$\sigma_1 = \sigma_2 = \sigma_3 = 2.11$ kg cm⁻²

$V_v'' = \underline{\quad 32.42 \quad}$ cm³

$V_s = \underline{\quad 59.2 \quad}$ cm³

$e_p = \underline{\quad 0.548 \quad}$

$D_d'' = \underline{\qquad}$ %

At Failure

$\sigma_1 = 11.60$ kg cm⁻²

$\sigma_2 = \sigma_3 = 2.11$

$\varepsilon_m = 4.36$ %

$\Delta e = 4.65$ %

$\dfrac{\sigma_1}{\sigma_3} = 5.5$

$\tan^2\left(45 + \dfrac{\varnothing}{2}\right) = \dfrac{\sigma_1}{\sigma_3}$

$\varnothing'' = 44°$

DRAINED TRIAXIAL TEST
ARKANSAS RIVER SAND

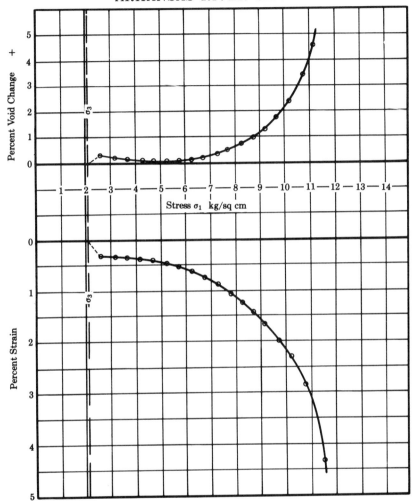

b. Cohesive Soils

(1) Preparation of Specimen

Prepare the test specimen in such a way that evaporative losses will be kept to a minimum. Perform the cutting and trimming operations as quickly as possible, preferably in a humid room. Compacted specimens and undisturbed specimens extruded from sampling tubes may often be tested in their original dimensions. If a 1.4 inch diameter specimen is to be taken from a hand-cut sample, use a wire saw to remove from the sample a block of soil $2'' \times 2'' \times 4''$. Square the ends of the block in a miter box and place the block in a vertical trimming lathe. Use a wire saw to trim vertical strips of soil from the block as the lathe is successively rotated a few degrees at a time. Collect some of the strips for a moisture content determination, recording the results on the Data and Computation Sheet. After the specimen has been smoothly trimmed to the proper diameter (the trimming guides on the lathe assure this), cut it to a length of 3 or $3\frac{1}{2}$ inches in a miter box having a semi-cylindrical bed. If an unconfined compression test is to be performed, record the dimensions of the specimen on the Unconfined Compression Test form and begin the test.

If a triaxial test is to be made, weigh a rubber membrane of the proper size and shape and record the weight on the Data and Computation Sheet. Place the membrane inside a metal cylinder which is about 1.6 inches in diameter and $3\frac{1}{2}$ inches long, into the side of which an evacuation port has been constructed, and fold the ends of the membrane over the cylinder. Expand the membrane by evacuating the air from the space between the membrane and the cylinder. Slip the membrane over the specimen, release the vacuum, and remove membrane and specimen from the cylinder. Record the weight of specimen + membrane, and the dimensions of the specimen, which is now ready for testing.

(2) Saturation of Apparatus

If an R test or an S test is to be performed, the porous stones and drainage lines must first be filled with water. This is done by the following procedure, for which Figure 11.20b applies schematically.

  (a) Close all valves on the apparatus.
  (b) Support cap in an inverted position by clamping it to a ring stand.
  (c) Place short sections of rubber tubing over the cap and pedestal

and fill both with (preferably) deaired water. The water must be replenished as necessary during subsequent steps.

(d) Apply a vacuum to top of burette and open valves A and B to fill line from pedestal to burette.

(e) Close valve B and open valve C to fill line from cap to burette.

(f) Close valve C, release vacuum, and open valves E and F to permit water from burette to fill reservoir.

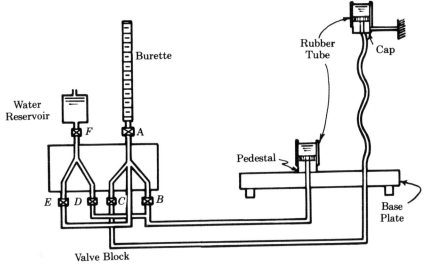

Figure 11.20b
Saturation of Triaxial Apparatus

(g) Close valve E and open valve D to fill line between reservoir and pedestal. Then again close all valves.

(h) Repeat steps (d) through (g) until no air bubbles remain in system.

(i) Remove rubber tubes from pedestal and cap.

(j) Open valves A, E, and F, and adjust water level in burette as desired for initial reading V_0. Close all valves.

(3) Assembly of Apparatus (Refer to Figure 11.20c)

(a) Wipe excess water from upper stone, place specimen in position on the inverted cap, and bind the membrane to the cap.

(b) Carefully remove cap and specimen from the ring stand and repeat step (a) with reference to the pedestal.

(c) Place the lucite chamber in position and screw the tie rods into the base.

(d) Set the head plate into position over the tie rods, being careful to introduce the loading piston gently into the socket in the specimen cap.   Draw the wing nuts down evenly on the upper ends of the tie rods.

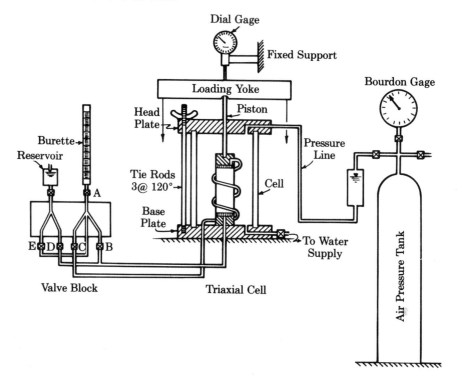

Figure 11.20c
Assembly for Triaxial Test

(e) Attach the pressure line to the head plate outlet and a water supply line to the base plate inlet.   Fill the chamber with water, allowing flow to continue until all air bubbles have been carried out through the pressure line.

(f) Set the assembly in the loading frame, centering it carefully, and adjust the loading yoke and dial gage.

(g) Connect pressure line to air pressure tank and measure difference in head between water in burette and water in pressure line.   Subtract this from gage reading on air tank to obtain the actual confining pressure for all tests or phases in which valves A and B or C are open.   For undrained loading the same correction may be applied as a reasonable approximation.

(*h*) Open valve to air tank, at the same time adding sufficient tare to the load hanger to balance the upward thrust on the piston.

(*i*) Open valves *B* and *C* (being sure that all other valves on the valve block are closed) and adjust dial reading to zero. The testing may now proceed.

(4) Test Procedure

The following procedures are for tests in which the rate of loading is controlled. If the rate of strain is controlled, these procedures would be slightly altered during the axial loading phase of the test. In controlled rate of loading tests some sort of stop (preferably adjustable in height) must be provided beneath the platform of the load hanger to prevent excessive movement when the specimen fails.

(*a*) U Test

A fairly common loading machine consists of a platform scale on which has been mounted a base plate. A loading yoke, having a cross-bar mounted above the platform on two vertical rods extending through the platform, is connected to a screw jack mounted on the frame beneath the platform. At the front of the machine a wheel is mounted for convenience in turning the jack screw for the purpose of positioning the cross-bar relative to the platform. The specimen is loaded by adding weights to the hanger on the scale beam. The multiplication factor of the lever system is customarily 100.

1. Center the specimen in the loading device and level the scale beam by adjusting the position of the beam rider.
2. Bring the loading head on the cross-bar into contact with the specimen, arrange a dial gage to measure the travel of the cross-bar relative to the platform, and record the initial dial reading.
3. If the test is not set up in a humid room, arrange moist towels loosely around the specimen to prevent evaporation of moisture during the loading.
4. Place load increments on the hanger gently, at one minute intervals, recording the load and corresponding dial reading on the data sheet (see example). The dial gage is read 55 seconds after application of the load. The magnitude of the load increments is so chosen that 5 to 10 increments will produce failure of the specimen.
5. Continue the loading process, while observing the specimen as closely as possible for indication of shear planes, until failure occurs. Sketch the specimen after failure.

6. Crank up the cross-bar, remove the specimen, and determine its water content. For this, use the entire specimen or take a full length vertical slice from the center of the specimen.
7. Perform all computations on the Data and Computation Sheet and plot the Stress-Strain curve.

(b) Q Test

Having set up the test to the point indicated in section (3)(i), load the specimen without drainage, as described for the U test. Record the data on the undrained loading Summary and Computation Sheet, perform all computations, and plot the strain as a function of the deviator stress.

(c) R Test

Having set up the test as in (3) above, record the initial dial and burette readings on the Consolidation Data form. Open valve A, recording the time on the data sheet. Read the dial and burette at intervals as in the consolidation test—i.e., at 6 sec, 15 sec, 30 sec, 1 min, 2 min, etc., doubling the elapsed time from start for each subsequent reading—and record the readings. Plot the dial readings and burette readings as functions of the log time, continuing this phase of the test until primary consolidation is complete. Close valve A and load the specimen to failure as in the Q test, recording the data on the Summary and Computation Sheet. Perform the indicated computations and plot the Stress-Strain curve.

(d) S Test

Proceed as in the R test with consolidation of the specimen. With valve A open, apply the axial load in increments, allowing complete consolidation under each increment before applying the next. Record the dial and burette readings for each increment on the Consolidation Data form. The first load increment may be fairly large, perhaps 25 per cent of the estimated failure load. Subsequent increments should be made increasingly smaller. The final increment, under which failure occurs, must be small (not over one per cent of the total axial load) so that the pore pressure at failure is small compared to the total deviator stress. Although time curves need not be plotted for each increment, frequent readings should be taken as failure approaches so that the dimensions at failure can be computed. The pressure in the air tank should also be observed regularly, as it fluctuates with temperature changes during tests of long duration. Failure of the specimen may possibly result from a decrease in confining pressure due to a drop in temperature.

(5) Example Data

The following pages show examples of data forms which have been found suitable for the tests described above. Data from actual laboratory tests have been placed on the forms for purposes of illustration. The U, Q, and R test data have been selected for specimens taken from a single hand-cut block sample. Stress-strain curves for the three tests have been plotted on the same graph to permit a direct comparison of the results.

SOIL MECHANICS LABORATORY
UNCONFINED COMPRESSION TEST
Data & Computation Sheet

Name___J. V. Parcher_____   Sample & Test No.__H 211 B - U1__

Date___Oct. 29, 1954_____   Sheet No.____1____

Description of Sample___Gray marine clay from Portland, Maine.___
_____Hand-cut, undisturbed sample.___

Dimensions at Start of Test: Height, $H =$____3.5____in. =____8.89____cm

Diameter =_____in. =_____cm

Area, $A =$_____in.^{2} =____10____cm^2

Rate of Loading___2 Kg/min.___

Elapsed Time	Load kg	Dial Reading	Strain	Corrected Area	Stress kg /sq cm	Water Content Determination	
0	0	2.000	0	10.00	0	Test No.	1
1 min.	2	1.986	0.00158	10.02	0.200	Tare No.	E-12
2	4	1.976	0.00270	10.03	0.399	Wt. Wet	213.93
3	6	1.966	0.00383	10.04	0.597	Wt. Dry	166.68
4	8	1.957	0.00484	10.05	0.796	Wt. Water	47.25
5	10	1.947	0.00596	10.07	0.994	Tare	58.89
6	12	1.936	0.00720	10.08	1.190	Wt. Solid	107.79
7	14	1.921	0.00888	10.10	1.385	$w\%$	43.9 %
8	16	1.890	0.0124	10.12	1.581 max.	Test No.	
—	11	1.690	0.0349	10.37	1.062	Tare No.	
—	9	1.630	0.0417	10.44	0.948	Wt. Wet	
						Wt. Dry	
						Wt. Water	
						Tare	
						Wt. Solid	
						$w\%$	

Remarks___Dial readings in cm.___

Note: Corrected Area $= \dfrac{A}{1 - \text{strain}}$

SOIL MECHANICS LABORATORY
TRIAXIAL COMPRESSION TEST
(Cohesive Soil)
Data & Computation Sheet

Name___J. V. Parcher___   Sample & Test No. H 211B - Q1

Date___Dec. 10, 1954___   Sheet No.___1___

Description of Sample___Gray marine clay from Portland, Me.___
___Hand-cut, Undisturbed sample.___

Before Test: Wt. Sample Wet + Rubber Membrane....___160.31___ gm

Wt. of Rubber Membrane..............___3.29___ gm

Wt. Sample Wet W.....................___157.02___ gm

	w from Sample Trimmings		
Thickness of Membrane ___0.025___ cm	Tare No.		
Height of Specimen, H ___8.89___ cm	Wet + Tare		
Diameter of Specimen ___–___ cm	Dry + Tare		
Area of Specimen, A ___10.0___ cm²	Wt. Water		
Volume of Specimen, V ___88.90___ cm³	Tare		
After Test:	Wt. Solids		
Tare No. ....___L-4___	$w\,\%$		

Wt. Sample Wet + Tare ....___–___

Wt. Sample Dry + Tare ....___288.46___ gm

Wt. Water W_w.............___–___

Tare.....................___178.15___ gm

Wt. Sample Dry W_s........___110.31___ gm

$w\,\%$ ....___–___

Initial Conditions:

Burette Reading V_o...........___–___ cc

G_s – Det. ___✓___ Est. _____... ___2.76___   Vol. Solids $V_s = W_s/G_s = $ ___40.0___ cc

Volume of Voids $V_v = V - V_s$..___48.90___ cc

Wt. Water in Sample
$\quad W_w = W - W_s$ ....___46.71___ gm   Vol. of Water $V_w = $ ___46.71___ cc

$w = W_w/W_s$.................___42.3___ %

Degree of Saturation V_w/V_v ...___95.6___ %   Vol. of Water $V_w = $ ___46.71___ cc

Void Ratio $e_o = V_v/V_s = $ ___1.222___

SOIL MECHANICS LABORATORY

TRIAXIAL COMPRESSION TEST

(Cohesive Soil)
Undrained Loading
Summary & Computation Sheet

Name J. V. Parcher Sample & Test No. H211B-Q1
Date Dec. 10, 1954 Sheet No. 2

Elapsed Time Min.	Piston Load kg	Dial Reading cm	Dial Change cm	Strain ε	Corrected Area cm ^{2}	Deviator Stress kg /sq cm	Vertical Stress kg /sq cm
0	0	0	Same as dial readings	0	10.00	0	7.0
1	2	0.0020		0.00023	10.00	0.200	7.2
2	4	0.0048		0.00054	10.00	0.400	7.4
3	6	0.0082		0.00092	10.01	0.600	7.6
4	8	0.0119		0.00134	10.01	0.800	7.8
5	10	0.0155		0.00174	10.02	0.999	7.999
6	12	0.0202		0.00227	10.02	1.198	8.198
7	14	0.0261		0.00293	10.03	1.397	8.397
8	16	0.0368		0.00414	10.03	1.594	8.594
9	18	0.0800		0.0090	10.09	1.785	8.785
9:10	19	0.1200		0.0135	10.14	1.872	8.872

Conditions at Start of Test

$\sigma_1 = \sigma_2 = \sigma_3 =$ 7.0 kg /sq cm

Void Ratio, $e' =$ 1.222

Conditions at Failure

$\sigma_1 =$ 8.872 kg /sq cm

$\sigma_2 = \sigma_3 =$ 7.0 kg /sq cm

Remarks

$$\text{Corrected Area} = \frac{A}{1 - \varepsilon}$$

SOIL MECHANICS LABORATORY
TRIAXIAL COMPRESSION TEST
(Cohesive Soil)
Data & Computation Sheet

Name _____ *J. V. Parcher* _____ Sample & Test No. _*H 211 B - R1*_

Date _____ *Dec. 8, 1954* _____ Sheet No. ____*1*____

Description of Sample _*Gray marine clay from Portland, Maine.*_
_____ *Hand-cut, undisturbed sample.* _____

Before Test: Wt. Sample Wet + Rubber Membrane _*161.48*_ gm

           Wt. of Rubber Membrane . . . . _*3.14*_ gm

           Wt. Sample Wet W . . . . _*158.34*_ gm

Thickness of Membrane _*0.025*_ cm

Height of Specimen, H _*8.89*_ cm

Diameter of Specimen _*—*_ cm

Area of Specimen, A _*10*_ cm^2

Volume of Specimen, V _*88.90*_ cm^3

w from Sample Trimmings		
Tare No.		
Wet + Tare		
Dry + Tare		
Wt. Water		
Tare		
Wt. Solids		
w %		

After Test:

Tare No. . . . . _*L-7*_

Wt. Sample Wet + Tare _*—*_

Wt. Sample Dry + Tare _*300.62*_ gm

Wt. Water W_w . . . . _*—*_

Tare . . . . _*186.45*_ gm

Wt. Sample Dry W_s . . . . _*114.17*_ gm

w % . . . . _*—*_

Initial Conditions:

Burette Reading V_0 . . . . _*0.00*_ cc

G_s—Det. _✓_ Est. _____ . . . . . . _*2.76*_   Vol. Solids $V_s = W_s/G_s =$ _*41.4*_ cc

Volume of Voids $V_v = V - V_s$ _*47.5*_ cc

Wt. Water in Sample
    $W_w = W - W_s$ . . . . _*44.17*_ gm   Void Ratio $e_0 = V_v/V_s =$ _*1.148*_

$w = W_w/W_s$ . . . . _*38.7*_ %

Degree of Saturation V_w/V_v . . . _*93*_ %   Vol. of Water $V_w =$ _*44.17*_ cc

SOIL MECHANICS LABORATORY
TRIAXIAL COMPRESSION TEST
(Cohesive Soil)
Consolidation Data

Name_____*J.V. Parcher*_____　　　Sample & Test No. _*H211B-R1*_

Date_____*Dec. 8, 1954*_____　　　Sheet No. _____2_____

Increment of Vertical Load: _____—_____ kg to _____—_____ kg

Increment of Liquid Pressure _____0_____ kg /sq cm to _____7.0_____ kg /sq cm

Date	Time	Elapsed Time Min.	Temp. °C	Dial Reading (inches)	Dial Change cm	Burette Reading cc	Remarks
12/8/54	1135	0	21	0.4757	—	0.00	
		0.25		0.4658	0.0099	0.80	
		0.50		0.4613	0.0144	0.88	
	1136	1		0.4554	0.0203	1.03	
	1137	2		0.4509	0.0248	1.26	
	1139	4		0.4326	0.0431	1.59	
	1143	8		0.4120	0.0637	2.09	
	1150	15		0.3894	0.0863	2.72	
	1205	30		0.3675	0.1082	3.37	
	1236	61		0.3465	0.1292	4.30	4 cc drained
	1335	120	21.5	0.3233	0.1524	1.38(5.38)	
	1535	240	22	0.2936	0.1821	2.38(6.38)	2 cc drained
	1815	400	21	0.2687	0.2070	0.86(6.86)	
12/9/54	0855	1280	20.5	0.2649	0.2108	1.23(7.23)	
	1316	1541	21.5	0.2629	0.2128	1.33(7.33)	
	1345	1570		0.1030	0.2140	1.34(7.34)	

From Sheet No. _1_ :　$H = $ _*8.89*_ cm　　$A = $ _*10*_ cm^2　$V = $ _*88.90*_ cc

$V_0 = $ _*0*_ cc　　$V_v = $ _*47.5*_ cc　　$V_s = $ _*41.4*_ cc

Elapsed Time, t_1, for 100% Consolidation .... = _____*800*_____ min.

Burette Reading Corresponding to $t_1 = V_1$ .... = _____*7.2*_____ cc

Dial Change Corresponding to $t_1 = \Delta H$ .... = _____*0.211*_____ cm

Corrected Height, $H' = H - \Delta H$ .... = _____*8.679*_____ cm

Volume Change from Initial Condition $\Delta V = V_0 - V_1$ .... = _____*7.2*_____ cc

Corrected Volume $V' = V - \Delta V$ .... = _____*81.70*_____ cc

Void Ratio, $e' = \dfrac{V_v - \Delta V}{V_s}$　Corrected Area, $A' = \dfrac{V - \Delta V}{H - \Delta H} = $ _____*9.42*_____ cm^2

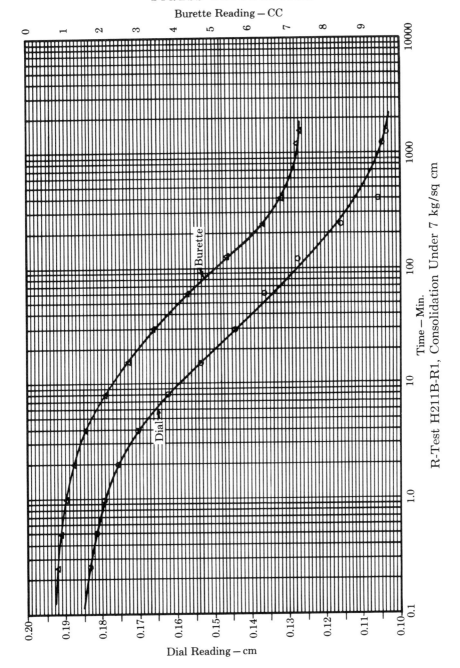

R-Test H211B-R1, Consolidation Under 7 kg/sq cm

SOIL MECHANICS LABORATORY
TRIAXIAL COMPRESSION TEST
(Cohesive Soil)
Undrained Loading
Summary & Computation Sheet

Name___*J. V. Parcher*_____ Sample & Test No. *H211B-R1*___

Date___**Dec. 9, 1954**_____ Sheet No._____*3*_____

Elapsed Time Min.	Piston Load kg	Dial Reading cm	Dial Change cm	Strain ε	Corrected Area cm^2	Deviator Stress kg/sq cm	Vertical Stress kg/sq cm
0	*0*	*0*	—	*0*	*9.42*	*0*	*7.0*
1	*4*	*0.0022*		*0.00025*	*9.42*	*0.425*	*7.425*
2	*8*	*0.0045*		*0.00051*	*9.42*	*0.850*	*7.850*
3	*12*	*0.0067*		*0.00075*	*9.43*	*1.272*	*8.272*
4	*16*	*0.0097*		*0.0011*	*9.43*	*1.695*	*8.695*
5	*20*	*0.0130*		*0.0015*	*9.44*	*2.120*	*9.120*
6	*24*	*0.0183*		*0.0021*	*9.44*	*2.540*	*9.540*
7	*28*	*0.0231*		*0.0026*	*9.45*	*2.960*	*9.960*
8	*32*	*0.0310*		*0.0035*	*9.46*	*3.380*	*10.380*
9	*36*	*0.0425*		*0.0048*	*9.47*	*3.800*	*10.800*
10	*40*	*0.0628*		*0.0071*	*9.50*	*4.210*	*11.210*
11	*44*	*0.1190*		*0.0134*	*9.55*	*4.610*	*11.610*
12	*48*	*0.4370*		*0.0492*	*9.94*	*4.840*	*11.840*
12:01	*52*	*Failure*			*10.5 (Est.)*	*4.950*	*11.950*

Dial Reading as and same (handwritten vertical annotation in Dial Change column)

Conditions at Start of Test

$\sigma_1 = \sigma_2 = \sigma_3 =$ ___*7.00*___ kg/sq cm

Void Ratio, $e' =$ ___*0.974*___

Remarks_____

Conditions at Failure

$\sigma_1 =$ ___*11.95*___ kg/sq cm

$\sigma_2 = \sigma_3 =$ ___*7.00*___ kg/sq cm

Corrected Area $= \dfrac{A}{1 - \varepsilon}$

Deviator Stress
kg/sq cm

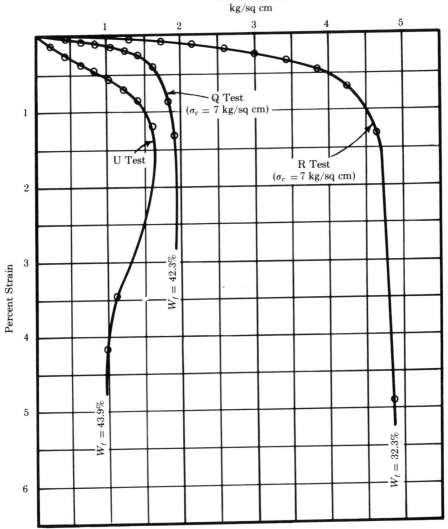

Percent Strain

Q Test
($\sigma_c = 7$ kg/sq cm)

U Test

R Test
($\sigma_c = 7$ kg/sq cm)

$W_l = 42.3\%$

$W_l = 43.9\%$

$W_l = 32.3\%$

SOIL MECHANICS LABORATORY
TRIAXIAL COMPRESSION TEST
(Cohesive Soil)
Data & Computation Sheet

Name___J.V. Parcher_____ Sample & Test No. 1C-56

Date___June 23, 1956_____ Sheet No. _____1

Description of Sample Gray silty clay from Cambridge, Mass.

Before Test: Wt. Sample Wet + Rubber Membrane.... 164.50 gm

Wt. of Rubber Membrane............... 2.86 gm

Wt. Sample Wet W.................... 161.64 gm

Thickness of Membrane 0.02 cm

Height of Specimen, H 8.88 cm

Diameter of Specimen ___—___ cm

Area of Specimen, A 9.85 cm²

Volume of Specimen, V 87.5 cm³

w from Sample Trimmings		
Tare No.	C-53	D-53
Wet + Tare	25.7253	28.7045
Dry + Tare	21.6239	23.6980
Wt. Water	4.1014	5.0065
Tare	11.6234	11.7503
Wt. Solids	10.0005	11.9477
w %	41.0%	41.8 %

After Test:

Tare No. ....____A____

Wt. Sample Wet + Tare ....___—___

Wt. Sample Dry + Tare .... 289.64 gm

Wt. Water W_w.............___—___

Tare..................... 175.79 gm

Wt. Sample Dry W_s........ 113.85 gm

w % ...._____

Initial Conditions:

Burette Reading V_0........... 0.00 cc

G_s – Det. ✔ Est. _____ ... 2.77   Vol. Solids $V_s = W_s/G_s = $ 41.05 cc

Volume of Voids $V_v = V - V_s$.. 46.45 cc

Wt. Water in Sample

$W_w = W - W_s$ .... 47.79 gm   Void Ratio $e_0 = V_v/V_s = $ 1.131

$w = W_w/W_s$................ 42.0 %

Degree of Saturation V_w/V_v ... 103 %   Vol. of Water $V_w = $ 47.79 cc

SOIL MECHANICS LABORATORY
TRIAXIAL COMPRESSION TEST
(Cohesive Soil)
Consolidation Data

Name _____J.V. Parcher_____   Sample & Test No. ___1C-56___

Date _____June 23, 1956_____   Sheet No. _____2_____

Increment of Vertical Load: _____—_____ kg to _____—_____ kg

Increment of Liquid Pressure _____0_____ kg /sq cm to ___6.21___ kg /sq cm

Date	Time	Elapsed Time Min.	Temp. °C	Dial Reading cm	Dial Change cm	Burette Reading cc	Remarks
6/23/56	1502	0	28.0	2.5000	—	0.00	
		0.1		2.4880	0.0120	0.50	
		0.25		2.4832	0.0168	0.60	
		0.50		2.4785	0.0215	0.75	
	1503	1		2.4716	0.0284	0.97	
	1504	2		2.4627	0.0373	1.27	
	1506	4		2.4504	0.0496	1.73	
	1511	9		2.4311	0.0689	2.53	
	1517	15		2.4165	0.0835	3.24	
	1540	38		2.3932	0.1068	4.96	5cc drained
	1628	86	28.5	2.3738	0.1262	1.92 (6.92)	
	1750	162	28.0	2.3538	0.1462	3.55(8.55)	3cc drained
6/24/56	1010	1148	27.0	2.2933	0.2067	2.57(10.57)	
	1922	1700	27.5	2.2908	0.2092	2.63(10.63)	
6/25/56	0958	2516	27.0	2.2878	0.2122	2.66(10.66)	

From Sheet No. _1_: $H =$ __8.88__ cm $A =$ __9.85__ cm^2 $V =$ __87.5__ cc

$V_0 =$ __0.00__ cc $V_v =$ __46.45__ cc $V_s =$ __41.05__ cc

Elapsed Time, t_1, for 100% Consolidation . . . . $=$ _Use Final Rdg._min.

Burette Reading Corresponding to $t_1 = V_1$ . . . . $=$ ___10.66___ cc

Dial Change Corresponding to $t_1 = \Delta H$ . . . . $=$ ___0.2122___ cm

Corrected Height, $H' = H - \Delta H$ . . . . $=$ ___8.67___ cm

Volume Change from Initial Condition $\Delta V = V_0 - V_1$ . . . . . $=$ ___10.66___ cc

Corrected Volume $V' = V - \Delta V$ . . . . $=$ ___76.84___ cc

Void Ratio, $e' = \dfrac{V_v - \Delta V}{V_s}$ Corrected Area, $A' = \dfrac{V - \Delta V}{H - \Delta H} =$ ___8.86___ cm^2

SOIL MECHANICS LABORATORY
TRIAXIAL COMPRESSION TEST
(Cohesive Soil)
Drained Loading
Summary & Computation Sheet

Name _____ *J.V. Parcher* _____   Sample & Test No. _ *IC-56* _

Date _____ *July 30, 1956* _____   Sheet No. _ *64 (see remarks)* _

Duration of Test _*35 days*_ .   Time Interval for Load Increments _*Varies*_

Elapsed Time for Computation _*Total*_ .   Time for 100% Consolidation _~350 min._

Initial Conditions for S—Loading: $H' =$ _*8.67*_ cm $A' =$ _*8.86*_ cm ^{2}

$V' =$ _*76.84*_ cc $V'_v =$ _*35.79*_ cc $V_s =$ _*41.05*_ cc $e' =$ _*0.871*_

Liquid Pressure = _____ lb /sq in $\times$ 0.0703 = _____ *6.21* _ kg /sq cm

Elapsed Time for Increment Min.	Piston Load kg	Dial Change cc cum.	Burette Reading cc	Volume Change cc ΔV	Void Change $\Delta e = \dfrac{\Delta V}{V_s}$	Void Ratio e	Strain ε	Correct Area cm ^{2}	Deviator Stress kg /cm ^{2}
—	0	—	2.66	—	—	0.871	—	8.86	0
1400	8	0.0622	3.11	0.45	0.011	0.860	0.0072	8.87	0.90
1436	16	0.1364	3.73	1.07	0.026	0.845	0.0157	8.89	1.80
796	24	0.2081	4.22	1.56	0.038	0.833	0.0240	8.90	2.70
943	28	0.2767	4.74	2.08	0.051	0.820	0.0319	8.91	3.14
1455	32	0.3395	5.17	2.51	0.061	0.810	0.0391	8.93	3.59

Considerable quantity of data omitted. During this period, dial was reset once and burette drained twice.

995	117.5	2.6294	11.47	8.81	0.214	0.657	0.303	11.27	10.43
480	118.0	2.6527	11.50	8.84	0.215	0.656	0.306	11.30	10.44
400	118.5	2.7200	11.55	8.89	0.217	0.654	0.314	11.42	10.38

Conditions at Failure: $\sigma_1 =$ _____ *16.63* _ kg /sq cm

$\sigma_2 = \sigma_3 =$ _ *6.25* _ kg /sq cm   Corrected Area $= \dfrac{V' - \Delta V}{H' - \Delta H}$

Remarks: _*Sheets 3 through 63 are same as sheet No. 2 — One*_ _____
_____ *sheet for each increment of vertical load.* _____

Index

461